BEYOND SPINOFF

BEYOND SPINOFF

Military and Commercial Technologies
in a Changing World

JOHN A. ALIC

LEWIS M. BRANSCOMB

HARVEY BROOKS

ASHTON B. CARTER

GERALD L. EPSTEIN

Science, Technology, and Public Policy Program
Center for Science and International Affairs
John F. Kennedy School of Government
Harvard University

Harvard Business School Press

BOSTON, MASSACHUSETTS

96 95 94 93 92   5 4 3 2 1

Library of Congress Cataloging-in-Publication Data

Beyond spinoff : military and commercial technologies in a changing
    world / John A. Alic . . . [et al.].
        p.   cm.
    Includes bibliographical references and index.
    ISBN 0-87584-318-2 (alk. paper)
    1. Technology.   2. Military art and science.   I. Alic, John A.
    T15.B48   1992   91-36175
    600—dc20   CIP

The recycled paper used in this publication meets the requirements of the American National Standard for Permanence of
Paper for Printed Library Materials Z39.49-1984.

Table of Contents

Acknowledgments

Any book credited to five authors listed alphabetically has a story behind it. In this case, the story is one of truly collaborative effort, involving many drafts discussed, debated, and passed back and forth over a time period longer than we perhaps should admit.

Nor were we the only people involved. Two of the chapters were written by others. J. Nicholas Ziegler, assistant professor at the Sloan School of Management, MIT, and former fellow at the Kennedy School's Center for Science and International Affairs (CSIA), wrote Chapter 7, "Cross-National Comparisons." Charles A. Zraket, CSIA senior scholar in residence, took on the task of Chapter 9, "Software: Productivity Puzzles, Policy Challenges." We are grateful not only for the contributions these chapters make to the book, but also for the many other ways in which Nick and Caz added value. We also owe a great deal to Teresa Johnson whose tireless and skillful editing deserves the credit for turning five voices into an approximation of one—and none of the blame for the infelicities that remain.

Many other people helped as well. Our colleagues in the Kennedy School's Science, Technology, and Public Policy Program (STPP), Paul Doty and Dorothy Zinberg, participated in the planning and initiation of the research leading to *Beyond Spinoff*. Joshua Lerner, with contributions from Todd Flynn and Craig Stetson, gathered much of the statistical data on industry structure presented in Chapter 6 and its appendices, and analyzed the data on mobility of engineers between defense and civilian sectors discussed in Chapter 4. Todd Watkins served as rapporteur at several workshops on dual use, and explored the role of subcontractors for Chapter 6.

Gary Rahl prepared several of the case examples and verified much source material, particularly in Chapters 3 and 6. Vincent Ruddy also provided research assistance. Susan Fox, then STPP's administrative director, provided essential support. Other individuals too numerous to mention aided us in thinking through the issues discussed in *Beyond*

Spinoff—in a three-year series of workshops and seminars, field interviews, and individual discussions.

Our undertaking began with a research project supported by the Alfred P. Sloan Foundation and the Carnegie Corporation of New York entitled *Research on Dual-Use Technologies: Balancing Economic and Security Interests in Federal R&D Investment Strategies*. Gerald Epstein was project director and Lewis Branscomb was principal investigator. John Alic participated while on leave during 1988–1989 from the congressional Office of Technology Assessment.[1] This research received important additional support from the MITRE Corporation, and from the EG&G, Harman International, Polaroid, Raytheon, and Rockwell International Corporations, together with a grant from the Bernard O'Keefe Charitable Foundation.

We are grateful to all who supported this project, tangibly and intangibly. Any errors, omissions, or misinterpretations in *Beyond Spinoff* are our responsibility alone.

—the authors
Cambridge, MA and Washington, DC
January 1992

[1] The views expressed in this book are those of the authors alone, not necessarily those of the Office of Technology Assessment.

PART ONE
The Problem at Hand

1

The Changing Role of Technology in Military and Economic Power

Technology has been a key component of both America's military security and its economic performance throughout the post–World War II period. In the future its importance can only grow. But American strategies for military and commercial technology have become unsuited to the times; they remain rooted in the postwar pattern of open-ended military competition with the Soviet Union and the comfortable dominance of world commerce. These strategies must be updated. The passing of the postwar military and economic order requires a fundamental reassessment of the wellsprings of American military and economic power. Given the importance of technology to both, it is crucial to rethink national strategies for bringing technology to bear on defense and commerce. This book aims to contribute to that reassessment by examining the nature of the U.S. technology base, by identifying how its military and commercial aspects interact, and by analyzing the effects of government policies on the development and application of new technology.

In the United States, sponsorship of technical effort is divided about evenly between the private and the public sector, with defense providing nearly two-thirds of the public funds. Understanding the relationship between these two efforts provides the necessary basis for designing appropriate and practical government technology policies. Those policies, in turn, will have a central place in a renewed strategy for enhancing U.S. commercial competitiveness and preserving U.S. military security.

3

THE PARADOX OF "DUAL-USE" TECHNOLOGIES

Throughout this book the term "dual-use technology" refers to technology that has both military and commercial applications, and the relationship between the military and commercial sectors is called the "dual-use relationship." Most technology is in fact multiuse. For example, the microchips that make precise missile guidance possible also turn up in children's toys, in automobiles, and in the manufacturing equipment used to make the chips. The theoretical knowledge and practical know-how that comprise the essence of technology are even more protean than the artifacts themselves. The same computational techniques used to design a blast-resistant missile silo, for example, can be used to design skyscrapers. The technologies for designing, making, and using jet engines, optical fiber cable, magnetrons, boron fiber composites, rocket propellants, ion implanters, three-dimensional hydrodynamic codes, and automated lathes are all dual-use technologies.

While some degree of dual-use potential is inherent in many technologies, most technical work sponsored by government agencies—and almost all sponsored by private companies—is not intentionally multiuse. Instead, most such investments are made with a more or less specific end use in mind. Private companies fund research and development (R&D) to further particular business goals. Government agencies fund research and development to further public missions: national defense, health, energy efficiency, environmental protection, space flight, and so on. Dual use is not the first priority of any government agency or of most private firms.[1]

Sometimes the pursuit of different goals leads to technologies that are truly single-purpose, not dual use: it is difficult to imagine a commercial use for an x-ray laser powered by a nuclear bomb, for example, or a "stealth" airplane. But even when the technical knowledge developed in connection with one intended use could in principle be applied directly to another use—such as for the composite materials used in both stealth planes and civilian aircraft—it is a separate challenge to facilitate the movement of ideas, tools, and experience from their originator to a new user, or from one user to another. Industrial managers testify to the difficulty and expense of transferring technology even between different parts of the same firm when both parties are eager

---

[1] An important exception is those suppliers of components, materials, instruments, tools, or technical services that define their businesses in technological terms. Such firms serve both defense prime contractors and commercial end-product manufacturers with similar dual-use products. See Chapter 6.

to do so. The process becomes far more difficult when the attempted transfer must bridge the gap between parties having goals and organizational cultures as different as those of military and commercial institutions.

Thus there is a disjuncture between the inherently multiuse nature of much of modern technology and the fact that technology is sponsored by organizations that have very different goals. This disjuncture and the paradoxes it presents to public policy form the central theme of this book.

DUAL-USE INTERACTIONS

The technological relationships between the world of defense and the world of commerce are important for both at this juncture in history, but for somewhat different reasons. Commercial technology policies must address whether, and how, the government should be involved in supporting the nation's technology base as the share of defense's contribution shrinks. Military planners, on the other hand, must recognize defense's increasing dependence on technologies from the commercial sector. Both factions must come to terms with the inevitable fading of U.S. technological dominance in an increasingly sophisticated world economy.

Defense Influence on Commercial Technology

Massive U.S. government spending on defense has had mixed effects on the commercial sector, as this book details. But the key factor in understanding its future influence is that defense's overall impact has become smaller and smaller—even before any post-cold war cutbacks—as the U.S. private sector and foreign nations have increased their technological efforts. One index of this relative decline is the fraction of total R&D spending provided by defense. Since 1960, defense R&D spending has increased about 50 percent in real terms, but spending by U.S. private industry has nearly quadrupled, and R&D spending by other Western industrialized nations has gone up by a factor greater than 5 (see Chapter 4). Thus in 1960 the U.S. Department of Defense (DoD) funded fully one-third of all R&D in the Western world, whereas today it funds less than one-seventh (Figure 1-1). This remains a significant fraction, but it will continue to decline.

At the same time, military and civilian technology development have grown increasingly isolated from one another as a result of techni-

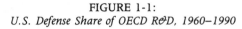

FIGURE 1-1:
U.S. Defense Share of OECD R&D, 1960–1990

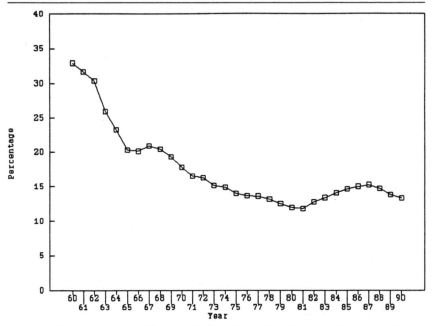

Source: Data were provided by the Division of Scientific, Technological, and Industrial Indicators, Directorate for Science, Technology, and Industry, Organization for Economic Co-operation and Development. See Chapter 4, especially Figure 4-2 and footnote 12.

cal specialization and the business-culture differences that arise in serving government and commercial markets (Chapter 5). It is likely that post-cold war reductions in overall defense spending will hasten this 30-year trend. In some sectors, such as electronics and computers, the defense share of the market and of R&D spending has declined even more dramatically, and DoD has been a relatively minor player in these fast-moving fields for many years.

Although it is declining steadily in relative terms, defense technology spending remains large. Close to one-third of all dollars spent on R&D in the United States still comes from defense agencies, and about 15 percent of the nation's technical professionals work on defense projects (Chapter 4). Defense's heavy involvement in technology makes its effects on the competitiveness of U.S. industry important. Defense has played significant roles in creating and maintaining the nation's technology base over the past five decades. Defense has funded R&D on a broad menu of generic and high-risk technologies, set enormously challenging goals for technical performance, and provided training and hands-on experience with advanced technology to many thousands of scientists, engineers, and technicians. These are roles that the smaller defense establishment of the future will not play to the same degree—yet they are all components of a technology policy that a government seeking to maintain its share of responsibility for the nation's technological infrastructure might wish to retain.

Because of defense's dominance for half a century, the defense-commercial relationship analyzed in this book is an important guide to the do's and don'ts of a future federal technology policy. However, the guidance that any retrospective examination can provide is incomplete, because the world has changed in ways that establish a new context for policy. Consequently, this book's larger purpose is to provide a revised intellectual framework for the relationship between government and the private sector in technology, and to define the proper scope of public policy in an era when the technology base that supports economic performance has become truly international.

Commercial Contributions to Defense Technology

The defense-commerce technology relationship is important from the standpoint of military security as well as that of economic security. While the commercial sector must address a diminishing contribution from defense, the trends for defense point to increasing dependence on commercial technology.

By 1991, the defense budget (excluding the costs for Operation Desert Storm) had already declined by 21 percent in real terms from its peak in 1985. It will probably continue to decline for several years at least, to between two-thirds and one-half its peak level of the mid-1980s. Although military threats to vital U.S. interests will remain, as the Gulf War of 1991 made clear, they will not command the share of national wealth formerly devoted to the cold war.

The portions of the defense budget that fund procurement of high-technology weapons and related R&D will take their share of cutbacks, perhaps disproportionately. The defense acquisition process that seeks to insert the latest technology into fielded military equipment is already in internal crisis. Defense systems are widely seen as costing too much, taking too long to develop and produce, and nonetheless frequently failing to perform to expectation.[2] These negative evaluations of high-tech weaponry have sometimes been overstated: U.S. equipment gave

---

[2] See, for example, Jacques S. Gansler, *Affording Defense* (Cambridge, MA: MIT Press, 1989).

a generally good performance in the Gulf War. Still, much of the criticism may be warranted, as the circumstances of that war were ideally suited to showing these weapons to best advantage. The cancellation of the Navy's A-12 fighter in early 1991, amidst allegations of continuing mismanagement, typifies the problems that have repeatedly surfaced in major weapons system acquisition.

We should expect the importance of applying technology to national security problems to increase, not decrease, with the winding down of the cold war. The U.S. defense establishment will be reorienting itself from the familiar framework of superpower confrontation to a more fluid and uncertain world of fractious nationalism, the eruption of regional conflicts suppressed in part by the rigidity of the cold war, proliferation of high-technology weaponry, terrorism, and drugs. Technology is always an important tool for meeting change and for hedging against an uncertain future.[3]

To meet its technology needs, the Defense Department will be forced to rely increasingly on the commercial sector. Defense's shrinking share of Western world R&D cannot support a state-of-the-art military by itself. Defense will need to learn to draw on the larger, dynamic, and increasingly global commercial technology base. Thus the military-commercial technology relationship is central to military security as well as to economic security.

SPINOFF: THE POSTWAR PARADIGM

Throughout the postwar period, most Americans understood that their economic and military security rested in important measure on the strength of the national technology base. But their view of that base featured a sharp divide between the proper roles of defense (or government in general) and the private sector, notwithstanding an implicit assumption of strong linkage between weapons work and economic prosperity. They assumed, in short, that there was no dual-use *policy*, but that there was a great deal of dual use.

According to this prevailing view, the nation's pursuit of economic and military security involved two separate spheres of technological activity. The first consisted of self-initiated actions by private firms aim-

[3] Harvey Brooks, "The Typology of Surprises in Technology, Institutions and Development," in William C. Clark and R.E. Munn, eds., *Sustainable Development of the Biosphere* (Cambridge: Cambridge University Press, 1986), Chapter 11, pp. 325–350. See also a report of the Carnegie Commission on Science, Technology, and Government, *New Thinking and American Defense Technology* (New York: Carnegie Commission, August 1990).

ing at profits and productivity in a market economy. The second effort, for national defense (and, in smaller measure, for other accepted missions such as health, spaceflight, and energy), was exclusively public and relied on federal funds to generate the technology needed by the Pentagon and other so-called mission agencies. It was understood that technology developed under defense auspices influenced commercial practice. But such "spinoff," as the name implied, did not require any deliberate coordinating action.

As viewed in this model, defense scientists working on a major DoD R&D project would make a discovery that in turn made possible some new product or process. Scientists and engineers working in commercial firms would learn of the discovery and conceive marketable goods or services based on it. In this way, the invention would "spin off" from the defense sector to the commercial sector, from research lab to office or factory floor. Since the new technology was not yet available elsewhere in the world, Americans would enjoy exclusive profits from selling the new products. This process was viewed as more or less automatic—unmanaged and cost free—and thus did not violate the principle that government should not deliberately intrude in matters affecting the civilian economy.

The reality was different, as the chapters that follow will show. Defense investments in technology were both more and less effective in generating economic gain than the simple spinoff view would predict. They were more effective, because the technical relationships between defense and commerce were richer and more complex than the spinoff model implies (Chapters 2 and 3). However, they were also less effective, since very few technologies proceeded effortlessly from defense conception to commercial application. The technology transfer process typically required substantial additional attention and investment.

Without DoD's broad involvement in the nation's technology base, key technologies such as semiconductors, computers, jet aircraft, and communications satellites would have developed much more slowly, and the early U.S. commercial leads in these technologies would not have been achieved so readily. Defense was not just "helpful" to these sectors, but thoroughly dominated them during their formative years through the sheer volume of Pentagon spending.[4] Moreover, the Pentagon did not confine its support to technologies of immediate relevance

---

[4]Richard R. Nelson, ed., *Government and Technical Progress: A Cross-Industry Analysis,* (Elmsford, NY: Pergamon Press, 1982); Kenneth Flamm, *Creating the Computer: Government, Industry, and High Technology* (Washington, DC: Brookings Institution, 1988).

to national defense. Through its basic and applied research programs, DoD pursued a great variety of technologies of broad applicability, often long before specific defense applications were clear (Chapters 4 and 5). DoD technology programs contributed greatly to the nation's supply of scientists and engineers. Other government agencies, such as the National Aeronautics and Space Administration (NASA; earlier the National Advisory Committee for Aeronautics) and the Department of Energy (DOE; earlier the Atomic Energy Commission), also pursued technology of commercial relevance, while the National Science Foundation and the National Institutes of Health provided support for basic research and for training scientists and engineers. (The life sciences are largely outside the scope of this book because defense involvement in these fields is relatively small.)

Paying a premium price for systems incorporating new technologies, the Pentagon frequently financed rapid movement down the learning curve until commercial markets took over. Defense procurement was just as important as R&D in influencing the course of development of computers or jet-propelled aircraft.

Despite its failure to capture the full breadth and depth of defense contributions to national technology, however, the spinoff paradigm in other ways *overestimates* the commercial impact of that contribution. Spinoff and other transfers of technology from defense to commercial projects were far from cost free (Chapter 3); they required additional expenditures for adapting defense technology and exacted opportunity costs as other uses for the resources devoted to the original defense project were forgone. Nor were spinoffs always beneficial. Military requirements can distort priorities toward complex, high-performance objectives with limited commercial applicability, although it is sometimes difficult to determine whether such distortions are attributable to defense support or simply to poor business judgment on the part of the affected industry (see Appendix 10-A).

The important point about the spinoff paradigm is not that it was a half truth at best, but that the unusual circumstances of the postwar world did not force Americans to question it. Few asked whether spinoff was an *efficient* way to link government efforts to commercial performance: for the first two decades or so after World War II, no Japanese or German competitors were breathing down the necks of American high-tech companies, threatening to snatch their markets through better production and distribution or more agile commercialization of new technology. Few asked whether some technologies, or some stages in the movement of technology from laboratory to application, were more important than others or more in need of government help. Into the 1970s America led in most if not all, and the imperative of cold war with the Soviet Union meant that defense had to dominate across the board. Few asked whether the reverse of spinoff—the use of commercial technology for defense purposes—might be important, since defense set the pace. Finally, few Americans questioned the premise that there were only two important creators of the technology base, the U.S. government and U.S. industry: foreign generators and users of technology, apart from the Soviet Union in the military sphere, were viewed as bit players.

As the prevailing—if unspoken—conception of the relationship between public and private sectors, the spinoff paradigm became deeply ingrained. It shaped and continues to shape the Washington climate. For most Americans, it remains the point of departure when they contemplate the relationship between defense and commerce or between public and private R&D. But it never corresponded very well to a more complex reality, and with the passage of time the correspondence has become ever fainter.

To the extent that a defense-driven technology strategy as implied in the spinoff paradigm is inappropriate for the contemporary world, the problem will be partially self-correcting: defense will continue to decline as a fraction of national (and international) technical effort. But there is serious danger that the nation will borrow defense technology paradigms uncritically, transposing them to nondefense policies as it seeks to compensate for the decline of DoD investment in the technology base. Evidence of this temptation can already be seen in the commitment of a disproportionate fraction of nondefense R&D funds to large and very costly science and engineering projects such as the space station, the national aerospace plane, and the Superconducting Super Collider, as well as in ideas for transporting defense institutional patterns (e.g., DARPA) to civil agencies. Meanwhile, American business, accustomed to letting DoD carry much of the burden, has been slow in responding to aggressive technological investments by Japanese firms, even as the latter outdistanced them first in process and then in product engineering. The cost to Americans of carrying around the wrong mental image of how the technological system works will be paid in terms of lost markets, overpriced weapons, and wasted resources.

THE OBSOLESCENCE OF THE POSTWAR PARADIGM: FOUR REVOLUTIONS

The urgent need to depart from the spinoff paradigm stems from four parallel changes in the world that have shattered the paradigm's foundation. These changes—so fundamental (even if not sudden) that they can be termed "revolutions"—stand poised to trigger a change in public policies toward science and technology fully as profound as those that followed World War II and the launch of the Soviet Sputnik.[5] With one exception—the collapse of communism in the Eastern bloc—these changes have been going on for many years, and it is the growing American awareness of them and their potential impact on technology policy that merit the term "revolutions."

What are these revolutions? First, the steady rise of Japan and much of Western Europe, particularly Germany, has brought other nations to technological parity with the United States. With highly competent overseas rivals, particularly in high-technology industries, the shortcomings of the U.S. technological system stand out as they did not in earlier years. American industry must begin viewing its technology base in worldwide terms, and so must DoD, given the extent to which defense systems depend on imported components and production equipment. Second, the end of the cold war has ushered in a revolution in national security affairs. A smaller military establishment with a structure different from the past will accelerate the ongoing decline in DoD's relative influence. Third, the recipe for industrial competitiveness through innovation has changed. Fast-paced, incremental improvement in productivity and product characteristics, always important, becomes even more so when the United States cannot count on the fundamental breakthroughs that once resulted from an overall lead in science and technology. Fourth, given a technological system that is truly international, national governments have lost much of their ability to control events through investments in technology or restrictions on its spread. Grandiose projects patterned on the Apollo moon landing or the Strategic Defense Initiative will be increasingly irrelevant

[5] World War II marked the beginning of an entirely new relationship of government to science and technology. At the war's end, Vannevar Bush called for continuing this relationship in *Science: The Endless Frontier* (Washington, DC: U.S. Government Printing Office [U.S. GPO], 1945). Before the war, federal R&D funding was meager and was earmarked mostly for agricultural research by civil servants. By 1950, however, federal R&D expenditures were relatively large, channeled to private institutions, and covered fields of great importance for all sectors of the economy. See Harvey Brooks, "National Science Policy and Technological Innovation," in Ralph E. Landau and Nathan Rosenberg, eds., *The Positive-Sum Strategy: Harnessing Technology for Economic Growth* (Washington, DC: National Academy Press, 1986), pp. 119–168. The Sputnik launch in 1957 set off another round of innovation in federal institutions for science and technology, including greatly strengthened White House science advisory mechanisms, creation of the Advanced Research Projects Agency within DoD (ARPA, later Defense ARPA or DARPA), and formation of the National Aeronautics and Space Administration.

to world technological competition. Nor can a top-down approach from Washington, without the active participation of private firms, succeed. Together, these four revolutions have made it clear to a growing number of Americans that the policies, institutions, and habits of mind that have governed the nation's science and technology system during the past half century are not suited to the present and must be rethought for the future.

The Revolution in the International Economy

The first revolution is the change in the international economy. The across-the-board technological dominance enjoyed by American industry in the postwar period is giving way to a position of first among equals.[6] This relative decline is the result (even if not fully intended) of the resounding success of U.S. postwar military and economic policies, which sought the recovery of both allies and enemies as successful free-market democracies while containing Soviet expansion within its 1946 sphere of influence. But there is no ignoring the fact that the trading partners the United States fostered are now its technology competitors. Any number of economic and trade indices reflect this equalization among the nations of the Western industrialized world (including Japan). Equalization occurred first in traditional sectors such as steel and automobiles but soon progressed to high-technology industries that Americans had considered their own. In 1970, 95 percent of the high-tech products purchased by Americans were made in the United States; by 1986, this fraction had fallen to 82 percent.[7] U.S. factories produced 51 percent of the total high-tech output of the advanced industrialized nations in 1970, but only 42 percent in 1986. U.S. goods comprised 28 percent of world high-tech exports in 1970, but only 22 percent in 1986. More dramatic has been the penetration of U.S. markets by imported automobiles, consumer electronics, machine tools, and semiconductor chips. The message is clear: American and foreign buyers alike will force American companies to compete for their business. When foreign goods offer superior price, performance, or quality, customers will choose them without thinking twice.

---

[6] Committee on Engineering as an International Enterprise, National Academy of Engineering (Thomas H. Lee, chairman), *National Interests in an Age of Global Technology* (Washington, DC: National Academy Press, 1990).

[7] These figures are based on OECD data compiled in *Science Indicators 1989*, NSB 89-1 (Washington, DC: U.S. GPO, 1989), Appendix Tables 7-1, 7-6, and 7-10, pp. 371, 374, and 377.

Competitiveness in the sense suggested above is a valid concern for Americans because they earn their wages in American enterprises; the nationality of the *owners* of the offices and factories that make up the enterprise is of secondary importance in this context.[8] There is no reason that American enterprises should have continued indefinitely to turn out goods of all sorts better and more cheaply than foreigners. But if the United States is to hold its position of first among peers in the world economy, its high-technology industries must excel in many product categories, even if they cannot dominate in all.

Foreign companies supply goods that compete effectively in U.S. and other markets for many reasons (including the relative openness of the American market), but an important one is their access to and effective use of technology. Foreign enterprises, especially in Japan and Germany, have for decades been increasing their investments in R&D at a higher rate than enterprises in the United States. Foreign companies have also learned to absorb technology developed elsewhere and to translate it into high-quality, low-cost, efficient production. These abilities depend only in part on the kind of organized R&D that is captured by comparative statistics on R&D spending. Like the trade statistics, trends in R&D spending reflect the natural convergence of the economies of the industrialized world nearly a half century following World War II. Yet the new competition highlights U.S. weaknesses and creates fear that Japan and Germany will not only catch up with America, but surpass her, and that America will go the way of Britain—clever but poor.

Success in the new international economy requires competitive costs, high quality, and the flexibility to respond quickly to new market demands and new technological or market opportunities. Competition is forcing American firms to shed the sloppiness that an unquestioned leader could afford. It is forcing companies to look beyond sheer technical virtuosity to the seemingly humbler—but actually very demanding—tasks of improving quality, controlling costs, moving quickly from an idea to its realization, and continuously improving product designs and production processes in light of feedback from the marketplace and the factory floor. Internationalization of the technology base makes it increasingly important for U.S. firms to learn to draw on technological advances made in other countries. This can be accomplished only by the private sector, not by government. But to the extent that government can and should play a helping role, government policies must adapt to the changing environment.

[8] Robert B. Reich, *The Work of Nations* (New York: Alfred A. Knopf, 1991).

The Revolution in National Security Affairs

The second revolution, in national security affairs, stems from Mikhail Gorbachev's "new thinking" and the discontinuity in world politics that followed in its wake, including literal revolutions in Eastern Europe. These changes amount to nothing less than the overthrowing of the assumptions on which U.S. defense planning has rested since the beginning of the cold war. The Warsaw Pact has disappeared and the Soviet Union fragmented. NATO has declared an end to the threat of short-warning massive attack on Western Europe, to which the large peacetime U.S. military establishment was chiefly directed. Pending arms control agreements covering both nuclear and conventional forces, if fully implemented, will further reduce the military danger to the United States and its allies.

At the same time, the 1991 Gulf War vividly reminded Americans that threats other than the Soviet Union can draw the United States into military confrontation. Indeed, regional tensions that would likely have remained suppressed during the height of the cold war standoff may now be more likely to erupt into armed conflicts that could involve the United States. Yet such military contingencies, absent the threat of East-West confrontation that has driven U.S. defense planning since World War II, will not be sufficient justification to maintain U.S. defense budgets and military forces at their cold war levels.

The revolution in military affairs will change the size and the character of DoD's technology investments, with profound implications for dual use. The overall defense budget is likely to decline by the mid-1990s to perhaps one-half to two-thirds of its mid-1980s peak. All four major elements of the defense budget—operations and maintenance, personnel, procurement, and research and development—are likely to share in the reductions, although probably not in equal proportions. Defense's technical activities are funded through both R&D and procurement accounts. Reductions will accelerate the trend that has made U.S. defense a smaller and smaller contributor (in relative terms) to the world's technology base (see Figure 1-1).

Despite these planned reductions, it is likely that the United States will maintain a substantial military establishment for some time to come. America and its NATO partners will need to preserve the option to reconstitute their heavy forces for the defense of Europe if political events there should turn sharply for the worse. Regional conflicts might involve high-technology weapons in the hands of possibly unstable governments locked in bitter, sometimes ancient, antagonisms. The U.S. military may also be drawn into unconventional wars against terrorism and illegal drug traffic. Finally, in the more distant future, one cannot rule out the emergence of new economic and military superpowers that the United States may wish to deter with its own military capabilities.

Thus one can argue that the importance of technology to defense will increase, not decrease, in coming years. Asked to identify the threat to which U.S. defense efforts should henceforth be directed, President Bush summed up the predicament by answering, "unpredictability, uncertainty, and instability." Technology is an important insurance policy against an uncertain strategic future. In coming years the Defense Department will be trying simultaneously to squeeze more military muscle from lower budgets, to reorient its efforts to a broader and less predictable spectrum of military threats, and to retain the option of returning to its current size on relatively short notice if necessary. Technology would be an indispensable tool in any of these efforts.[9]

If previous downturns in defense spending are any guide, however, Pentagon spending on new technology is likely to take a disproportionate share of the budget cuts. Broad, ambitious, and generic work without immediate payoff—funded in what DoD calls its "technology base"—could be the hardest hit. Although it did not share in the defense buildup that began in the late 1970s and continued through the Reagan years, technology-base funding could well share in reductions now under way. Moreover, a significant amount of dual-use work takes place outside the formal technology base, within the large weapons-development programs that absorb the lion's share of the R&D budget, and as part of the manufacturing that is funded through the procurement budget. This work will also suffer as overall DoD spending declines.

Impending budget cuts could therefore have a profound effect on investments in dual-use technologies. Reductions in defense spending will hasten the retreat from a half-century of DoD involvement as a source of broad support and as an explorer of new directions in any and all fields of technology that could conceivably have military applications. Under severe funding pressure, defense technology managers normally turn inward, using scarce resources to cover a narrower front of technical fields of direct and immediate relevance to national security. Defense's technology efforts have tended to become less relevant outside fields unique to military applications, even at current spending

---

[9] Carnegie Commission on Science, Technology, and Government, *New Thinking and American Defense Technology*.

levels. An abrupt and unplanned withdrawal from technical fields that defense has long supported could leave gaps in the national effort at a time they can ill be afforded from the standpoint of U.S. competitiveness.

DoD's internal management of technology has also been in crisis. The Pentagon's problems in getting technology from drawing board to field mirror those in civilian industry, and are probably worse. The weapons acquisition process is widely viewed as suffering from cost, schedule, and performance shortfalls. Numerous "reforms" have been announced, but all have stalled when faced with the vastness of the task.[10] Much of DoD's acquisition problem involves management more than technology. High cost, lengthy schedules, and poor technical performance have many nontechnical causes, including ponderous government contracting procedures and the political twists and turns of an ever more arcane budgetary process. Difficulties in fielding new or upgraded systems in a timely manner limit DoD's ability to capitalize on technical progress.

Defense's way of doing business provides little guidance for coping with the pressures of the new international economy. Defense technologists take their cues from government "requirements," not from a competitive market. DoD emphasizes functional performance objectives over schedule and cost; one consequence is that it spends five times more on R&D, as a fraction of total system costs, than commercial firms do. Major defense projects extend over a decade or more, much longer than in civilian industry. Defense programs tend to follow a "pipeline" progression, in which a separately funded and managed R&D phase precedes production. In contrast, commercial businesses, constantly improving their products, pursue R&D in parallel with production and feed in new technology incrementally. The low volume of defense production means that less attention is given to efficient manufacturing than in the commercial world.

If DoD buys fewer planes, ships, and tanks in the years to come, instead spending more of its money ensuring against an uncertain future, it will probably pay even less attention to manufacturing. Insurance would take the form of maintaining small but sophisticated forces,

---

[10] The best-known study of defense acquisition is that of the Packard Commission (1986). Among its predecessors over the past two decades are the Fitzhugh Commission (1970), the Commission on Government Procurement (1972), and the Grace Commission (1983). See *Defense Acquisition: Major U.S. Commission Reports (1949–1988)*, prepared for the Defense Policy Panel and Acquisition Policy Panel of the House Committee on Armed Services, 100th Cong. 2d sess., November 1, 1988, Committee Print No. 26.

preparing to mobilize quickly, and relying on technology to increase the flexibility and responsiveness of a smaller military establishment. In this scenario, the defense R&D effort will be a forum for technology assessment, rather than the first stage in large-scale production of weapons.

The distinctive nature of the defense business, including the multitude of government bookkeeping and auditing rules, unique specifications and standards, and intellectual property restrictions, causes many private firms to specialize in either defense or commercial markets; if they pursue both, they do so in separate divisions insulated from one another except at the highest management levels. Thus the military and commercial sectors have evolved two different technical and business cultures. Influences from the defense world may not be of much help to the commercial world, and they might even be harmful (see Chapter 2, especially Table 2-3).

Two conclusions emerge from observing the shrinking relative size and peculiar character of DoD's technology investments. First, the Pentagon cannot provide the best technology to the military services unless it breaks out of its enclave, learning to use technology spawned in the commercial sector and attracting capable high-technology firms willing to supply its needs. Since the best technology will not always originate in the United States, using the newest and best will unavoidably entail dependence on foreign suppliers for components critical to defense systems.[11] Second, the decreasing size of defense's contribution to the world store of technology, together with the idiosyncrasies of its technical culture, means that it cannot continue to play the part cast for it in the spinoff paradigm—that of sparkplug for high-technology industry.

The Revolution in the Process of Technological Innovation

The third revolution is in the nature of technological innovation itself. Fast-paced international competition places higher premiums on certain features of the technological process now than in the past. Some of these features are new, while others have always been present but have become more critical for competitive outcomes. Together they suggest that the nature of the technological process is dramatically at odds with much of the conventional wisdom, especially in Washing-

[11] See Theodore H. Moran, "The Globalization of America's Defense Industries: Managing the Threat of Foreign Dependence," *International Security*, vol. 15, no. 1 (Summer 1990), pp. 57–99.

ton, about how technology is made and used. The blunt realities of international competition attach great importance to features of the technological process that the United States could safely downplay in the early postwar years, when Washington was forging its role in science and technology. Competitiveness, for example, depends on product engineering and on manufacturing capabilities, activities well outside the accepted scope of government involvement in the United States.

Four closely related themes characterize a proper understanding of the technological process: the importance of "downstream" or non-R&D technical activities, the pertinence of alternatives to the so-called pipeline view of technology generation and application, the convergence of science and engineering, and the importance of borrowing and adapting technology as well as generating it.

Upstream versus downstream.   The prevalent pipeline metaphor sees new technology emerging from successive steps of basic research, applied research, exploratory development, engineering, and manufacturing. "Upstream" activities refer to organized science and engineering R&D focused on new techniques and new products. "Downstream" refers to engineering design and development for efficient manufacturing, including controlling cost and quality, and for meeting market demand, closely matching products to the needs of the buyers who pay for them. In earlier years the American penchant for inventiveness and American domination of the upstream process allowed U.S. firms to neglect the downstream processes, although they have always been important to commercial success. But U.S. preeminence at the upstream end of the pipeline no longer results in automatic success in the marketplace because foreigners, with easy access to the fruits of basic and applied research, sometimes have stronger downstream capabilities to absorb and apply research results.

Alternatives to the pipeline view.   The pipeline model is not a bad approximation for radical innovations in which new science makes possible unprecedented technological capabilities: discovery of the laser, for example, or the transistor. But such breakthroughs are rare exceptions even in high-technology sectors.

Alternatives to the pipeline model, such as the "cycle" view,[12] recognize that companies compete on the basis of existing products, not those yet to be created. For these products, companies engage in a very

[12] Ralph Gomory, "Of Ladders, Cycles, and Economic Growth," *Scientific American* (June 1990), p. 140. Gomory refers to the pipeline as the "ladder."

different process of innovation in which they seek constant, incremental improvements—typically in both product features and manufacturing processes. In the cycle model, speed is critical, because the longer the improvement cycle, the more likely it is that competitors will get their own improvements into the marketplace first. This view gives equal weight to technical virtuosity in all of the functions—research, design, production, and marketing—and it stresses the importance of close coupling among them. It also emphasizes the importance of complementary assets (not always technological) that often determine competitive success: supplier firms, distribution and service networks, and so on.[13]

In reality, neither the pipeline view nor the cycle picture of incremental development is adequate by itself. Commonly, long periods of incremental innovation with gradually declining benefit/cost ratios are punctuated by bursts of more radical technological change, usually traceable to upstream research, that remove fundamental barriers to further incremental improvement. The important point is that no simple view of innovation is rich enough to capture the variety of processes that contribute to technological advance. Government policy should not be based on a caricature of that complexity.

The convergence of science and engineering. The period since World War II has witnessed an accelerating transformation in the relationship of science to engineering and thus to production. In the nineteenth century, much of engineering remained a craft, empirically based and with little direct connection to science. Over the past hundred years, products and processes found everywhere in the economy have come to depend on science and on theoretically based engineering methods. By now, most of the products of industry have strong underpinnings in science and systematic theory. Through improved instrumentation and use of custom-designed materials unknown to nature, industrial processes now press more closely on the limits of scientific knowledge.

Since the 1960s, computer models incorporating complex mathematical constructs and quantitative data from the new instrumentation have permitted the design of products, simulation of manufacturing processes, and calculation of the performance of both product and process with ever closer fidelity to nature. Computer-aided engineering techniques allow rapid comparisons of product and process alterna-

---

[13] David J. Teece, "Capturing Value from Technological Innovation: Integration, Strategic Partnering, and Licensing Decisions," in Bruce R. Guile and Harvey Brooks, eds., *Technology and Global Industry: Companies and Nations in the World Economy* (Washington, DC: National Academy Press, 1987), pp. 65–95.

tives, particularly in the chemical and electronics industries; design and production issues can be addressed concurrently with rapid feedback from one to the other. Moreover, when precise design and production data is in computer-readable form, it can be communicated electronically between technical groups. Older and slower methods of transferring technology, through apprenticing and learning-by-doing, are thus partly replaced. Together with close control of every step of the manufacturing process, modern production methods—which depend as much on management and organization of plant and workforce as on sophisticated computer tools—are becoming indispensable in commercial competition, enabling shortened product cycles, greater flexibility in responding to market conditions, and low-cost high-quality production.

Diffusion versus generation. For the first several decades after 1945, most foreign nations took it for granted that major inventions and discoveries would come from the United States. Foreign firms sought to borrow technology from American innovators and make better, cheaper products with it, or to produce standardized goods for which new technology was less important. Foreign governments helped through policies such as requiring transfers of American technology as a condition for investment, protecting domestic markets, and promoting diffusion of best-practice know-how. American innovators accepted this pattern of borrowing because they wanted access to those foreign markets and because they planned to stay one step ahead. According to the product-cycle theory associated with Raymond Vernon, by the time foreign firms mastered a technology and began to apply it better or at lower cost, U.S. innovators would abandon it for new technologies, thereby staying ahead of the imitators.[14] The United States emphasized the generation of new knowledge, other nations its application and diffusion.

With other countries again technological peers as in the prewar years, the postwar division of labor has begun to work against U.S. interests. In its new role as first among equals, the United States must learn to borrow also. American businesses, less practiced as hunters and gatherers of technology, need help with diffusion. They must explore a wider set of technological niches and learn to adopt best-practice know-how, regardless of source. The very slow spread of Japanese quality control and production management techniques—some

[14] Raymond Vernon, "International Investment and International Trade in the Product Cycle," *Quarterly Journal of Economics*, vol. 80 (May 1966), pp. 190–207.

of them originally from the United States—shows how big an adjustment is now required.

The Revolution in the Scope of National Government Action

The fourth revolution is the diminishing scope for national government action in a technology system that is increasingly international. In the broadest sense, this book is about the influence of national policies on the world of commercial technology: federal policies have had and will continue to have a profound influence on the technical capabilities of American firms. But it is necessary to admit at the outset that crucial determinants of the technological future are escaping from Washington's control, and indeed escaping from the control of governments everywhere. Still beginning to make itself felt, this trend will limit the effectiveness of unilateral national technology policies. These constraints reflect the patterns of international business activity that have emerged over the past two decades and the nature of modern technological innovation. Policy design must proceed with these constraints in mind.

The first constraint stems from the globalization of technology. With international trade and investment having expanded steadily for more than four decades as a share of gross national products (GNPs), and with firms based in other parts of the world catching up in technology and managerial expertise, U.S.-based multinational corporations have been forced to rethink their corporate strategies. During the 1950s and 1960s, the overseas activities of American multinationals centered on exports and direct investment. Newer technology went abroad embodied in goods. Overseas affiliates typically got older technology. Multinational corporations kept the latest knowledge at home, where it could be protected more easily.

Companies still prefer to maintain a tight hold over their technical know-how, using it only to produce for export or transferring it to controlled subsidiaries abroad. But these choices are less practical today than in the past. By the 1970s, U.S. multinationals were already transferring technology to their affiliates much earlier in the product cycle—that is, they were transferring newer, advanced products and processes.[15] By the 1980s, U.S.-based companies, struggling to keep

---

[15] Raymond Vernon, "The Product Cycle Hypothesis in a New International Environment," *Oxford Bulletin of Economics and Statistics*, vol. 41, no. 4 (November 1977), pp. 255–267; Edwin Mansfield and Anthony Romeo, "Technology Transfer to Overseas Subsidiaries by U.S.-Based Firms," *Quarterly Journal of Economics*, vol. 95 (1980), p. 739.

up with ever more capable rivals, found it necessary to accelerate the entire process. Many markets, in many parts of the world, demanded the latest and best. Instead of first developing products for home markets and then moving them abroad if they proved successful, companies began to design products for world markets, modifying them only slightly for different countries. With globalization of this kind, new technologies are introduced simultaneously in Japan, Europe, the United States, and elsewhere.

Pursuing such strategies, corporations based in the industrialized nations are extending and integrating themselves globally, and at the same time they are decentralizing—farming out more production, contracting for services once provided internally, and pursuing joint ventures and other intercorporate linkages and alliances such as sharing in technology development projects. As multinationals seek to rationalize their operations on a worldwide basis, they disperse their design, development, production, distribution, and marketing operations worldwide, clustering them within major markets rather than concentrating them at home. In so doing, they come to rely on complex global networks of wholly and partly owned affiliates, suppliers, and strategic partners (who may include putative rivals).

In this complex international web of commercial life, national policies have less reach and force. Globalization means that many companies headquartered in the United States are no longer "American" in any simple sense, and thus the "American interest" that Washington policies are supposed to advance is not readily identifiable. Governments everywhere have increasing difficulty distinguishing "our" firms from "their" firms. Well-intended Washington policies might not have much effect on the global operations of U.S.-based multinationals and could do inadvertent harm; conversely, the policies of foreign governments might cause plans carefully laid in Washington to unravel. The world is not a global village, and the technology and products made in the United States are still largely sold in the United States.[16] But the trend is toward a worldwide technology base that will not respond to policies crafted in a narrowly national framework.

Even in domains of science and technology that have traditionally been the preserve of national governments, such as security, environment, and "big science," international considerations are becoming paramount. High-tech weapons depend on ball bearings and electronic

---

[16] Pari Patel and Keith Pavitt, "Large Firms in the Production of the World's Technology: An Important Case of Non-Globalization," *The Journal of International Business Studies,* vol. 22, no. 1 (1991), p. 1.

components no longer manufactured in the United States. Environmental problems, such as greenhouse warming, acid precipitation, and ozone depletion, are global in cause and effect. Pure science is becoming internationalized as the costs of investigation—space exploration and particle accelerators, for example—exceed the willingness of individual countries to pay for them.

While national governments must come to grips with loss of influence, local, state, and regional governments are growing presences in technology. U.S. state governments, German *Länder* governments, Japanese prefectures, and local entities around the world are unabashed practitioners of industrial policy.[17] Local governments have some advantages over national governments in this respect. They generally have a closer acquaintance with the needs and problems of their high-tech firms. They are accustomed to competing among themselves for jobs, and many have become active promoters of technology diffusion. And for better or worse, state capitals harbor fewer ideological qualms than Washington does about "interfering" in the free market. Federal technology policies should be fashioned in partnership with, and certainly not in ignorance of, the activities of state governments.

A final constraint on policy arises from the nature of technology itself. The technology base is a complex sociotechnical system that operates through myriad private initiatives and the expert judgment of many thousands of scientists, engineers, and managers. The technological enterprise does not lend itself to top-down central planning or to decision making at a distance by bureaucrats who regard all technologies as mysterious or interchangeable "black boxes." Policies must be tailored to the details of each situation to be effective. This inevitably puts a premium on sophisticated judgments by government officials and on their understanding of technical and market realities, an understanding that in turn requires a close and collegial relationship between government officials and business executives, groups that have often stayed at arm's length to avoid the appearance or the reality of conflicts of interest.

BEYOND SPINOFF: TOWARD A NEW FEDERAL TECHNOLOGY POLICY

The spinoff paradigm that underlies much of U.S. technology policy is in key respects at variance with the four revolutions we have described.

[17]David Osborne, *Laboratories of Democracy* (Boston: Harvard Business School Press, 1988).

Originating in an era when the United States dominated world technology and national defense dominated U.S. technology, it provides little guidance to American firms seeking superior technology abroad, or to defense planners who increasingly must rely on commercial know-how.

Spinoff emphasizes revolutionary developments that create entirely new markets, rather than the processes of incremental improvement and rapid response on which commercial competitiveness more typically depends. Whereas competitive success requires concentrated attention to downstream activities such as production engineering and manufacturing, the spinoff paradigm focuses on R&D, an activity that in defense acquisition is kept distinct from production. The paradigm gives priority to technical novelty originating in big national projects like aerospace planes, supercomputers, and moon voyages, as opposed to the equally challenging but less visible tasks of designing consumer products, improving their quality, and cutting costs. Diffusion is not ignored in the spinoff paradigm, but it is portrayed as easy, almost automatic: potentially useful technology from defense R&D or big weapons projects is assumed to be readily recognized by potential borrowers and applied by them with little extra effort.

Policymakers rarely articulate the spinoff paradigm, but it continues to influence decisions in Washington and the government actions that Americans view as legitimate regardless of their political persuasion. To avoid controversy, government funds R&D for socially approved missions like defense, space, and health; everyone also accepts public investments in basic science, since in the pipeline view research is safely separated from the free market. These accepted federal roles are mirrored in the agencies most active in technology: DoD, the Department of Energy, NASA, and the National Institutes of Health for "mission" work and the National Science Foundation for basic research. The four revolutions have increased the mismatch between the familiar and politically accepted government policies and agency missions, on the one hand, and the day-to-day processes of technology development on the other. In the future, government will either have to content itself with a smaller role in the national technology base, or adapt itself to a newer understanding of the technological process.

In other words, the federal government must look beyond spinoff if it hopes to formulate an effective response to the challenges facing American industry in today's highly competitive environment. New policies must be found that are compatible with the realities of interest-group politics, American style, and of a fragmented and intentionally weak federal government that offers a host of opportunities for advocates of almost any position to make themselves heard. New policies must also be compatible with the constraints defined by processes of technological innovation, and by the humbling realities of the new international competition. We draw out the implications of these themes for federal technology policy in Chapter 12.

Just as the spinoff paradigm fails to illuminate the range of future federal technology policies, it also fails to do justice to the breadth and diversity of the existing technical relationships between defense and commerce. Simple spinoff has always been just one mechanism through which the two interact—and by no means the most frequent or important. But the broader record of dual-use interactions examined in this book is not robust enough to serve as the basis for technology policy either. Today, government contributions to the commercial technology base are typically the unintended by-products of government actions taken for other reasons. But defense contributions to national R&D funding, to the training and job experience of scientists and engineers, and to the technological infrastructure of facilities, technical tools, and instrumentation, have had and continue to have important impacts, as we show in later chapters.

Analysis of dual-use relationships touches the heart of the technological process itself: how knowledge is generated, how it is diffused and applied, and how the motives of its sponsors determine its character and its usefulness. Sound technology policy is not possible without a solid understanding of the technological process. "Dual use" is thus a window on the contribution of technology to economic and military security, on the changing relationship between public and private investments in the technology base, and on the nature of the technological process itself. Through this window we will consider appropriate goals for policy in a changing world.

ORGANIZATION OF THIS BOOK

This book is organized in four parts. Part I (Chapters 1–3) explores the nature of the issues we are dealing with: the nation's technology base, government's influence on it, and the interactions between its commercial and military portions. Part II provides greater detail about the U.S. science and technology system: measures of R&D budgets and personnel, government policies and procedures that influence (and mostly impede) dual-use synergies, and corporate structures in the defense and commercial worlds (Chapters 4–6). Providing a counterpoint to the U.S. discussion is an overview (Chapter 7) of the institutional and industrial structures of France, Sweden, Germany, and Japan. Part III (Chapters 8–10) provides three case studies illustrating significant aspects of dual use: microelectronics, computer software, and manufacturing technologies.

We present our findings and conclusions in the two chapters of Part IV. Chapter 11 sums up what the United States can expect to achieve through dual-use technology policies—using defense investments to promote both military and economic goals. Seeking dual-use synergies is important for the commercial world, and it will be a necessity for defense. However, as Chapter 11 explains, defense policies will not be sufficient to keep U.S. industry neck and neck with technologically sophisticated foreign competitors. In Chapter 12, therefore, we examine the need, the constraints, and the opportunities for new technology policies that are not based on defense.

2

Technology Policy and the Technology Base

Processes of technology development and diffusion are complex. In the United States they involve dozens of government agencies, hundreds of federal laboratories and universities, thousands of industrial corporations, and hundreds of thousands of engineers, technicians, managers, and scientists. The technological enterprise is huge in financial terms as well. The official figures show that U.S. R&D spending exceeds $145 billion annually (Chapter 4).[1]

To understand how government policies affect technology, it is necessary to understand the activities of the people and institutions that develop and apply technical knowledge. This chapter outlines the many types of technical knowledge and the varied processes by which that knowledge is generated and transmitted. Exploring the dual-use paradox introduced in Chapter 1, it shows that although technical knowledge itself cannot be divided along military and commercial lines, the institutions and innovation processes that apply it to achieve particular military or commercial ends are largely distinct. The causes and implications of that separation will be explored in later chapters. This chapter concludes by examining government's role, establishing a framework for analyzing U.S. technology policy, and distinguishing it from the more contentious subject of industrial policy.

[1] Such figures understate the total because official statistics do not capture much design and applications engineering in smaller firms, and much manufacturing engineering in firms of all sizes. Keith Pavitt and Pari Patel, "The International Distribution and Determinants of Technological Activities," *Oxford Review of Economic Policy*, vol. 4, no. 4 (Winter 1988), pp. 35–55.

The picture that emerges suggests that the U.S. science and technology system works reasonably well for generating new knowledge, particularly the knowledge needed to support well-defined missions of federal agencies, whether national defense or seeking a cure for AIDS. This is what the system was designed to do. But our current policies are not well suited for supporting an internationally competitive array of industries. The great bulk of federal R&D money goes to electronics, aerospace, and the biomedical sciences. That distribution is not broad enough to support an economy as vast and diverse as that of the United States. Moreover, federal R&D is focused on the generation—rather than the diffusion and broadest possible utilization—of knowledge. The nation does not yet have effective policy mechanisms for bridging the gaps between public and private sectors, or between research and downstream engineering.

TECHNOLOGY AND TECHNICAL KNOWLEDGE

Technology consists of purposeful human artifacts and the knowledge needed to create and use them. Development of an artifact depends on knowledge; for example, a fly-by-wire aircraft control system requires mathematical models of flight stability, statistical data for the failure rates of the control system's components, and much more. We can visualize the technology base as consisting of both artifacts and knowledge. The technology base is far more than just the science base.

Technical Knowledge

At the core of the technology base lies the widely known and accepted knowledge found in textbooks and other reference works. Table 2-1 lists these and other forms of technical knowledge, from established facts and principles to tacit know-how. Publicly available knowledge, such as basic research results (e.g., Maxwell's equations for electromagnetism) or well-documented engineering methods in widespread use (such as empirical equations used in predicting heat transfer) can be exploited by anyone with technical understanding sufficient to evaluate it critically and apply it correctly.[2] A great deal of technical information,

---

[2] Although some technical papers, and much of the review literature, summarize well-accepted results or describe current practices, others contain speculative results (e.g., cold fusion) that may turn out to be wrong. The insights needed to steer clear of incorrect or useless information are found in the tacit realm. On patterns of literature use, see Thomas J. Allen, *Managing the Flow of Technology: Technology Transfer and the Dissemination of Technological Information within the R&D Organization* (Cambridge, MA: MIT Press, 1977).

TABLE 2-1:
Types and Characteristics of Technical Knowledge

Description	Characteristics
Text and reference books; monographs; review papers; handbooks.[a] Journals; conference proceedings. Patents.	"Glass box": transparent, readily accessible.[b] Often available in English irrespective of origin. Typically quantitative, based on mathematical models, algorithms, empirical results. Well-developed quality controls, particularly for the reference literature.
Unpublished technical documents (conference papers; government reports; handouts for academic and training courses; internal company reports; consultants' reports).	Gray literature; scattered, hard to find; may not be available through publishers, bookstores, or libraries; may be restricted to citizens or classified secret; may be available only in foreign languages; may have had little or no reviewing for accuracy; may be incomplete or require other, unspecified information for use.
Trade and business press. Computer programs and documentation; physical and mathematical models; computerized databases. Military and industrial standards; military design handbooks.	
Engineering drawings, blueprints; bills of materials; process sheets; instruction and repair manuals.	Often proprietary.
Experimental methods and techniques.	Some are well codified (e.g., standard test methods sanctioned by professional or trade associations), others are matters of experience and tacit know-how. Quality control tests may be proprietary.
Organizational routines; company operations and training manuals.	May be embodied in people (tacit); can sometimes be learned and transferred only through personal experience.
Bid and proposal packages for DoD and other contracts.	Normally confidential.
Trade secrets.	Proprietary by definition.
Tacit knowledge (experience-based judgment, heuristics, intuition), both individual and organizational.	"Black box": may be difficult to state with precision; often hard to explain; typically nonquantitative.

[a] As the practice of technology relies ever more heavily on up-to-date information, computerized databases will gradually replace handbooks and other data sources.

[b] The glass-box/black-box distinction follows that applied to design and analysis techniques by J. Christopher Jones, *Design Methods: Seeds of Human Futures* (London: Wiley-Interscience, 1970), pp. 45–58. Glass-box literature is (or should be) transparent and rigorous: one can locate it and see into and understand the logic.

including test results and other empirical data, exists only as unpublished or scantily distributed "gray literature"—for example, in the reports of research conducted under contract to government agencies. Aircraft designers, for instance, still depend on reports issued in the 1930s by the National Advisory Committee for Aeronautics (NACA), which contain extensive tabulations of wind tunnel results for differing airfoil cross-sections. Generic technical data such as materials properties may fall in this category, as may nonproprietary process design procedures (e.g., time-temperature relationships for chemical reactions).

Technical knowledge also exists in forms that are not easily or willingly transmitted, either because they are uncodified (see Box 2-A on tacit knowledge) or because they are protected by firms seeking competitive advantage or by governments for reasons of national security. These relatively immobile forms of technical knowledge can include management skills and organizational techniques, such as just-in-time (JIT) production methods, as well as rules of thumb and tacit know-how (e.g., understanding the appropriate dimensional tolerance for machined parts, or technical and managerial judgment). Much of this knowledge migrates slowly and uncertainly not because it is secret but because learning and implementation are difficult when the rules are not clear. Americans struggled for years to grasp the principles embodied in JIT production, largely missing the point of a simple system built around self-initiated actions by many individuals rather than around top-down planning.

Whether a firm will be able to tap a given body of knowledge depends on the education and training of its employees, along with skills and abilities that come only through experience. Engineering development, for example, depends on the interpretation of often-ambiguous test results. And simply knowing that something is possible—for example, that an airplane can pass through the "sound barrier" without disaster—may have great impact in stimulating technological development. Thus technical knowledge includes not only the well-codified information found in text and reference books, but the black-box categories at the bottom of Table 2-1.

In industry, tacit know-how has always been important for proprietary technology and competitive advantage—especially when it is organizational rather than individual, as reflected, for example, in learning curves representing improvements in productivity, cost, or quality.[3]

[3] Linda Argote and Dennis Epple, "Learning Curves in Manufacturing," *Science*, vol. 247, no. 4945 (February 23, 1990), pp. 920–924.

BOX 2-A: TACIT KNOWLEDGE IN THE PRACTICE OF TECHNOLOGY

The practice of technology hinges on tacit knowledge and skill—in Polanyi's telling phrase, what people know but cannot tell (can do but not explain).[1] Expertise comes only with experience. This is true especially in design and manufacturing, the heart of the technical enterprise. Individuals and groups make decisions—choice of research strategies, selection of design parameters, process details—based on what they know and can articulate, combined with tacit know-how, instinct, and intuition. Some of this tacit knowledge—associated with craft skills as well as with the professions—people will not even be aware of calling on.

Software engineers designing a large program, for instance, face thousands of choices in arranging instructions, branches, and subroutines (Chapter 9). Some of these decisions they can rationalize in logical (glass-box) fashion. Others they will be able to make only on a black-box basis and can justify only by recourse to judgment or intuition—the decision process remains opaque. Some people pick up these tacit black-box skills—which may have little in common with the skills required for writing nominally error-free software on a line-by-line basis—more readily than others. Some seem born with good heuristics; others learn fast; some never do. Thus the productivity of programmers with comparable experience varies by factors of 10, 20, or even more. Much the same is true when it comes to using complex computer programs for solving engineering problems in structural mechanics or fluid flow.

Other forms of tacit knowledge have their origins on the factory floor. Toyota's famous kanban or just-in-time production system evolved through day-to-day, case-by-case solution of shopfloor problems. Many people contributed. When Toyota decided to replace this informal system with computer algorithms and a network, it took the firm several years to understand the logic embodied in the system's actual functioning. Japanese manufacturers have achieved advantages in shopfloor productivity through kanban and other techniques, in part because they do a better job of tapping and documenting the know-

---

[1] Michael Polanyi, *Personal Knowledge* (Chicago: University of Chicago Press, 1958). Also see Herbert A. Simon, *The Sciences of the Artificial* (Cambridge, MA: MIT Press, 1969); and, among the writings of J. Christopher Jones, *Essays in Design* (Chichester, UK: Wiley, 1984), "Designing Designing," pp. 125–142, and "Continuous Design and Redesign," pp. 191–216. Joseph Badaracco, *The Knowledge Link: How Firms Compete through Strategic Alliances* (Boston: Harvard Business School Press, 1991), distinguishes between "migratory" and "embedded" knowledge, with the latter very similar to the concept of tacit knowledge as it is presented here. For a discussion of the economic theory, see Joseph E. Stiglitz, "Learning to Learn, Localized Learning and Technological Progress," in Partha Dasgupta and Paul Stoneman, eds., *Economic Policy and Technological Performance* (Cambridge: Cambridge University Press, 1987), pp. 125–153.

how of blue-collar employees. By the same token, the high mobility of technical employees characteristic of the United States may tend to break up teams and destroy the benefits of group learning that are mainly embodied as tacit or embedded knowledge.

As the kanban example illustrates, tacit knowledge can be organizational as well as individual. Organizational knowledge—something more than the bits and pieces of know-how embodied in a company's workforce—reflects group learning, history and tradition, institutional style and habit.[2] When DoD or some other federal agency funds large R&D projects or contracts for the production of a weapons system, one result will be organizational learning within the contracting firms. In working together over a period of years, engineers, scientists, and managers who have shaped a new airplane or computer system learn to draw on one another's strengths while compensating for blind spots and weaknesses. Trite as the comparison may seem, such groups resemble athletic teams in at least the following respect: no matter how capable the individuals may be, it takes time and the experience of both success and failure before the group performs at its full potential. Once broken up, engineering groups cannot be easily reconstituted, any more than gifted athletes can be quickly melded into a winning football team. Thus, for example, national security concerns will require that the United States retain at least some experienced weapons design groups in the years ahead, no matter how much the DoD budget may decline.

[2]Japanese firms do a much better job documenting the learning experience of innovation teams so that lessons learned from past mistakes are not repeated in subsequent development cycles. See Phillip Barkan, "Productivity in the Process of Product Development," in Gerald I. Susman, ed., *Integrating Design and Manufacturing for Competitive Advantage* (New York: Oxford University Press, to be published 1992). Such lessons tend to reside only as tacit knowledge within organizations; the public literature usually deals only with what worked.

In other ways too, the intangible character of tacit knowledge gives it special significance for policy, even though it has often been overlooked by analysts and policymakers. Tacit knowledge is a route for maintaining a technological edge in military systems: what cannot be written down can hardly be stolen. And when the federal government supports education and training in science and engineering through research grants, graduate student stipends, and postdoctoral fellowships, it is helping build the nation's store of tacit know-how as well as book learning. (Some of this tacit knowledge migrates overseas when foreign graduate students return home.)

Knowledge flows into the technology base from tacit know-how, from R&D, and from experience accumulated and codified during design, development, and testing. Knowledge flows out when it becomes obsolete (superseded by newer results or proved wrong). Results published in the open technical literature enter immediately. Proprietary knowledge seeps in gradually, despite the efforts of companies to hold onto it.

Technology Diffusion

Flows of technical knowledge help drive innovation and economic growth. Knowledge diffuses in part through networks of engineers and scientists, linked in technological communities that exist independently of corporate organizations and government agencies.[4] Table 2-2 lists the institutions that play a role in technology diffusion. Diffusion can take place from one nation to another or from government projects to commercial projects. It can take place horizontally, from one firm to another in the same business; or vertically, from laboratory to factory floor; or between end-product manufacturers and their suppliers. Access to diffusion networks is especially important for smaller firms that do not engage in R&D. Even in large firms, a principal function of research scientists and engineers is to recognize and help interpret R&D results from elsewhere.[5]

Technology does not flow easily, as gas flows in a pipeline. Learning is difficult; so is unlearning. Ideas and know-how remain embedded in laboratories, design teams, and factories. Crucial insights may never be written down. People working in one area may not realize that their insights could solve problems in another area. Others needing technical information may not know whether it exists or where to look for it, especially if it is tacit knowledge or if it can be found only in the gray literature. Security classification rules, proprietary restrictions, export control laws, and other formal barriers expressly discourage individual transfers and slow diffusion processes, desirable as well as undesirable.

There is no simple, easy prescription for improving diffusion. Most

[4] The diffusion of innovations, technological or social, can be pictured something like chemical diffusion at the atomic level: the processes are highly random individually, while responding on average to gradients of knowledge (for the spread of new engineering methods) or to the driving force of economic returns to investment (for the spread of, say, numerically controlled machine tools). See, in general, Everett M. Rogers, *Diffusion of Innovations*, 3d ed. (New York: Macmillan, 1982). Technology transfer carries a narrower range of meanings: one-way transfers from a source (e.g., an R&D laboratory) to one or a few users—the point-to-point movement of prepackaged knowledge.

[5] Harvey Brooks, "National Science Policy and Technology Transfer," in Harvey Brooks, *The Government of Science* (Cambridge, MA: MIT Press, 1968), Chapter 10, pp. 254–278.

TABLE 2-2:
Institutional Actors in Technology Diffusion

Organizations	Comments and Examples
Industrial corporations	Explicit mechanisms include technical licensing agreements, cross-licensing of patents, and vertical links with suppliers. Linkages among American firms tend to be weak both horizontally and vertically, although the number of strategic alliances is growing rapidly.
	Manufacturers of laboratory instruments and experimental apparatus play particularly important roles, along with companies that develop and supply production equipment (e.g., machine tools) and with capital goods suppliers generally. Even advertising by these companies helps diffuse knowledge.
Labor unions	Primary vehicles include apprenticeships and other training programs.
Technical services and consulting firms	Suppliers of specialized technical services include companies that design equipment and facilities (e.g., clean rooms for semiconductor production), contract engineering firms, and organizations (including universities) that operate laboratories under contract to government. Both profit-seeking R&D and consulting firms (e.g., SRI) and not-for-profits (e.g., MITRE) diffuse technology through contract research as well as other technical service activities. The consulting industry would be much smaller if technology diffused through the U.S. economy rapidly, inexpensively, and otherwise without friction.
Industrial consortia	Examples include Sematech, Computer-Aided Manufacturing International (CAM-I), and Microelectronics & Computer Technologies Corporation (MCC).
Trade and industry associations	The Semiconductor Industry Association provides financial support for the Semiconductor Research Corporation, which funds university R&D and graduate student training.
Standards-setting bodies	American National Standards Institute (ANSI), American Society for Testing and Materials (ASTM), and other professional and technical societies.
Publishers	Several have introduced computerized bibliographic information services and on-line technical databases.
Professional and technical societies	Most oral and published transmission of new scientific information is organized by these societies through means such as books, journals, conferences, short courses, and tutorials. They also play a major role in defining and systematizing standards of quality in science and engineering.

TABLE 2-2 (*continued*):

Organizations	Comments and Examples
Federal agencies	The National Technical Information Service (NTIS) distributes technical reports on federally funded R&D conducted in government labs, universities, and industrial firms. By indexing, advertising, and disseminating these reports, it makes some of the gray literature more available. During the 1980s, Congress and the administration explicitly addressed transfers of publicly funded R&D results to industry (see Chapter 3).
	INTERNET, a collection of more than 2,000 computer networks, links engineers and scientists throughout the country. It had its origin in ARPANET, developed with DoD support.
State governments	A number of states, seeking to promote economic development, have adopted diffusion-oriented technical assistance programs—e.g., industrial extension services aimed at helping small manufacturers find and use technology successfully.
Colleges and universities	Many schools offer continuing education programs. Telecommunications as a delivery mechanism continues to spread. Community colleges and vocational-technical schools organize education and training programs intended to meet the needs of local industry.

companies have modest technical capabilities compared with IBM or General Dynamics. How can the government help them tap the know-how that might enable them to improve their product technologies, quality levels, and productivity? While policymakers have focused on the tangible outcomes of federal R&D spending—technical reports, patents, licensing agreements—informal channels for technology diffusion are at least as important. And even when the knowledge is well codified, people must absorb it and learn to use it if transfer is to take place. The multiplicity of vehicles and channels listed in Table 2-2—overlapping networks of people and organizations, and differing but complementary procedures for validating and codifying know-how, packaging it, and transferring it—suggests the need for a multiplicity of policy initiatives. In Chapter 12 we discuss the actions that are needed.

THE U.S. TECHNOLOGY BASE: DUAL USE AND MORE

The Common Base of Technical Knowledge

Technical knowledge—in contrast to physical objects or artifacts—must be presumed to have an inherently dual-use or multiple-use character, until and unless analysis shows otherwise. For centuries, many if not most major technological innovations were intrinsically dual use—steam-powered ships, the motor truck, the airplane. Since World War II, however, much military hardware has diverged from its nearest civilian analogs—or, perhaps more appropriately, the civilian sector has diverged from defense, as the growth of private R&D over the past few decades swamped that of government-funded R&D (see Chapter 4). When it comes to tanks, aircraft, and electronic command and control systems, similarities remain at component and subsystem levels. But whereas early World War I tanks were built from farm tractors, today few parallels can be found at the system level between the Army's M-1 tank and civilian trucks, tractors, or off-road construction equipment.[6] Ballistic missiles incorporate components such as ball bearings and integrated circuits (ICs) that may be similar or occasionally identical to parts found in civilian products, but no commercial products resemble the missiles themselves. With exceptions such as general-purpose digital computers or machine tools, only aircraft engines and transport planes, among major technological systems, remain similar in their military and civilian versions. Even then, although they may share a common core, the fan-jets for commercial transports differ in many ways from military engines.

At the same time, engineers call on a common body of design and analysis methods in, say, analyzing aircraft structures (Box 2-B). The example of structural integrity technology discussed in Box 2-B illustrates how different fields of knowledge interact, how research results find their way into the practice of technology, and how both public and private institutions—government agencies, companies, professional societies—contribute to the development of new knowledge and its diffusion. In these respects, there is nothing special about structural integrity technology; it has evolved and diffused much as any multidisciplinary technical subject today.

---

[6] Even so, Caterpillar's unsuccessful contract bid for rubber-tracked MX missile launchers during the 1970s led, a few years later, to a new line of farm equipment. Dave Fusaro, "Rubber-Tread Farm Machine Market Escalates," *Metalworking News* (September 25, 1989), pp. 1, 41.

BOX 2-B: STRUCTURAL INTEGRITY TECHNOLOGY: A DUAL-USE EXAMPLE

Military and civilian aircraft differ in many ways—design lifetimes, duty cycles, the magnitude of maneuver loads—but the essential requirements of structural integrity are the same: no structural failures that could endanger pilot and passengers. Only when it comes to battle damage and its consequences do structural integrity requirements for military and civilian aircraft diverge markedly. Even then, many of the analytical problems—predicting the weakening effects of damage created by shell fragments—can be tackled with the same methods used in analyzing cracks left after fabrication or created by fatigue stress during routine service.[1]

Structural design techniques for aircraft remained relatively crude until the 1930s, when all-metal construction became the rule. World War II brought a rapid spurt in aircraft technology: greater performance meant greater structural loads. At the same time, widespread failures caused by brittle fracture in Navy ships spurred research on a parallel but unique set of problems (ship steel behaves differently than aircraft aluminum). Efforts to understand and prevent wartime ship fractures took two paths: (1) experimental studies aimed at developing test procedures for ranking or qualifying materials, and at design guidelines; and (2) mathematical models of the behavior of cracks under load. Different groups at the Naval Research Laboratory (NRL) pursued each path. Metallurgists guided the experimental work, following a traditional engineering approach—the loop marked ''Service Experience, Failure Analysis, Test Development, 1940s–Present'' in Figure 2-1. The other group, led by physicists, pursued research associated with the loop marked ''Sharp Crack Models, 1950s–Present.''

Both these research paths had their own genealogies going back decades (as summarized in the figure) and drew on both practical engineering experience and basic science. During the 1840s, for instance, W.J.M. Rankine found that fatigue failures in railway car axles stemmed from gradual cracking. Crystallization as an explanation became even less tenable when x-ray crystallography—a basic research tool—revealed all metals to be inherently crystalline. Many years later, knowledge of crystal structure from x-ray studies helped scientists develop dislocation models of cracks—hence the feedback loop at the upper left in Figure 2-1. Other pieces of knowledge came at different times from different fields. A.A. Griffith's studies of fracture in glass during the 1920s depended heavily on earlier work in applied mathematics and elasticity theory, yielding the Griffith crack model noted in the figure. Despite its idealized nature—Griffith chose glass as a model

---

[1] This discussion is based on John A. Alic, ''The Federal Role in Commercial Technology Development,'' *Technovation*, vol. 4 (1986), pp. 253–267.

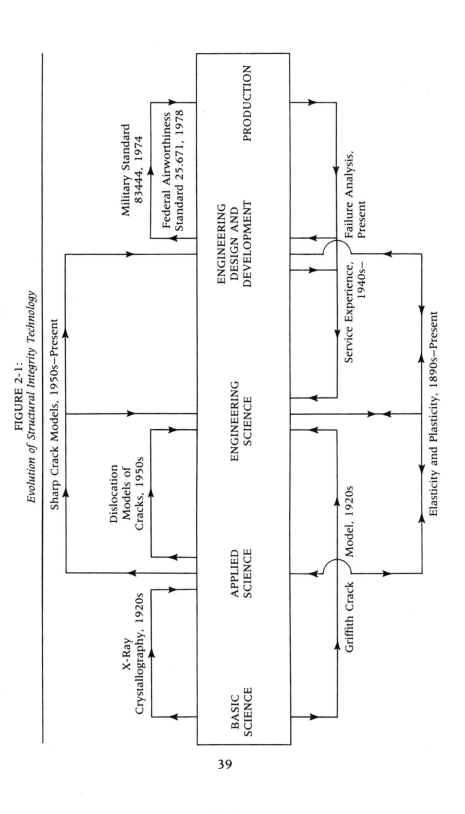

FIGURE 2-1:
Evolution of Structural Integrity Technology

39

substance close to the perfectly elastic behavior required by his mathematics—this model continues to underlie the methods in use today, hence the loop at the lower left in Figure 2-1.

By the 1960s, research in applied mathematics and continuum mechanics, solid-state physics, and materials science had advanced to the point that idealized cracks of many types could be modeled. Computers were put to work analyzing crack extension processes. Engineers developed design procedures permitting, for instance, prediction of crack growth under service loadings. For aircraft, these were validated through laboratory experiments, full-scale testing, and extensive fleet tracking programs (for gathering data from planes in service). Over this same period, related but distinct groups of engineers and scientists were working on closely related fracture problems in boilers and pressure vessels—particularly important for the growing nuclear power industry. By the middle 1970s, as noted in the figure, the Air Force was ready to promulgate a military standard covering design/analysis methods for aircraft structural integrity. The Federal Aviation Administration (FAA) later adopted similar requirements as part of its airworthiness standards.

Starting about 1920, and at a greatly increasing pace since 1950, theory and application interacted in the complicated fashion summarized in the figure. R&D groups in Europe and Japan made their contributions, but most of the currently accepted methods originated in the United States, frequently under DoD sponsorship. The process of diffusion through the broader technical community continues, particularly into consumer products industries, civil engineering, and electronics (e.g., failures of IC lead wires and interconnects under thermal cycling).

Much of the technology discussed in Box 2-B stems from R&D sponsored by defense agencies. Indeed, a great many of the tools used on a routine basis in the modern practice of engineering have their origins, wholly or in part, in mission-relevant R&D funded by federal agencies. Many of these methods are practical only because of the digital computer. Examples illustrating the scope and impact of these methods:

- The most widely used of all computer-based engineering tools, finite-element-method (FEM) programs enable engineers to pare down load-carrying structural members everywhere, whereas in earlier days calculations could be undertaken only at locations thought critical. As a consequence, structures can be made lighter, consuming less material (and energy). NASTRAN (NASA

Structural Analysis), the finite-element program in most common use, stems directly from federal R&D (see Chapter 3). It is perhaps the best example of spinoff from government R&D that takes the form of technical knowledge itself.

- Computational fluid dynamics (CFD), supported and used by many federal agencies, simulates flow over airfoils—and indeed over an entire airplane—with much better accuracy than was possible 20 years ago. CFD calculations, consuming many hours of supercomputer time, are perhaps the most elaborate and expensive analytical techniques in widespread use, but they save money by cutting down on the need for wind-tunnel testing. The applications spread well beyond aerospace: for example, CFD has brought lower drag coefficients, hence greater fuel economy, for passenger cars and trucks.
- Closed-loop control systems, using digital processors, have found widespread applications in industrial process control, fly-by-wire aircraft, and now in drive-by-wire automobiles. For military airplanes, the new technology opens the door to artificial stability and high maneuverability through active control; commercial transports ride more smoothly in turbulent air and burn less fuel overcoming aerodynamic drag. Fly-by-wire technology stems from work done for both space vehicles and military aircraft.

In a rather different example, Air Force funding during the 1950s led to the development of numerically controlled machine tools and laid the foundations for the ongoing improvement of factory production methods through flexible automation (Chapter 10). Ironically, lack of incentives for cost reduction in defense production (because of procurement procedures that prevent the contractor from capturing much of the savings) meant that DoD captured less of the benefits of its investments in this area than did civilian industries, U.S. and foreign.

The private sector, of course, also contributes to the technology base, while defense contractors and civilian firms alike draw on it. They do so in ad hoc fashion over the course of design and development projects that sometimes continue for a decade or more; typically, it is impossible to predict in advance that knowledge of a particular type will prove indispensable to a successful outcome.

Given the multiuse character of so much knowledge, there is little point in trying to define a defense technology base that might be distinguished in any usefully detailed way from a commercial technology base. The terms retain a commonsense distinction: the graphite fiber technology used in golf clubs differs from that used for Air Force jets, and the more applied aspects of phased-array radars, ballistic missile guidance and control, or submarine design have little obvious relevance for commercial industry. Other parts of the technology base may be mostly the preserve of commercial firms. But even for highly specialized products, quite generic knowledge provides essential underpinnings. Equations for electromagnetic imaging are the same regardless of the type of radar system; ballistic missile guidance begins with notions of feedback control going back to James Watt's flyball governor; the laws of hydrodynamics hold for submarines and sailboats. There are simply too many disciplines feeding the technology base, too many hierarchies of analytical techniques and design methods, all in a more or less continuous state of flux, for rigorous and reliable distinctions between military and commercial technologies.

The tools of technology, then, are much the same on both sides of the economy, even though military systems differ in a multitude of respects from the familiar products of civilian industry (see Chapter 5). Leaving aside mass production, the equipment found in manufacturing plants—machine tools, forging presses, heat treating furnaces—will be similar if not identical whether the plant serves military or civilian markets. So are the computers that engineers use for design calculations and managers use to control work-in-process inventories. When output volumes are small, as they are for most military systems and in civilian capital goods industries, the parallels extend to the organization and management of production. Finally, engineers, scientists, and managers—and technicians and blue-collar workers—get their training in many of the same institutions, even though their job histories may give them quite different skills at later stages in their careers.

All this says nothing about whether the military and commercial worlds are drawing further apart from the viewpoint of end products or artifacts. In fact, this question probably has no general answer. In aircraft design, stealth technology has little to offer the civilian market directly. But fiber-reinforced composite structures, part of the stealth package, do. Beech's composite Starship aircraft, designed primarily for the corporate market, won certification before the first B-2 bomber left the ground. The knowledge going into the Starship came in large part from several decades of DoD-sponsored R&D on composite materials and composite structures: micromechanics of fiber-matrix bonding; aeroelastic behavior as a function of ply orientation; environmental degradation (epoxy absorbs moisture over time, losing stiffness). Recent case studies of fiber optics, computer software, and polymer-matrix composites show little evidence of substantial divergence between military and civilian requirements.[7] Commonality of many sorts has always existed, and probably always will.

Military and Commercial Innovation: Two Distinct Systems

As the preceding section has shown, both military and commercial projects contribute to technical knowledge through processes that are often complex, idiosyncratic, and interrelated. Yet as we pointed out in Chapter 1, the systems, institutions, and motivations for applying technical knowledge—to produce marketed goods and services, to defend the nation, or to satisfy other public and private needs—do differ, often markedly. Most technology investments are made with a particular purpose in mind, by institutions with a specific mission. In Chapter 6, for instance, we will see that most defense production is performed by business units that sell primarily to the Defense Department. We can therefore represent military and commercial innovation as proceeding via two coupled but largely distinct systems—one financed and managed by government, the other funded by and responsive to private markets.[8]

Both systems draw on the common base of technical knowledge, and as described in the following chapter, they interact in important ways. Nevertheless, the two involve different sets of institutions and in general operate quite differently—the result of differences in goals and technical requirements, as well as in the managerial arrangements accompanying defense production in particular and all government activities in general (see Chapter 5). For instance, commercial firms constantly seek to improve their products and production processes in response to market signals, a feedback mechanism having only a few parallels in military innovation. As a result, in the majority of cases, military and commercial innovation have evolved distinctive technical "cultures," as summarized schematically in Table 2-3.

To be sure, intermediate cases can easily be cited. Large, complex, and specialized capital equipment for manufacturing or for regulated service industries (e.g., telecommunications or utilities) share some of the characteristics of major weapon systems. Conversely, mass-

---

[7] Office of Technology Assessment, *Holding the Edge: Maintaining the Defense Technology Base,* vol. 2: Appendixes (Washington, DC: Office of Technology Assessment, January 1990), pp. 61–113.

[8] An *innovation* is the application of technical and/or marketing ingenuity to the creation of a new or improved product, process, or service that is successfully introduced into the market (for commercial innovation) or into the field (for defense). Relatively few *inventions* ever become successful *innovations*.

TABLE 2-3:
Two Cultures—Civil and Military Innovation

	Civil	Military
Impetus for design	Market driven, opportunistic introduction of new products	Dictated by military "requirements"
Nature of response	Rapid, incremental improvements, punctuated by more fundamental redesigns	"Big leap" improvements
Product cycle	Measured in years	Measured in decades
Priorities	Process technology for low-cost manufacturing, high quality, and flexibility	Product technology for functional performance and long shelf life
Production	High rates and volumes (in consumer product industries)	Low production rates and unit volumes
Linkage of R&D and production	Integrated management of R&D, production, and customer service	R&D and production separately contracted and sequential
Technology sharing	Success based on proprietary technological advantage	Success may require sharing know-how with second-source contractor

produced weapons such as ordnance share some characteristics with mass-produced civilian goods. Nevertheless, notwithstanding the occasional similarities and interactions between the two systems, opportunities for synergy are limited. For the same reason, government technology policies—to which we turn next—will not automatically or efficiently promote commercial innovation if they rely too heavily on defense as a vehicle.

U.S. TECHNOLOGY POLICY

In the United States, government investments in R&D constitute nearly half the nation's total R&D spending. This contribution stocks the technical knowledge base in ways that cannot avoid influencing commercial innovation as well. Governments in all the advanced industrialized economies also pursue policies intended to stimulate the application of technologies resident in the base. They do so—some governments more than that of the United States—by encouraging the transfer of scientific knowledge from university or government laboratories to industry, by helping companies make use of best-practice know-how, and by establishing technical standards such as those covering screw threads or computer languages. Many policy tools, of course, seek to encourage both the generation of knowledge and its application. One of the earliest examples of conscious technology policy is the granting of patents on condition that the new knowledge be made publicly available.

What Is Technology Policy?

U.S. technology policy has never been very clearly defined. In contrast to national security policy—the subject of extensive debate involving parties inside and outside government—technology policy has normally been a derivative category, shaped by decisions made on other grounds. Year-by-year, case-by-case budgetary decisions by DoD, NASA, and other mission agencies, by the Office of Management and Budget, and by Congress drive much of U.S. technology policy.

Superimposed on the annual budgetary decisions have been a series of larger issues for which technology has been primarily a means to an end: the cold war, the U.S. venture into space, medical research, energy, agriculture, and so forth. Only for basic research—about 17 percent of federal R&D funding in 1990—has the output of R&D been viewed as good in itself, rather than good because it helps solve a specific problem such as disease or because it deals with some policy concern such as whether the Soviet Union would beat the United States to the moon.[9]

One result of the mission focus is that federal agencies have never paid much attention to industrial technologies other than those employed by defense contractors. Rather, the view has been that private firms should commercialize the results of scientific research and

---

[9]Federal agencies have had a policy of contributing to the national stock of technology and trained scientists and engineers as a kind of "central overhead" charged to each mission. Harvey Brooks, "Can Science Be Planned?" in Harvey Brooks, *The Government of Science* (Cambridge, MA: MIT Press, 1968), pp. 54–80. Implicit in the funding of universities by the Office of Naval Research immediately after World War II, such a policy was made explicit by President Eisenhower's Executive Order 10521, March 17, 1954, which states that "The conduct and support by other Federal agencies [besides the NSF] of basic research in areas which are closely related to their missions is recognized as important and desirable especially in response to current national needs, and shall continue." Quoted in Bruce L.R. Smith, *American Science Policy Since World War II* (Washington, DC: Brookings Institution, 1990), p. 199, n. 18.

mission-oriented R&D on their own, and that government should encourage this, but not play a direct role. Most Americans have believed that the proper role for government is a limited one: paying for basic research and mission-oriented R&D, and creating a climate conducive to entrepreneurship and industrial innovation through macroeconomic and regulatory policies (e.g., antitrust). Other acceptable indirect measures range from protection for intellectual property, strong in the United States for two centuries, to such relatively recent initiatives as the R&D tax credits in place since 1981, and the National Cooperative Research Act, passed in 1984 to signal industry that the Justice Department would take a tolerant view of collaborative R&D.

Government policies affecting commercial technology can be divided into four categories, as shown in Table 2-4, of which the first three arouse little debate. The fourth category in Table 2-4—direct government actions to promote the technological competence of commercial industry—has been viewed as out of bounds in the United States (although not in other countries).[10] Only with growing concern over the competitiveness of U.S. industries—and in particular, the startling declines in such high-technology sectors as microelectronics—have departures from the traditional view of an appropriate technology policy been seriously considered. From the 1950s into the 1970s, technology seemed to many observers to be a secondary factor—compared, say, to labor costs or to alleged unfair trade practices—in the competitive decline of traditional industries like steel and textiles. But during the 1980s, as U.S. high-tech industries also began to slip, technology moved to the forefront of the competitiveness debate.

Technology Policy and Industrial Policy

Before technology policy moved to the forefront in the mid- to late 1980s, "industrial policy"—a much broader and more contentious notion—was the subject of a lengthy and politicized debate. Industrial policy has often been taken to mean a portfolio of sector-specific measures such as subsidies or import restrictions that give one industry advantages over others, domestic as well as foreign. Many such "tar-

---

[10] For general discussion of government technology policies, see Harvey Brooks, "Towards an Efficient Public Technology Policy: Criteria and Evidence," in Herbert Giersch, ed., *Emerging Technologies: Consequences for Economic Growth, Structural Change, and Employment* (Tubingen: Institut fur Weltwirtschaft an der Universistat Kiel, 1982), pp. 329–378; Richard R. Nelson, "Government Stimulus of Technological Progress: Lessons from American History," in Richard Nelson, ed., *Governmental and Technical Progress: A Cross-Industry Analysis* (New York: Pergamon, 1982), Chapter 9, pp. 451–482.

TABLE 2-4:
Federal Government Policies Affecting Commercial Technology

Modes of government involvement agreed to be appropriate:
- basic research
- science and engineering education
- public missions: defense, health, space, energy, environment, agriculture, agricultural extension, and so forth
- public infrastructure: roads, airports and air traffic control, launch vehicles for communications satellites, and so forth

Modes of government involvement agreed to be inappropriate:
- commercial product development
- short term, moderate risk research with benefits to identifiable firms

Modes of unavoidable government involvement:
- antitrust
- tax policy
- intellectual property protection
- health, occupational safety, and environmental regulation
- trade policies, including national security export controls

Modes under dispute and discussion:
- direct investment in commercially relevant technology[a]
  —pathbreaking technology
  —infrastructural technology
  —strategic technology
- diffusion of technology for economic benefit, industrial extension
- government-industry partnerships
- international partnerships

---

[a] We define and describe these three categories of investment in Chapter 12.

geting" measures have little or nothing to do with technology. As policy tools for rescuing industries threatened by international competition, sectoral measures have been attacked in the United States as likely to exact high economic costs—both directly, as measured by prices paid by consumers, and indirectly, through inefficiencies in allocation of such resources as investment capital.[11] To the opponents of such measures, U.S. competitive strength has been built on free, hence efficient, markets. The cures for U.S. economic ills are to be found in

---

[11] Such measures are not absent in the United States. But they are typically justified in the name of national security (e.g., "Buy American" provisions to protect industries deemed critical to defense) or economic adjustment (aid for firms, workers, or localities harmed by sharp increases in imports). In other cases, the political process has led to a tradition of public support for certain economic sectors—agriculture being the paramount example. Many American policymakers would prefer to view such measures as exceptions to the general U.S. predisposition against sectoral policies, justified by special circumstances of one sort or another, but not part of an overall industrial strategy.

controlling inflation and federal spending, changes in the tax code, or an appropriate value for the dollar against foreign currencies, not in policies that subsidize some economic sectors at the expense of others.

In this view, shared by recent administrations, indirect incentives for private industry should suffice for commercialization of results from basic science and mission-oriented R&D. Accordingly, government involvement in commercially relevant technology beyond basic research has often been seen as a subsidy—unjustified and likely to create harmful economic distortions, as well as to violate the principles of multilateral trade that the United States has consistently espoused in international fora such as the General Agreements on Tariffs and Trade (GATT).[12]

But technology policies need not be viewed in this light. Public intervention for the creation and diffusion of industrially relevant technology can be considered more legitimate than other possible targets of public policy because technology is in part a public good, and government subsidization of industrial technology provides benefits to many firms. It is unlike most other forms of intervention, such as trade protection or direct subsidy, where the game becomes zero-sum in that one industry's or country's advantage becomes another's disadvantage.

Technology policy and the industrial policy debate. In the latter part of the 1980s, technology entered the industrial policy debate as a means to an end—in this case, the end being productivity and international competitiveness. Thus Congress, the administration, and representatives of the private sector argued, among other issues, the wisdom of federal R&D funding as a support for commercial industry. Sematech became the most visible case in point (see Chapter 8).

Sematech differs from other federal policies with similar intent— creating advantages for American firms—in that the instrument is money for R&D, rather than import restrictions (put in place to help U.S. machine tool firms), pressure for market-opening steps abroad (the U.S.-Japan Semiconductor Arrangement), or international negotiations intended to secure stronger protection for U.S. intellectual property (e.g., computer software). Still, it is not really much of a departure. Precursors can be traced back at least to the 1901 founding of the National Bureau of Standards (NBS), now the National Institute of Standards and Technology (NIST).[13] Established in part as a response to

---

[12] For alternative views, stressing dynamics, see the literature on strategic trade theory, as summarized, for example, in Paul R. Krugman, "Is Free Trade Passe?," *Journal of Economic Perspectives*, vol. 1, no. 2 (1987), pp. 131–144.

[13] According to its charter, the bureau was to determine "physical constants and the properties of materials, when such data are of great importance to scientific or manufacturing interests and are not to be obtained of sufficient accuracy elsewhere." Quoted in A. Hunter

the creation of similar laboratories in Europe, NBS, like the European laboratories, was intended to create advantages for national industry through the establishment of standards that could serve as barriers against imports.[14]

In a similar vein, NASA's forerunner, the National Advisory Committee for Aeronautics, was established in 1915 to conduct research and testing on aircraft. NACA pursued R&D on behalf of the civilian aircraft industry and in support of military aviation, where World War I found the United States behind.[15] While NACA developed notable expertise in wind-tunnel testing and could boast successes in engine cowling development and low-drag airfoils, it eventually devolved into an organization content to help meet the short-term needs of aircraft companies for design information. Conservative leadership, the product of years of warding off attacks from opponents in both government (often on grounds of waste) and the private sector (including allegations that NACA helped insulate the major U.S. aircraft manufacturers from competition), helped seal NACA's fate as an independent agency.

In effect, NACA's history proves the rule: the United States has not succeeded in developing a durable national consensus around institutional mechanisms for supporting commercial (or explicitly dual-use) technology development. The troubled history of energy R&D—not only civilian nuclear power, but also solar power, fossil fuels, and synthetic fuels—reinforces the point.

The scope of technology policy. Any comprehensive view of technology policy must see it as encompassing the entire range of government measures that affect knowledge generation, diffusion, and use—from "pure" research, to fellowships for engineering students, to civil works projects (which, like procurement of defense matériel, lead to learning through experience). In other countries, governments often fund commercially oriented R&D as one ingredient in technology (or industrial) policies. Examples range from subsidies for development of the successive generations of aircraft introduced by Europe's Airbus Industrie, to visionary R&D on fifth- and now sixth-generation computer sys-

---

Dupree, *Science in the Federal Government: A History of Policies and Activities to 1940* (Cambridge, MA: Belknap Press/Harvard University Press, 1957), pp. 272–273.

[14] For example, different voltages and hundreds of different designs for electrical plugs and sockets are used around the world. National standards-setting bodies often exclude foreign firms, with the result that appliance manufacturers typically have a small cost advantage in their home market. Despite many years of international discussions, there has been no progress toward agreement on a universal standard. J. Callcott, "A World-Wide Plug Faces Disconnection After 74-Year Effort," *The Wall Street Journal*, April 1, 1982, p. 1.

[15] Alex Roland, *Model Research: The National Advisory Committee for Aeronautics, 1915–1958*, vols. 1 and 2, NASA SP-4103 (Washington, DC: National Aeronautics and Space Administration, 1985).

tems financed by Japan's Ministry of International Trade and Industry. But government funding of commercial technologies—like industrial policy itself—has never been considered legitimate in the United States. Nor have U.S. policymakers viewed R&D spending by the mission agencies as a conscious tool for leveraging the productivity and competitiveness of U.S. industry, their optimism about the value of spinoff notwithstanding.

Nevertheless, federal expenditures over many decades have supported not only R&D, but also the infrastructure for technology, by paying directly and indirectly for the development of scientific apparatus and equipment, along with techniques for using them. Wind tunnels exemplify the experimental facilities, often elaborate and expensive, used to generate and verify technical knowledge. The computer itself, now indispensable for both experimental work and for design calculations, is perhaps the best example of how federal expenditures for both R&D and procurement can stimulate engineering tools and generic technologies. Government support of graduate education in engineering and science, primarily through research funding, is vitally important for diffusion as well as knowledge generation. Technology policy encompasses all these activities and more.

R&D Is Not Enough

It may seem paradoxical that a nation can spend as much on R&D as the United States and still suffer from technological inadequacies. The first problem is one of allocation: in many fields of technology and science neither the government nor industry has spent enough, particularly in fields that are less than glamorous and not of overriding importance for agency missions. Manufacturing processes, to take an obvious example, suffer on both counts. Few physicists or chemists put their skills to work attacking problems of metal flow in forging or solidification in casting. The practical nature of these problems limits their attraction for university researchers, who tend to see in them little scope for advancing theoretical and conceptual frontiers. Because DoD buys mostly complex systems in small numbers, many essentially handmade, it does not have strong incentives to support manufacturing R&D. Industry underinvests because the work is generic: rarely can a single company expect to reap significant advantages over its rivals. (Proprietary processes, especially those that can be protected by patent coverage, are exceptions.)

Manufacturing processes are hardly alone: the argument can be made that the United States invests too little in cases that range from chemistry and mathematics to technologies that are important for the public infrastructure but lack well-placed advocates within government (e.g., highway durability). Nor have indirect measures stimulated enough private spending to fill the gaps in federal support. With the exception of the corporate laboratories of a few leading multinational firms, American industry, under pressure to produce short-term profits, pays for little R&D with intermediate or longer time horizons.

R&D spending by industry, moreover, lags when compared with our major international competitors, Japan and Germany (measured as a fraction of corporate revenues or gross national product). Currently, the United States spends about 1.9 percent of GNP on civilian (i.e., nondefense) R&D; Japan spends 2.6 percent of GNP, and Germany 2.7 percent (see Chapter 4). Three points concerning such comparisons need mentioning. First, in all three countries, most of the money for civilian R&D comes from industry, not from government. Neither the Japanese nor the German government heavily subsidizes industrial R&D. This does not mean that government spending for civilian R&D cannot have very substantial impacts; it can, particularly when it is channeled to projects where private firms cannot expect to capture high rewards, but where paybacks to the economy as a whole might be large. This is, of course, the familiar justification for government funding of basic research, but it applies much more broadly.

Second, inputs to R&D—money spent, number of engineers and scientists at work—tell nothing about the outputs of R&D. Those outputs are hard to measure, as every research manager knows. Third, although neither Japan's nor Germany's government spends lavishly on commercial R&D, both countries have well-developed support systems for commercial technology, while the United States does not. For example, Japan's fifth-generation computer project and its earlier Very-Large-Scale Integration (VLSI) microelectronics project were both intended to diffuse technology as well as develop it. Germany relies on a broad range of measures that include extensive public support for technical education and training, exemplified by its apprenticeship system.[16]

In sharp contrast to these two countries, industrial technology has never had a strong advocate in the U.S. government; indeed, commercial considerations have historically been almost invisible in U.S. technology policy. When the United States was broadly superior to other

---

[16] The German "dual-track" system of training for skilled workers, along with the training/apprenticeship systems of several other countries, is described in William E. Nothdurft, *Schoolworks: Reinventing Public Schools to Create the Workforce of the Future* (Washington, DC: Brookings Institution, 1989).

nations in technology, neither the gaps in the nation's knowledge base nor the time lags associated with diffusion from defense to commercial sectors did much immediate harm. But today we have a system that is still geared to lead, rather than to race neck and neck with highly capable rivals. Competitive advantages now and in the future will be fleeting without continuous incremental innovation in products, processes, and management techniques.

The new environment poses particular difficulties for the United States. Competitive outcomes hinge on organizational learning. Time is critical, with individual companies working to shave months or even weeks off product development cycles. But the U.S. economy is huge and diverse. It takes a long time for knowledge to make its way to industries that do little R&D of their own, or to smaller firms in any industry, particularly because mechanisms for technology diffusion are poorly developed compared with those in countries like Germany or Japan. It will take several decades to make substantial improvements in the average quality of a labor force exceeding 125 million people, even if Congress, the administration, and the states and localities that control the public schools agree on measures for improved education and training—including not only formal schooling, but also retraining and continuing education of the existing workforce. Moreover, the U.S. policymaking system is decentralized, slow moving, and reactive, with Congress and the executive debating major decisions for years.

Needed: New Policies

By concentrating on mission-relevant R&D plus basic science, and relying on indirect incentives for commercialization (some of these of debatable effectiveness—e.g., tax incentives), U.S. government policies contribute to underinvestments in technology of three types. First, the United States is increasingly hesitant to use public funds to explore major, long-term opportunities of the sort that led in earlier years to entirely new industries such as computers. Cuts in defense spending may greatly aggravate this problem. Second, the nation spends too little on R&D relevant to industrial needs, such as manufacturing processes. Third, and perhaps most serious, the United States underinvests in the diffusion and application of both new and existing knowledge.

Washington is not unaware of these problems, and agreement seems to be growing that the federal government should do more to address them. Yet there is no consensus on how to proceed, or even the grounds for debate. One step in the right direction was the establishment within the National Institutes of Standards and Technology of new programs with an explicit focus on the generation and diffusion of commercially relevant technology, as called for in the 1988 Omnibus Trade and Competitiveness Act. Another was the Bush administration's September 1990 statement on technology policy—a document that could be seen as preparing the way, or perhaps as testing the waters, for greater governmental emphasis on the needs of the U.S. commercial sector.[17] In addition to efforts to ensure a high-quality technical workforce and the dissemination of best practices to help U.S. firms remain competitive, the document endorses federal participation "with the private sector in precompetitive research on generic, enabling technologies that have the potential to contribute to a broad range of government and commercial applications."

Such initiatives pose few of the risks that make industrial policy contentious. Just as the government supports basic research, it may need to bear part of the burden of funding industrially relevant R&D that has a "public good" nature, and to sponsor pathbreaking technologies that will pay off only in the long run. In both cases, private rewards may be small, difficult to appropriate, uncertain, or many years off, but the social returns on the investments may be large. Furthermore, public investments in the diffusion of existing knowledge arouse few of the political inhibitions that attach to direct subsidy of industrial R&D for commercial purposes. We return to these points in Chapter 12.

---

[17] Executive Office of the President, Office of Science and Technology Policy, *U.S. Technology Policy*, Washington, DC, September 26, 1990.

3

Military-Commercial Technology Linkages

Public and private actions in research, development, production, and acquisition interact in many different ways to influence the success of both government and commercial innovations. This chapter explores the nature of these technological linkages. It begins with an examination of the traditional idea of spinoff: thought to be an automatic process by which the civil economy benefits, at little incremental expense, from the government's large R&D investments. However, although spinoffs from government R&D can indeed be identified, they do not happen easily, automatically, or without cost. Nor are they typical of government-commercial technology linkages.

After presenting a more realistic picture of spinoff, this chapter describes a variety of other interactions between military and commercial technical activities. Many of these linkages do not result from direct government funding of R&D at all. Purchases of high-technology products agencies need for their own use, for example, have had substantial impact. So, too, have government investments in infrastructure and in the development of technical tools. Nevertheless, the U.S. government has rarely sought to influence commercial technology directly through such means.

What it has done is seek economic benefits from expenditures made in the course of fulfilling preexisting missions. The last part of this chapter, therefore, discusses technology transfer policies governing the rights to federally developed or federally funded R&D results. These policies have been evolving rapidly over the past dozen years toward promoting private sector commercialization—even the granting of exclusive licenses—instead of insisting that publicly financed technologies remain in the public domain, where they typically had gone unexploited.

THE MYTH OF SPINOFF

A spinoff is the commercial application of a product or technology originally developed for a particular government mission. It can refer to commercial products stemming from government R&D—a *vertical* transfer of technology—and also to *horizontal* transfers in which military devices, tools, or technology are adopted in new civil applications.[1] In both cases, the implication is that technical effort on the part of government is responsible for civilian applications.

Government agencies—particularly those such as the National Aeronautics and Space Administration that are continually pressured to justify their activities—tout the spinoff value of their investments in sometimes quite extravagant claims. NASA actively pursues its congressional mandate to "promote expansion of spinoff in the public interest."[2] It publishes an annual report called *Spinoff* to chronicle each year's transfer of NASA technology to other sectors of the economy, and sponsors touring exhibits at science and technology museums that feature examples drawn from the "more than 30,000 items, including Velcro and the Jarvik artificial heart . . . developed from technology associated with the nation's space program."[3]

Department of Defense officials, too, are quick to remind the Congress of the indirect economic benefits of defense technical activity. However, DoD's mission—if not its budget—is unchallenged, and with the exception of the controversial Strategic Defense Initiative Organization, DoD devotes less attention than NASA to claiming credit for and advertising its spinoffs.

However, there have always been skeptics of spinoff. In the late

---

[1] "Horizontal" and "vertical" technology transfers are distinguished by Harvey Brooks in "National Science Policy and Technology Transfer," Chapter 10 of *The Government of Science* (Cambridge, MA: MIT Press, 1968), pp. 254–278.

[2] National Aeronautics and Space Administration, Technology Transfer Division, Office of Space and Terrestrial Applications, *Spinoff 1980* (Washington, DC: U.S. Government Printing Office [GPO], April 1980), p. 3.

[3] Lindsey Tanner, "Astronaut Donates Outer Space Flag to Unusual Exhibit," Associated Press, June 23, 1989.

1950s, annoyed by exaggerated claims for spinoff benefits from Atomic Energy Commission research, Dr. Ralph Lapp called it "drip-off" to convey his opinion that very little momentum accompanied the transfer of defense technology to commercial companies.[4] Samuel Doctors concluded that NASA's Technology Utilization Program "was founded primarily in response to political pressures and has continued to be used as a device for partial justification of NASA R&D funding, rather than as a technical project in its own right."[5] And popular examples of spinoff products that are cited in press reports as deriving from space technology—Velcro, Tang, and Teflon—are spurious (see Box 3-A). Although marketing of each of these products benefited from favorable NASA publicity, their technical origins were not military- or space-related.

Serendipitous spinoff, indeed, is not a typical form of commercial innovation. Nevertheless, although less common than claimed by spinoff advocates, there are many valid examples of products (and more frequently components, technical tools, and materials) developed for space or defense use that have found their way into commercial markets. It is instructive to examine one such case in detail: Raytheon's household microwave oven.

A Classic Case of Spinoff: The Raytheon Radarange

The name Raytheon gave its microwave oven, "Radarange," tells its technological ancestry.[6] But its evolution from military to commercial product was neither automatic nor cost free. This example—successful though it was—illustrates an important shortcoming of promoting spinoff as a cost-effective way to stimulate commercial activity: no one would suggest that the nation should spend billions on the development of military radars in order to get a head start in the world microwave oven market. Even if spinoff were an effective diffusion mechanism, the value of efforts to promote it must be subject to scrutiny.

In 1946, Raytheon emerged from World War II as a 15,000-

[4]Ralph Lapp used this metaphor in testimony opposing Atomic Energy Commission activities in the late 1950s (interview with LMB, Spring 1989).

[5]Samuel Doctors, *The Role of Federal Agencies in Technology Transfer* (Cambridge, MA: MIT Press, 1969), p. 161.

[6]The Radarange story is told in Otto J. Scott, *The Creative Ordeal: The Story of Raytheon* (New York: Atheneum, 1974). Additional information was provided in interviews with Raytheon's chief executive officer, Tom Phillips (HB and LMB, 1988) and with D. Brainerd Holmes (LMB, May 1990), who joined Raytheon as senior vice president in 1963 and became president in 1975.

BOX 3-A: "SPINOFFS: A MYTH IS BORN"

Many newspaper accounts have attributed Teflon, Velcro, and Tang to the space program, no doubt reflecting a belief widely held among their readership.[1] However, on the twentieth anniversary of the first moon landing, a *Boston Globe* article set the record straight.

What do Tang, Velcro and Teflon have in common? They all were NOT spinoffs of the Apollo space program.

In the public mind all three are inextricably linked to the moon landing program, but all were developed long before the National Aeronautics and Space Administration ever heard of them.

It's true that the astronauts did wear space suits coated with Teflon and some drank powdered Tang orange drink while sitting in spacecraft that had patches of Velcro on the walls. But the belief that the three products were fruits of the space program is a testament to the myth-making power of Madison Avenue, not to the technology-driving power of the race' to the moon.

"It was a myth that the public bought," chuckles Thomas O. Paine, NASA administrator between 1968 and 1970, who oversaw seven Apollo missions. What Apollo did, said Paine, was to popularize Velcro and Teflon and their less-known innovations by finding new uses for them. That is not to say that Apollo did not have an enormous impact on technology. In fact, Paine said in a telephone interview, the hallmark of the program was encouraging scientists and engineers to explore ideas at the forefront of technology. Over time, many of those ideas turned into industrial and consumer products. . . .

Putting a dollar value on the spinoffs from the Apollo program is almost impossible, but NASA officials have tried. They claim that direct benefits alone—sales of items transferred directly to the private sector—total well over $1 billion a year. More than 30,000 secondary applications of space technology have emerged in the past three decades as well, said James T. Rose, NASA's assistant administrator for commercial programs.

But Tang, Velcro and Teflon were not among them.

The breakfast drink was developed in 1957, 12 years before the first lunar landing. In 1965, Tang was listed as one of the optional drink selections for the Gemini 4 astronauts. . . .

Velcro dates to the late 1940s when George de Maestral, a Swiss researcher, studied the cocklebur plant under a microscope to see what made it so adhesive. Based on what he saw, he developed fiber patches containing tiny hooks or eyelets that clung to each other when pressed together. The astronauts used them to hold small tools in place in a weightless environment. . . .

And Teflon, the coating used to make non-stick cookware, can be traced back to 1938. That's when DuPont researcher Roy J. Plunkett noticed that

[1]"The spinoffs from the space program are legion, and they extend far beyond Teflon, Velcro and Tang," wrote Lee Dembart, in "Worth the Expense? A New Era Dawning for U.S. in Space," *Los Angeles Times*, April 21, 1985, p. 1. "Teflon and 'Tang,' the orange powdered drink], are the classic examples [of technology spinoffs from the U.S. space program]"; Louise Kehoe, "How Down to Earth Research Has Gained Ground At NASA," *Financial Times*, November 10, 1987, p. 34.

certain combinations of fluorocarbons in his refrigerant gas experiments produced an inert waxlike substance.

It wasn't until 1956 that a French company sold the first Teflon-coated frying pans. Five years later, just as Kennedy was declaring that the United States would go to the moon, they were making their way into American households.[2]

Note that even if the $1 billion estimate of annual direct benefits claimed by NASA is correct, it is not a net benefit. See discussion of evaluation of spinoff claims, later in this chapter.

[2]Ronald Rosenberg, "Spinoffs: A Myth Is Born," *Boston Globe*, July 17, 1989, p. 21. Reprinted courtesy of The Boston Globe.

employee defense contractor, having grown from a small Massachusetts vacuum tube manufacturer. That year Washington canceled $150 million of $200 million in outstanding orders. Management laid off thousands and turned its attention to diversification into the civilian market.

One day late in 1945, Percy Spencer, a Raytheon engineer of unusual ability (138 patents), placed a bag of popcorn in front of the waveguide driven by the magnetron from a military radar set. The kernels began to pop inside the bag. The following day, Spencer exploded an egg over a group of assembled onlookers, who were nevertheless impressed. Laurence Marshall, then Raytheon's president, became very enthusiastic about the possibility of a microwave oven for restaurants. Many problems, such as radiation safety, had to be solved. But by 1947, Raytheon introduced a restaurant oven. The home appliance market promised to be far larger, but Raytheon managers knew they would have to reduce costs greatly to sell the Radarange even as a luxury product, much less a mass-market item. They would also need distribution and marketing channels.

In 1965, Raytheon undertook development of a much smaller, less expensive Radarange for home use. In the same year, by serendipitous good fortune, Raytheon merged with the Amana Refrigeration Corporation, which solved the problem of consumer product design knowledge and provided a ready-made sales and distribution network. The home oven could not meet its cost target unless Raytheon could manufacture the magnetron for $35 each, whereas the actual cost when manufacturing began was $125. Tom Phillips, Raytheon's president at the time, agreed to a corporate subsidy of the $90 difference for one year. In effect, the corporation subsidized the early technological learning so it could introduce its home Radarange at a realistic price. In 1967, Raytheon launched the product with great fanfare at a Chicago home appliance show, and sales exceeded the forecast of 50,000 units a year. The microwave oven was a great commercial success. By 1970, the year inventor Percy Spencer died at the age of 76, commercial businesses accounted for half of Raytheon's revenue.

Raytheon pursued two additional spinoffs from military radar. One, a microwave diathermy machine that used the radar's magnetron to generate heat inside the body for medical purposes, was never very successful, and the business was later sold. The other, Raytheon's commercial marine navigational radar, has been a good and steady business for many years.

Consider these elements of the Radarange case:

1. Percy Spencer made fundamental contributions to both the magnetron and the microwave oven. He was the primary inventor of both the military product and its commercial spinoff. Thus the primary "transfer" of the technological idea occurred within the brain of one man.

2. The product had an early executive champion with enthusiasm for its prospects—Laurence K. Marshall, president of Raytheon from its formation in 1928 through 1948. His successors, especially Tom Phillips, shared the enthusiasm and showed uncommon patience. (Twenty years were to pass between the launching of the first restaurant Radarange and the first home appliance.)

3. The company positioned the first Radarange models as industrial products, with high costs and low volumes. However, it recognized the possibility of a much larger market if it could bring costs down. Raytheon then undertook this second wave of innovation itself.

4. The company realized that it did not have the right skills either for product design or for advertising, sales, and distribution. Raytheon did not purchase Amana for the purpose of exploiting the Radarange, but the acquisition occurred at the right time to provide the necessary capabilities.

5. Management took the risk of introducing an initially unprofitable product, anticipating and achieving a fourfold reduction in manufacturing cost in little more than a year.

6. Raytheon licensed the microwave oven technology to Japanese companies, but was able to compete against them effectively throughout the product's years of most rapid growth.

In this case, as in others, investment, good management, and patience were needed to commercialize a defense technology. This recognition clashes with the political appeal of spinoff as a straightforward, low-cost, serendipitous process.

Spinoff as a "Free," Nontargeted Economic Benefit

Washington has sometimes fed expectations that military technology will yield spinoff dividends in order to sugar-coat the burden on taxpayers of huge defense budgets. At the same time, spinoff serves for many as an acceptable surrogate for a national technology policy. Compared to direct government investments in industrial technology, spinoff seems to have two important virtues. First, it appears to cost nothing, as long as the original defense investments were justified entirely on national security grounds. Second, spinoff appears not to need managing, targeting, or any other form of "interference" by government in the marketplace, since the underlying defense programs' technical content and goals are governed only by their national security objectives. Private firms may choose to take advantage of the "windfall" technology, or they may ignore it. It is politically acceptable for them to benefit from spinoff because the government takes no responsibility for selecting the beneficiary.

The attraction of spinoff as a free, nontargeted alternative to technology policy follows from the belief that no direct government effort is needed to ensure its success. This may explain why, except for agriculture, relatively little public effort has been devoted to measuring or enhancing the effectiveness of spinoff.[7] Until recently, federal technology transfer policies were limited to the offer of information to the private sector, without any explicit program to facilitate transfer. But despite the myth, spinoff is not free. Companies seeking to exploit the results of technology developed through government R&D and procurement must usually modify and adapt it for commercial use.

---

[7] Doctors, *The Role of Federal Agencies in Technology Transfer*, p. 21. In 1960, the Department of Agriculture spent almost as much money on disseminating the fruits of agricultural research to farmers ($142 million for the Cooperative Extension Service) as it did in generating new knowledge in the first place ($167.5 million in federal funding for agricultural research; this was, however, matched by another $112 million in state funds). For the rest of the government, however, the technology transfer component was far smaller. Out of a total federal R&D budget of $16.2 billion in 1967, expenditures on collection, evaluation, and transfer of nonagricultural scientific and technical information came to less than $25 million, or about 0.15 percent of the funds spent acquiring information. The NASA Office of Technology Utilization spent $7.5 million of this (about 0.1 percent of the NASA budget).

These adaptation costs will be smaller if military and civil versions of a product are developed by the same team and manufactured in the same (or similar) production facilities. But this is rarely the case, as we explain in Chapter 5. The majority of defense prime contractors keep their government and civil business units at arm's length. Thus the incremental costs for spinoff within these firms may be almost as high as between separate defense and civil firms.

The political benefits of spinoff are therefore offset by four factors. First, the substantial separation of the defense and commercial sectors of U.S. industry limits opportunities and raises the costs of spinoff. Second, proper accounting of the investments required to adapt a defense innovation to commercial use belies the notion of spinoff as a cost-free benefit of government technical effort. Third, in refusing to target technology investments based on their commercial importance or potential, a policy depending on spinoff ends up, by default, addressing only those technologies relevant to the mission needs of federal agencies— primarily aerospace and electronics in the case of defense (see Chapter 4, especially Table 4-8, and Chapter 6). And finally, to the extent that commercial spinoffs are seen as part of the Defense Department's rationale for funding new technology investments, the efficiency of such spinoff investments in producing commercial technology needs to be compared to alternative types of federal investment. We now turn to this comparison.

ASSESSING THE BENEFITS OF SPINOFF: COMPARED TO WHAT?

The net benefits of spinoffs from defense investments can be addressed only through the counterfactual. How might the original investment of public funds and technical talent otherwise have been spent? The analysis must ask, "Compared to what?" At least nine different standards of comparison are used in assessments of the indirect effects of defense R&D on the commercial economy, and in polemics for and against defense technology spending. But rarely are the standards explicitly stated. The "efficiency" of spinoff appears very different when judged by the different standards summarized in the following paragraphs.

1. *A comparable dollar volume of private sector R&D.* This comparison is made most frequently by critics of the defense effort, who compare the economic value of a dollar of defense R&D to the economic

value of a dollar's worth of R&D performed by commercial companies with their own funds. It is not surprising that the defense dollar comes up short. But this comparison is hardly relevant in the absence of a policy that would result in the shift of a defense dollar to a dollar of private R&D.

One way to approximate such a policy would be to cut defense R&D and simultaneously extend tax credits to private companies to increase their own R&D investment in proportion. Through R&D tax credits such as those offered in current law, the government in effect shares some of the cost of private R&D while leaving the firm free to choose which projects will be funded. However, this does not mean that government can count on tax credits to boost private R&D.[8]

2. *No research at all.*   This standard of comparison is the favorite of spinoff proponents. They display all the commercial artifacts that, they imply, would never exist if not for the programs they are championing. NASA is notorious for statements of this kind, as is the Strategic Defense Initiative Organization (SDIO). Since the implied alternative to the R&D in question is no R&D at all—that is, the standard against which spinoff is judged is no spinoff—the results are self-evident. They are also not helpful for policy purposes.

3. *Defense procurement.*   Champions of spinoff also like to claim that if the money were not spent for defense R&D, it would be spent to buy more hardware—one more bomber or aircraft carrier. This is the "bigger pie" argument, holding that defense R&D does not displace commercial R&D, but rather it displaces non-R&D and thus makes for a bigger overall pie of science and technology than would otherwise exist. This argument does not address the relative efficiency of defense R&D when compared to other R&D in achieving commercial objectives, nor does it take account of the fact that procurement contracts also have significant effects on commercial technology.

4. *Different defense R&D investments.*   This approach contrasts current defense R&D spending with that resulting from some hypothesized policy change: for example, shifting resources from the development of specific military systems to technology base activities. Since one kind of defense R&D is being compared to another, the conceptual leap between the two cases is reasonable. It should be possible to predict

---

[8] Tax credits are notoriously difficult to target. The research and experimentation tax credit introduced as part of the Economic Recovery Tax Act of 1981, and modified several times since, proved no exception. Analyses have not shown conclusively that the credits have raised corporate R&D spending substantially. See, for example, Edwin Mansfield, "The R&D Tax Credit and Other Technology Policy Issues," *AEA Papers and Proceedings,* vol. 76 (May 1986), p. 190.

the consequences of such a shift. Such a policy change would also be among the easiest to implement.

5. *R&D in other federal agencies.*   How does defense R&D influence economic performance compared to R&D in agencies such as the National Institutes of Health, the National Science Foundation, or the National Institute of Standards and Technology? This question is important when policymakers must decide which agency should support technologies, such as semiconductors, advanced materials, and super-conductors, that have direct relevance to both government missions and commercial applications. The answers depend on institutional and managerial capabilities, which differ widely among agencies.

6. *Basic research support.*   Basic research in nondefense agencies is the alternative advocated by those who believe government overemphasizes mission-oriented R&D at the expense of investigator-initiated technical activity with no externally imposed agenda. However, funds for basic research at NSF and NIH are found in an entirely different part of the federal budget from DoD R&D, and they are affected by different agency procedures, interest groups, and Congressional committees. Changes in DoD R&D funding levels are not simply translated into corresponding changes in nondefense R&D.

7. *Federally funded civilian technology initiatives.*   From time to time, the government has sponsored programs intended to contribute directly to the economic performance of a particular industry. Examples include the Civilian Industrial Technology Program of the Kennedy administration, and the Carter administration's proposed Cooperative Automotive Research Program.[9] These programs generally appear to have failed, but the causes are various: indifference (or even hostility) on the part of the industry being "helped," a poor technical strategy, structural problems in the industry, or a hostile political environment.

8. *Other government expenditures.*   Another variant of the bigger-pie argument maintains that federal dollars not spent on defense R&D would end up diffused among other federal spending programs that contribute much less to economic performance.

9. *Uncollected taxes.*   Those of both liberal and conservative bent sometimes suggest that the money would have a greater ultimate effect on economic performance if left in the pockets of citizens and firms, rather than spent on defense R&D.

[9]Lewis M. Branscomb, "Opportunities for Cooperation Between Government, Industry and the University," *Annals of the New York Academy of Sciences*, vol. 334 (December 14, 1979), pp. 221–227. For a close examination of an early case, see Dorothy Nelkin, *The Politics of Housing Innovation* (Ithaca, NY: Cornell University Press, 1971). For an analysis of large commercialization demonstrations, see Linda R. Cohen and Roger Noll, *The Technology Pork Barrel* (Washington, DC: Brookings Institution, 1991).

Which of the above standards of comparison is the "right" one? It is the one that matches the public policy alternative being proposed. Analysis and policy must go together: blanket assessments of the "value" of spinoff, pro or con, are rarely relevant to actual policy choices. Even when a specific approach, hence a particular standard, can be identified, evaluation is not straightforward. Many of the comparisons summarized here have been debated at least implicitly in the literature, but the results have remained inconclusive. Policies have not been tested on a scale sufficient for valid conclusions, and testing by means of analytic models has remained generally beyond the capability of modern econometric techniques (see the section "Can We Measure Dual-Use Interactions Quantitatively?" in Chapter 4).

MILITARY-CIVIL TECHNOLOGY RELATIONSHIPS

Not only is the spinoff model grossly oversimplified, it is only one in a spectrum of interactions between defense and commercial technological activities. Historically, government actions have affected private innovation at all of its stages, from sponsoring pure research to being a creative customer for industrial goods. Table 3-1 abstracts, from the

TABLE 3-1:
Defense-Commercial Technology Relationships

	Mode	Example
1.	Direct product conversion (true spinoff)	Microwave oven
2.	Defense procurement pull and commercial "learning"	Supercomputers
3.	Concurrent development of civil and military applications of a common technology	Jet engines, jet transports
4.	Shared infrastructure for defense programs and emerging commercial industry	Nuclear power, satellite communications
5.	Development of engineering techniques and tools to meet government needs	NASTRAN
6.	Dual-use technology developed from defense agency support of basic or generic research	Artificial intelligence, lasers
7.	Reverse spinoff ("spin-on") from civil to military	MILVAX, CMOS semiconductors
8.	Forced diffusion through demonstration programs	VHSIC

U.S. experience over the past four decades, the variety of ways in which military and civil technologies have influenced one another. Most of the linkages were indirect and perhaps even unintended consequences of policies pursued for other reasons. In the typical case, several types of interaction will be found together. This section of the chapter will explain and illustrate each of the modes in Table 3-1.

Mode 1. Direct Product Conversion (True Spinoff)

Occasionally a civil application is found for an innovative military product previously developed at government expense. Alternatively, a firm may develop two related product families, one civil and one military, at approximately the same time. Thus the relationship between military R&D and civil application may be either sequential or concurrent. Raytheon's microwave oven was a classical sequential spinoff within a single firm: the company had been a major manufacturer of magnetrons for military radar systems. The commercial version required substantial additional investment and technical work—a complete redesign of the magnetron microwave power source for ease of manufacture, a fivefold reduction in magnetron manufacturing cost through learning and scale, a patented safety seal to prevent the escape of microwave radiation out the oven door, and corporate acquisition of a distribution channel.

Mode 2. Defense Procurement Pull and Commercial "Learning"

Government purchases can help create a market and reduce business risks through early acquisition of new technology by technically sophisticated users. DoD's willingness to pay almost any price for compact, lightweight electronics for its missile programs was much more significant in stimulating the infant semiconductor industry than DoD's programs (Chapter 8). This early, cost-insensitive purchasing by DoD helped the companies pioneering this technology move down the learning curve, reducing their costs to the point where commercial customers could afford the new chips.

Defense procurement also made critical contributions to the early development of commercial computers. Thomas J. Watson, Jr., the son of IBM's founder, describes how he looked to the Defense Department for the market to drive IBM's entry into the computer business when the Korean War broke out in 1950.[10] His father telephoned President Truman "putting the resources of IBM at the government's disposal," and naming Tom Jr. as the man to contact. "It seemed to me," the latter wrote, "that if we could build a couple of one-of-a-kind machines under government contracts, we'd have a way of getting our feet wet." IBM studied computational needs across a wide range of defense programs and decided to invest $3 million in the development of a general purpose scientific computer, which it called the Defense Calculator. Shipped in December 1952 as Model 701, and sold primarily to government contractors, it stimulated strong demand from IBM's commercial customers for a business computer. That machine, the IBM 702, was shipped two years later, and thirteen were sold.

Supercomputers bought by weapons laboratories in the 1960s and 1970s illustrate the continuing impacts of purchases of sophisticated government users. The Los Alamos and Livermore laboratories of the Atomic Energy Commission placed the first orders for new supercomputers such as the IBM Stretch (1956, see Box 3-B), the CDC 6600 (1962) and 7600 (1968), and the Cray 1 (1976).[11] Many of these orders were placed before the first machine had been built, often at a price substantially below the firm's ultimate costs. The labs' scientific staffs helped debug the machines and created operating systems for them, as well as software essential for application development. The computer firms recovered some of their development costs through these early sales, better positioning them to price to commercial users in, for example, the oil exploration industry. They also benefited from the labs' software developments.

Even a failed government procurement can have a stimulating effect on a firm's civilian products. Capitalizing on design work for its unsuccessful bid on the C-5 military transport, Boeing went on to develop the 747 wide-body passenger jet, fortuitously in time for the rapid run-up in aviation fuel prices after the OPEC embargo. Lockheed, which won the C-5 contract, developed the L-1011 commercial wide-body concurrently. But difficulties in the C-5 program pressed Lockheed hard, technically and financially. At the same time, Rolls-Royce, engine supplier for the L-1011, was headed for bankruptcy, compounding Lockheed's problems and setting the stage for an elaborate U.S.-U.K. bailout of both firms. With Boeing's 747 well established

[10] Thomas J. Watson, Jr., and Peter Petre, *Father, Son & Co.: My Life at IBM and Beyond* (New York: Bantam Books, 1990), pp. 203, 230, 233.

[11] Emilio Gonzalez, *Government Policy and Supercomputer Development*, Thesis in the Field of Government for the Degree of Master of Liberal Arts in Extension Studies, Harvard University, November 1990.

BOX 3-B: IBM'S STRETCH COMPUTER

In 1955, the Atomic Energy Commission's Lawrence Livermore Laboratory asked IBM and Remington Rand to bid on a large, all-transistor computer to be called LARC (Livermore Automatic Research Computer).[1] IBM responded with a design five times faster than the Livermore specification that incorporated new proprietary transistor technology then in its early development stage. However, IBM's three-year schedule exceeded the 29 months proposed by its competitor (renamed Sperry Rand). Since early delivery was important to Livermore, it awarded the contract to Sperry.

IBM then offered its machine to the National Security Agency (NSA). NSA did not order one, but was interested enough to invest $1 million for development of a high-performance memory and another $250,000 for a computer design. Later in 1955, IBM learned that the Los Alamos Scientific Laboratory (LASL), Livermore's sister and rival, was interested in buying the machine turned down by Livermore. At that, IBM began development of the Stretch computer.

A joint steering committee of eight scientists—four from LASL, four from IBM—worked for eight months during 1956 to develop the technical specifications for the machine's architecture. LASL was a sophisticated customer. Seeking very high performance, its representatives proposed extensions to the capabilities of the IBM 704—IBM's top-of-the-line computer at the time—and influenced the design of Stretch in major ways.

In November 1956, IBM signed a contract with Los Alamos for delivery of a Stretch machine 42 months later. Although the contract did not specify the computing speed to be achieved, IBM discussed a goal of 100 times the speed of the IBM 704. The industry had never seen a commercial product jump in performance by a factor of 100 in only four years, and indeed never would. IBM's price had risen to $4.3 million, which still left an estimated $9.3 million in additional development costs to be recovered through other sales. The AEC allowed IBM to own or have free use of all the resulting patents, so that IBM could hope to use the technology in future commercial products.

Stretch remained in the IBM Research Laboratory for a full year before product development took over responsibility in January 1957. With the addition of a special processor called "Harvest" to meet its intelligence collection needs, NSA also ordered a Stretch. NSA representatives joined LASL and IBM on the joint steering committee, which became the "Three-in-One" committee.

There followed a series of very difficult technical problems, over-

---

[1]This narrative is based on Emerson W. Pugh, *Memories That Shaped an Industry* (Cambridge, MA: MIT Press, 1984), pp. 160–185, and Charles J. Bashe et al., *IBM's Early Computers* (Cambridge, MA: MIT Press, 1986), pp. 449–458. Supplemented by interviews (LMB, June 18, 1990) with members of the Stretch development team: Steve Dunwell, Harwood Kolsky, John Cocke, Paul Herwitz, and Frederick Brooks.

come by a series of inventions. In June 1958, as the Stretch technical team struggled, IBM saw an opportunity to capitalize on the new transistor technology being developed for Stretch. As the transistor logic matured, IBM decided to remap the successful 709 machine from vacuum tubes to transistors, calling it the 7090. In December 1959, only a year and a half later, IBM delivered the first two 7090s for DoD's Ballistic Missile Early Warning System. This was possible because of the decision to make no architectural changes in the 709, and because of a coincidence in memory word sizes that permitted the Stretch memory—developed for a new architecture—to be used in the 7090 without change.

IBM missed the scheduled May 1960 delivery date for LASL's Stretch. Mounting costs, diversion of effort to the 7090, and a great variety of technical difficulties all contributed to the delay. But in April 1961, Stretch was finally shipped to LASL. The NSA's Harvest machine was completed in January 1962.

Although it was the fastest machine available, Stretch's speed on typical tasks was only 60 times that of the 704, rather than the much-discussed goal of 100 times faster. Embarrassed at the shortfall, IBM Chairman Thomas J. Watson, Jr. ordered that its price to Livermore be cut to 60 percent of the original figure. With $25 million sunk in development costs, Stretch could not possibly be profitable at that price. Nevertheless, Watson allowed a total of eight Stretch computers to be built and sold at the reduced price. (In September 1980, the last Stretch machine in active use, at Brigham Young University, was shut down.)

This history illustrates the interplay of revolutionary and evolutionary technology. Stretch itself was a money-loser. But its technology broke ground for an extraordinary number of "firsts" and made possible the sale of some 200 IBM 7090s, a very profitable program. A conventional business plan for the 7090 would never have been based on such radical new technology. Stretch also became the starting point for the IBM System/360, in its day the most successful computer product in the industry's history.

Stretch benefited from a degree of risk-sharing among IBM, the AEC laboratories, and NSA that would be unlikely today. Indeed, those working on the project felt that even two or three years later, the AEC would not have allowed LASL or Livermore to commit $4.3 million for a "best efforts" computer, most of which had yet to be invented. Key members of the IBM team had originally acquired their technical expertise in government defense work; one member left LASL to join IBM midstream, moving from one side of the joint steering committee table to the other. The growth of IBM and the U.S. computer industry as a whole was accelerated by the close partnership between IBM and its government clients—a relationship that would not be tolerated today, with greatly heightened sensitivity to the appearance of conflicts of interest in defense procurement.

in the market, Lockheed never turned a profit on the L-1011, and later withdrew from the commercial air transport market.[12]

Mode 3. Concurrent Development of Civil and Military Applications of a Common Technology

Parallel development of both military and civil applications of a new technology can be mutually beneficial. AEC and Navy investments in the development of nuclear reactors for electric power production and submarine propulsion set design points for the dominant forms of the technology in the United States. The light-water reactors selected (because of their compact size) for submarine propulsion also came to dominate for power generation, although the fuel systems differed.

Instructive examples can also be found in the jet engine business. Both General Electric and Pratt & Whitney (now a division of United Technologies) design and manufacture related families of engines for both military and commercial aircraft. Another classic example is the relationship between two very successful Boeing aircraft: the KC-135 Air Force Stratotanker, which appeared in August 1956, and the Model 707 Stratoliner, which appeared 18 months later and ushered in a new era in passenger aviation. Although the 707 appeared after the KC-135, it was not a sequential spinoff from the KC-135, but rather a sibling from a common parent—the Dash-80 jet transport prototype (see Box 3-C).[13]

Mode 4. Shared Infrastructure for Defense Programs and Emerging Commercial Industry

The commercial communications-satellite industry was made possible because government paid for development of launch vehicles, launch facilities and services, and the tracking stations required to monitor satellites in orbit and link them with terrestrial communications networks. This infrastructure was just as necessary as the satellite technology itself, most of which came from DoD and NASA programs. Given this infrastructure—which can be thought of as a complementary

---

[12] John Newhouse, *The Sporty Game* (New York: Alfred A. Knopf, 1982), pp. 112–115, 175–184.

[13] A.T. Lloyd, "Thirty Years Young," *Boeing Airliner* (July–September 1986), pp. 18–22; William H. Cook, *The Road to the 707: The Inside Story of Designing the 707* (privately published, 1991).

BOX 3-C: A FAMILY TREE OF MILITARY AND CIVIL AIRCRAFT

When the Strategic Air Command began to purchase large numbers of intercontinental jet bombers in the early 1950s, the KC-97 propeller-driven aerial refueling tanker, of which 814 were in service, had to be replaced.[1] Boeing designed and built, at its own expense, a four-jet swept-wing aircraft. After the Dash-80, as it was called, first flew in July 1954, and even before it had demonstrated aerial refueling capability, the Air Force ordered 29. In less than two years the resulting tanker, labeled the KC-135, was flight-tested, and 18 months after that the 707, a commercial airliner also based on Dash-80, made its maiden flight.

While the 707 and KC-135 had a common parent and were in development at the same time, they were separate projects, and they have a number of significant differences, although similar in external appearance. Their wingspans are identical, but the 707-100 is 4.5 inches higher, 10 feet longer, and 4 inches wider than the KC-135. These differences emphasize the extent to which design optimization is practiced in this industry. Materials and structure differ because service utilization rates differ by a factor of almost 10: the KC-135 was designed to fly 375 hours a year, a 707 up to 3,500.

Thus it is reasonable to say that the very successful Boeing 707 was a spinoff from the Dash-80 prototype, which led to both the KC-135 and the 707. Boeing's willingness to invest in the Dash-80, in turn, followed from the prospect of both military and commercial sales. Boeing's corporate strategy successfully leveraged the demands of both defense and air transportation markets.

[1]This box is based on the sources listed in footnote 13.

asset—a satellite manufacturer can find a market without the need to offer all the ground-based system components or transport to orbit.[14]

Another example of shared infrastructure can be seen in the early years of the nuclear medicine industry. To promote the peaceful uses of atomic energy, the AEC funded research on diagnostic, therapeutic, and research applications of radioactive substances, as well as protection against harmful effects. With a knowledge base, clinical experi-

[14]David Teece, "Capturing Value from Technological Innovation: Integration, Strategic Partnering, and Licensing Decisions," in Bruce R. Guile and Harvey Brooks, eds., *Technology and Global Industry: Companies and Nations in the World Economy* (Washington, DC: National Academy Press, 1987), pp. 65–95. Similarly, a firm like Scientific Atlanta, which manufactures earth stations, can view government support for satellite development as a complementary asset for its business.

ence, and safety standards in place, private firms found the cost of entry into nuclear medicine acceptable.

If the government can provide complementary assets, it can take them away. Although previously insisting that the satellite communications industry use the Space Shuttle for launch services, after the *Challenger* loss NASA displaced commercial customers in favor of military payloads. Government aggravated the industry's problems by discouraging the use of foreign launch vehicles, ostensibly because of concern about transfer of the technology embodied in the satellites.

Infrastructure sharing may require compatible technical standards. For example, DoD has been frustrated for many years by the diversity of computer languages used in weapons systems, which impedes military interoperability (see Chapter 9). In other cases, military and commercial standards have been harmonized, creating a single market with cost advantages for both government and private customers. Civilian aircraft frequently make use of lubricants, hydraulic fittings and fluids, and other components supplied to military standards.

Federal and state governments also invest directly in technical infrastructure and services. Government agencies and private firms both use major government facilities, such as synchrotron light sources and supercomputers, which offer private industry—particularly smaller firms—the ability to conduct research that they otherwise could not afford. For proprietary work, time on these national facilities is typically offered on a cost-recovery basis. If the research is to be published in the open literature, time on the apparatus is free, and only the incremental costs associated with the user's equipment are paid by the user.

Substantial federal investments have been authorized to expand and rationalize the Internet computer communications system, forming a National Research and Education Network (NREN) that would serve a very broad community of scientists, engineers, and educators in the private sector as well as government.[15] Internet itself grew out of DoD's ARPANET, which pioneered both network applications (electronic mail) and technology (packet switching) later commercialized by private firms.

---

[15] *Grand Challenges: High-Performance Computing and Communications, The FY 1992 U.S. Research and Development Program,* A Report by the Committee on Physical, Mathematical, and Engineering Sciences, Federal Coordinating Council for Science, Engineering, and Technology, Office of Science and Technology Policy (Washington, DC: National Science Foundation, 1991). The High Performance Computing Act of 1991 was signed by the president on December 9, 1991.

Mode 5. Development of Engineering Techniques and Tools to Meet Government Needs

Federal agencies have paid for the development of many engineering tools, techniques, and methods having widespread applicability. Some of these tools have come from university research, others from DoD contracts with industry. Many have found their way into general industrial usage. An example is NASTRAN, standing for NASA Structural Analysis, which was originally developed by several firms under contract to NASA starting in the middle 1960s. A number of aerospace firms had been proceeding with independent, proprietary efforts aimed at the same or similar software for computer-based structural design and analysis.[16] NASA's objective was an integrated set of tools, portable among computers from different manufacturers and near-universal in applicability. Although the original software developed for NASA is in the public domain, NASTRAN and many variants are now available in proprietary form from private firms that supply updates as well as support.[17] Originally applied to space systems, this technology is now used in almost every area of mechanical and structural design.

Mode 6. Dual-Use Technology Developed from Defense Agency Research Support

Defense agencies—e.g., the Office of Naval Research (ONR), the Air Force Office of Scientific Research (AFOSR), and the Army Research Office (ARO)—fund basic research in universities and government laboratories in a wide range of subjects of potential military interest. Much of the work is sufficiently fundamental that there is little to distinguish military from civil potential. Other research, for example in materials science and engineering, promises short-term payoffs and may be equally valuable to civilian industry. This work can be called generic

---

[16]Thomas G. Butler and Douglas Michel, *NASTRAN: A Summary of the Functions and Capabilities of the NASA Structural Analysis Computer System*, NASA SP-260 (Washington, DC: National Aeronautics and Space Administration, 1971); "Quantifying the Benefits to the National Economy from Secondary Applications of NASA Technology," Report No. CR-2673 (Princeton, NJ: Mathematica, Inc., March 1976), pp. 111–126.

[17]Because of contractual limitations on the firms that had developed and maintained NASTRAN, NASA disputed the right of MacNeal Schwendler—the dominant vendor—to sell NASTRAN commercially without paying royalties to NASA. The matter was finally settled for a payment of $125,000 to NASA. Richard H. MacNeal, *The MacNeal Schwendler Corporation: The First Twenty Years*, privately published 1988, pp. 122–138.

research, or, as we term it in Chapter 12, "infrastructural" technology development; by funding it, government provides essential, if largely unguided, support to the industrial knowledge base.

Artificial intelligence (AI), for example, has been the subject of academic research funded by DARPA, NIH, NSF, and other agencies for three decades. A growing industry now specializes in AI software for both civil and government applications. The laser industry also owes much of its growth over the past 30 years to DoD research, as described in Box 3-D.

Mode 7. Reverse Spinoff or "Spin-on": Civil to Military

Technologies developed entirely in the commercial sector are increasingly being used, or adapted for use, by defense (reverse spinoff, or "spin-on"). For example, CMOS (Complementary Metal-Oxide Semiconductor) circuits were perfected in Japan for electronic wristwatches. Although initially complex to manufacture, CMOS chips had the great advantage of very low power consumption, so that electronic watches could run for a full year before battery replacement. With demand for CMOS circuits stimulated by the growth of the consumer electronics and low-end computer markets, CMOS has become the industry's dominant cost-performance chip technology. They are now used in many military applications, especially where weight and power are a concern. Government funding made little contribution until the later stages of the Defense Department's Very High Speed Integrated Circuits (VHSIC) program (see Chapter 8).

When commercial products are purchased for incorporation in defense systems without adaptation, they are referred to as "Commercial Off-the-Shelf" (COTS) acquisitions. The broader term "Nondevelopmental Item" (NDI) covers anything DoD buys without having paid for R&D, including products developed for sale to the military at corporate expense—e.g., Raytheon's line of MILVAX computers, manufactured under license from the Digital Equipment Corporation. MILVAX versions of Digital's popular VAX computers meet military specifications for packaging, environment, and so on. DoD not only saves the time and expense of developing a similar computer, but also benefits from the wealth of commercially available VAX software.

Mode 8. Forced Diffusion through Demonstration Programs

Military-funded technology demonstrations aim to stimulate demand by the services for technology not yet in use in weapons systems, or

BOX 3-D: MILITARY SUPPORT FOR LASER R&D

In the early years of the computer and semiconductor industries, DoD dollars had their greatest impact through procurement. Federal agencies needed the capabilities, and paid for them. The development of laser technology illustrates a different pattern, in which DoD spent heavily on research, but for many years found little to buy.[1]

The laser, based on a theoretical proposal of Charles Townes and Arthur Schawlow and first demonstrated in the laboratory by Theodore Maiman at Hughes Aircraft in 1960, represents one of the purest cases in which science—in this case quantum electronics—led quickly and directly to technological innovation. Theory and experiment interacted in classic fashion as research groups in several parts of the world competed to understand the underlying phenomena and to build new devices. Maiman and others immediately recognized the potential applications of the laser, both military and civilian: at the press briefing announcing his demonstration, Maiman stressed the communications possibilities—multichannel capability, low cost per channel. It took many more years to realize this potential.

DoD provided much of the early support for laser R&D—twice the industry's own spending in the early years—and today continues to pay for most work on high-power devices. As the customer, DoD guided R&D, evaluated alternative approaches, and indeed conducted much of the fundamental research in its own laboratories. From the beginning, the Pentagon wanted lasers of high power for such applications as battlefield weapons and ballistic missile defense. The Army also sought laser rangefinders for tanks, which in fact turned out to be the first significant product on the defense side. Civilian applications, meanwhile, began with eye surgery—another use immediately anticipated upon the laser's invention. Concentrated energy, finally, offered potentially useful applications in manufacturing, such as cutting, welding, and heat treatment. But although the military continually expressed interest in the use of lasers for manufacturing, most of these applications were eventually reduced to practice to meet civilian needs.

Some fifteen years after Maiman's breakthrough, truly dramatic impacts in communications began to appear, just as he had expected. But it took the solid-state laser in combination with low-loss optical fibers—an independent and equally important innovation depending heavily on steady, incremental improvement—to make fiber-optic communications systems a reality. By themselves, neither the laser nor optical fibers would have had remotely comparable significance. To-

[1]M. Bertolotti, *Masers and Lasers: An Historical Approach* (Bristol: Adam Hilger, 1983); Robert W. Seidel, "From Glow to Flow: A History of Military Laser Research and Development," *Historical Studies in the Physical and Biological Sciences,* vol. 18 (1987), pp. 111–145; Anthony J. DeMaria, "Lasers in Modern Industries," in John R. Whinnery, ed., *Lasers: Invention to Application* (Washington, DC: National Academy Press, 1987), pp. 17–44.

gether, they made possible a new family of telecommunications technologies, fully as important as satellite communications. And today, despite extensive continuing support for laser R&D, the military finds itself well behind the civil sector in applications of fiber optics and opto-electronic technology in general.

to provide incentives to commercial industry to make their technology available for government use. On the civil side, federal agencies have sponsored initial market-entry demonstrations of new products or technologies in order to stimulate the interest of either potential users or manufacturers. The Pentagon's VHSIC program provides a recent and largely unsuccessful example (Chapter 8). DoD funded "brassboard" development of specialized chips and circuits, and subsidized R&D needed to help commercial semiconductor manufacturers adapt or extend their manufacturing processes to meet military specifications, in the expectation that they would bid on military procurements of such parts. Few did so.

Civil technology demonstration programs were popular with Congress during the 1960s and 1970s, but began falling out of favor during the later Carter administration years.[18] Operation BREAKTHROUGH, for example, a Nixon-era demonstration of factory-produced housing based on new materials and performance-based building codes, failed in terms of building-code reform and consumer acceptance, even though its technologies met their targets.[19] Congress became reluctant to gamble on multimillion-dollar demonstration projects, either in testing unproven technology or in exploring new markets or the strength of complementary assets.

FEDERAL TECHNOLOGY TRANSFER POLICY

The linkage mechanisms described in this chapter show that government influence extends far beyond funding the R&D from which com-

---

[18] Walter S. Baer, Leland L. Johnson, and Edward W. Merrow, "Government-Supported Demonstrations of New Technologies," *Science*, vol. 196 (May 27, 1977), pp. 950–957; Cohen and Noll, *The Technology Pork Barrel*.

[19] See case study of BREAKTHROUGH in Baer et al., *Analysis of Federally Funded Demonstration Projects: Supporting Case Studies*, ibid.; also "Operation BREAKTHROUGH—Lessons Learned About Demonstrating New Technology," GAO Report PSAD-76-173 (Washington, DC: U.S. GAO, November 2, 1976).

mercial products sometimes arise.[20] By and large, however, the federal government has pursued only one set of policies explicitly intended to exploit military-civil technology interactions, and these have focused on research results. Certainly the sheer scope of the federal R&D effort motivates such a focus. The government funds almost half of the nation's annual R&D bill, spending a third of its share—over $20 billion annually—in its own labs or in federally funded research and development centers (see Chapter 4, especially Table 4-2). Much of the R&D conducted in federal laboratories may be peripheral or irrelevant to commercial applications, but its magnitude attracts attention. Moreover, it is politically expedient to view the simple spinoff linkage, in which research in government laboratories or in industry under government contract leads to commercial products, as the most important one. Policymakers seek to transfer technology out of the federal laboratories because this seems more straightforward—and is certainly less controversial—than figuring out a set of incentives and policies that would address the technological dimensions of competitiveness directly.

But giving federal laboratories the new objective of technology transfer is easier to state than accomplish. Without major change in the structure of federal R&D funding and in the roles of agencies and departments, most of the R&D will remain targeted on the particular missions of particular agencies. Even if those missions involve the development of dual-use technologies, pursuing the commercial applications of those technologies will generally require more money and a redirection of effort. Spinoff is not automatic or cost-free, as we have seen.

Issues of Equity and Fairness

Questions of equity and fairness complicate matters, a theme that also underlies much of federal procurement policy (see Chapter 5). The U.S. government has traditionally sought to ensure that the fruits of publicly funded R&D are available to all. However, a private firm has little

---

[20] Information in this section is drawn from the specified pieces of legislation and Wendy H. Schacht, "Technology Transfer: Utilization of Federally Funded Research and Development," Congressional Research Service Issue Brief IB85031, May 11, 1989; U.S. General Accounting Office, *Technology Transfer: Implementation Status of the Federal Technology Transfer Act of 1986*, GAO/RCED-89-154 (Washington, DC: GAO, May 1989); and "Technology Policy and Its Effect on the National Economy," Report of the Committee on Science, Space, and Technology, U.S. House of Representatives, House Report 100-1093, October 19, 1988 (Washington, DC: U.S. GPO, 1988).

incentive to invest the additional resources needed to exploit a federally developed technology unless it can protect its investment from free-riding imitators. Tensions between disseminating the results of federal R&D as a public good, and granting some form of monopoly rights to encourage commercial exploitation, thus pose a fundamental conflict.

Similarly, allowing federal researchers to benefit financially from commercialization of their work may motivate them to recognize and develop promising applications, but it can also create potential conflicts of interest or otherwise distort their research agendas. These ambiguities are especially troubling for national laboratories without a clear-cut mission, such as the Energy Department's civilian multipurpose facilities. DOE's nuclear weapons laboratories will also lose some of their mission focus as they seek new roles to preserve their staffs and budgets in the years ahead.

Evolution of Policy

Before 1980, the agencies had no consistent policy with respect to commercialization of federal R&D. Many retained title to patents and offered royalty-free licenses to any interested party, which typically inhibited commercialization. The Stevenson-Wydler Technology Innovation Act of 1980 for the first time made technology transfer a formal part of all R&D agency missions:

> It is the continuing responsibility of the Federal Government to ensure the full use of the results of the Nation's Federal investment in research and development. To this end the Federal Government shall strive where appropriate to transfer federally owned or originated technology to State and local governments and to the private sector.[21]

Stevenson-Wydler called for the establishment of Offices of Research and Technology Applications (ORTAs) within most federal laboratories to identify technologies with commercial potential and facilitate their transfer to the private sector.

Stevenson-Wydler also called for more explicit federal involvement in developing and disseminating commercially relevant technology, directing the Department of Commerce and the National Science Foundation to set up "Centers for Industrial Technology" that would support "technological and industrial innovation," particularly by small businesses and individuals. However, the Reagan administration did

---

[21] Public Law 96-480, Section 11(a), October 26, 1980.

not implement these provisions. Commerce declined to establish any centers, although NSF continued to fund its existing Industry/University Cooperative Research Centers, a program it had begun during the early 1970s.[22]

Patents and Licenses

The year 1980 also brought changes in patent law. New legislation that year permitted agencies to grant exclusive or partially exclusive licenses to private parties. It also gave universities, small businesses, and nonprofit organizations contracting with government the right to patent, license, and market for themselves any technologies they developed with government funds.[23] Later legislation and executive orders extended this policy to other federal contractors, including those operating federal laboratories (who had explicitly been excluded from the 1980 legislation).[24]

These changes to patent and licensing policies recognize that federal R&D is much more likely to be commercialized if contractors have a financial incentive to do so. But tensions remain between promoting commercialization on the one hand and recovering the taxpayer's investment in R&D on the other. In service of the latter, defense contractors are required to pay royalties to the government on the sale or license of products or technologies developed in part with government funds.[25] Some argue that this requirement, too, should be eliminated.[26]

---

[22] *Implementation of P.L. 96-480, Stevenson-Wydler Technology Innovation Act of 1980,* hearings, Subcommittee on Science, Research, and Technology, Committee on Science and Technology, U.S. House of Representatives, July 14, 15, 16, 1981. This section of the Stevenson-Wydler Act was amended by the Federal Technology Transfer Act of 1986 (Public Law 99-502, October 20, 1986), which called instead for "Cooperative Research Centers" with a somewhat changed emphasis.

[23] 1980 Amendments to the Patent and Trademark Laws (The Bayh-Dole Act), Public Law 96-517, December 12, 1980.

[24] Public Law 98-620, November 8, 1984; Presidential Memorandum on Government Patent Policy, February 18, 1983. However, national laboratories that are run by large for-profit corporations (such as Oak Ridge National Laboratory, operated by Martin Marietta), and nuclear weapons and naval nuclear reactors programs at the weapons laboratories remain excluded.

[25] Recovery of Nonrecurring Costs on Commercial Sales of Defense Products and Technology and of Royalty Fees for Use of DoD Technical Data," DoD Federal Acquisition Regulation (FAR) Supplement, Part 271; "Recoupment of Nonrecurring Costs on Sales of U.S. Products and Technology," DoD Directive 2140.2.

[26] Center for Strategic and International Studies (CSIS), *Integrating Commercial and Military Technologies for National Strength: An Agenda for Change,* Report of the CSIS Steering Committee on Security and Technology (Washington, DC: CSIS, March 1991), p. 62. Provisions that royalties be paid on "derivative" products having as few as 10 percent of their parts in common with the original product are viewed as particularly troublesome.

Cooperative Research and Development Agreements

Before 1986, federal agencies varied widely in their authority to enter into cooperative arrangements with outside parties. The National Aeronautics and Space Act of 1958[27] permitted NASA to enter into such agreements, but other agencies—even some deeply engaged in technology transfer such as the Department of Agriculture—had little or no statutory authority to do so. The Federal Technology Transfer Act of 1986 made policy uniform across agencies, authorizing government-operated laboratories (extended in 1989 to contractor-operated laboratories[28]) to enter into cooperative R&D agreements (CRADAs) with other agencies, nonprofit organizations, state and local governments, and private firms. Consistent with their missions, agencies were permitted to provide personnel, services, facilities, equipment, or other resources (except funds) to outside parties who contributed their own funds and resources.

The 1986 act permitted agencies to award title to any patents resulting from a CRADA to the outside party. It provided for cash awards and royalty sharing to encourage federal employees to promote technology transfer. Furthermore, it provided statutory authority for a Federal Laboratory Consortium and required federal laboratories to join, strengthening what until then had been an informal network linking federal laboratory representatives with potential users of federal technology. CRADAs represent a significant step beyond previous patent and licensing policies. By encouraging government and industry employees to work together, by permitting agencies to contribute resources and personnel, and perhaps most important, by involving industry from the outset, CRADAs have a better chance of generating commercially significant results than do programs aiming simply to transfer information. The September 1990 White House statement, "U.S. Technology Policy," calls for the federal laboratories to give greater consideration to potential commercial applications "in the planning and conduct of R&D," not merely after the fact.[29]

Yet tensions over deriving private gain from public funds will pose increasing difficulties as federal agencies and laboratories make greater use of their collaborative authority. The National Institutes of Health, for example, is struggling to define the boundaries of permissible behavior for researchers who consult for or hold stock in companies

---

[27] Public Law 85-568, July 29, 1958.

[28] National Competitiveness Technology Transfer Act of 1989, Public Law 101-189, Part C, Sections 3131–3133.

[29] Office of Science and Technology Policy, "United States Technology Policy," September 26, 1990, p. 6.

that participate in CRADAs.[30] Officials at DOE's Lawrence Livermore National Laboratory have been accused of favoritism in awarding exclusive patent rights to a firm founded by three laboratory employees.[31] In this case the charges, which had been made by a competing firm, were not ultimately supported. Even so, it is not likely to be the only such accusation. The laboratories will have to respond to the new directives from Washington in ways that do not compromise their primary missions.

Other Mechanisms

One more policy seeking to couple federal technology investments to the commercial sector deserves mention—the Small Business Innovation Research (SBIR) program, established by the Small Business Innovation Development Act of 1982.[32] The product of enthusiasm over the innovative potential and entrepreneurial behavior of small high-technology firms, and an extensive lobbying effort by small business interests, the SBIR program requires federal agencies with R&D budgets above $100 million to spend 1.25 percent of their contract R&D dollars in firms employing fewer than 500 people.[33] Eleven federal agencies participate in this program, with 55 percent of SBIR funds in recent years provided by DoD.

The SBIR program appears to have met with some success. Of a sample of SBIR projects, about one-quarter evidently led to the commercialization of products that would "definitely not" or "probably not" have been pursued in the program's absence.[34] DoD projects showed a lower percentage of such "unique commercializations," probably because the sample included immature projects that had al-

---

[30] See for example Barbara J. Culliton, "NIH, Inc.: The CRADA Boom" and "CRADAs Raise Conflict Issues," *Science*, vol. 245 (September 8, 1989), pp. 1034–1036; J. Mervis, "NIH Rebuffed, Rethinks New Ethics Regulations," *The Scientist*, vol. 4, no. 3 (February 5, 1990), p. 1.

[31] "Tech Transfer Triggers Protest," *Science*, vol. 246 (October 20, 1989), p. 329.

[32] Public Law 97-219, July 22, 1982.

[33] Total SBIR funding for fiscal year 1989, the most recent year for which data are available, was $432 million; cumulative spending had by then reached $1.8 billion. U.S. Small Business Administration, *Small Business Innovation Development Act: Seventh Year Results*, Office of Innovation, Research, and Technology, 7th Annual Report, July 1990.

[34] Joshua Lerner, "The Small High-Technology Company, the Government, and the Marketplace: Evidence from the SBIR Program," STPP Discussion Paper 89-11, Kennedy School of Government, Harvard University, September 1989. Not surprisingly, projects that were more likely to have been undertaken even without SBIR support were commercialized more frequently.

ready produced items of military value but had not advanced to commercialization. When these projects were excluded from the sample, leaving only projects that had come to completion or that had been commercialized before coming to completion, DoD's "unique commercialization" rate approached that of other agencies.

SPINOFF: TOO SIMPLE A VIEW

Spinoff from government technology projects is neither automatic nor cost-free. A simple view of spinoff is attractive to policymakers because it suggests that the United States can reap serendipitous benefits without deliberate policies. But by ignoring costly downstream technical effort and the complementary assets that are necessary to create a viable commercial product, such a picture misleads. Even a more accurate conception of spinoff cannot capture the diversity of government/civil technological relationships. We have described eight such relationships in this chapter, many of them involving activities quite distinct from R&D. This perspective makes it clear that leveraging government technology investments to improve productivity and competitiveness will require substantial effort and deliberate policy attention.

If technology policy is to be based in part on exploiting spinoff, policymakers must also recognize the opportunity costs: the alternative uses to which the funds and human resources could otherwise have been put. No analysis of the net value of spinoff to the economy is meaningful unless it posits how those investments would otherwise have been made—alternatives the spinoff paradigm does not address.

Much of existing technology transfer policy is based on the spinoff paradigm. Even so, these policies have been evolving in ways that are likely to promote a greater variety of dual-use interactions—e.g., by allowing federal labs and agencies to collaborate with private firms, and to do so at early stages in the evolution of new technologies. Concerns regarding the propriety of allowing federal investments to benefit private parties complicate technology transfer policies. Yet such tensions and tradeoffs will inevitably surround any policies that seek to promote competitiveness, since it is private firms that compete in the international economy. Indeed, the political acceptance of granting firms exclusive licenses to government technology acknowledges that the public interest can be served even when private interests are served as well. Equitable processes and successful experience with licensing federal technology would make it easier to contemplate more far-reaching policy changes.

PART TWO

The Technology Base: What Have We Built Since 1945?

Part II takes a closer look at the nation's dual-use technology base. In Chapter 4, we show how government (especially defense) investments in R&D and in the development of human resources are distributed through the economy, and we identify some of the implications of that distribution. Although quantitative evaluation of the net impacts of defense-related technical investments on productivity and economic growth would be useful for policy purposes, Chapter 4 explains why existing estimates entail too many assumptions and approximations to be of value.

Chapter 5 shows how technology development fits within the Defense Department acquisition system and explains the factors that impede exchange of people and technology between military and civilian sectors of the economy. Chapter 6 looks at dual-use relationships in industry, showing that the factors described in Chapter 5 have served to divorce military from civilian production in many industrial sectors, even when a given firm serves both markets. It looks at the ways that a firm's organization affects its ability to relate its defense business with its commercial activities. Chapter 7 broadens the perspective from the United States to a number of other countries—France, Sweden, Germany, and Japan—showing how government and corporate efforts to relate military and commercial technologies play out under quite different defense policies.

The examination in Part II reveals the complexity in dual-use interactions. Defense spending absorbs resources—people, dollars—but it also helps educate and train scientists, engineers, and technicians, and it creates knowledge and teaches lessons in the design and development of complex systems. Defense may not be efficient at this, but so long as defense absorbs many billions of dollars in public funds it makes sense to seek ways of capturing the benefits of this spending. In Part II, we show why this is difficult. In Parts III and IV, we offer policy guidance for improving the nation's ability to capture those benefits.

4

Patterns of Investment in Dual-Use Resources: People and Technology

To understand the relationship between commercial and military technologies, we need a picture of the nature and size of the activities undertaken by commercial and government institutions. We must also understand their relative contributions to the nation's human capital stock. Because it is difficult to quantify and evaluate the *results* of research and development activities, we look instead to the *pattern* of public and private R&D investments.

But analyses of spending patterns can easily be incomplete or even misleading. Research and development are by no means the only inputs to the innovation process, and funding is not the only input to R&D. Accounting conventions and costs differ across industrial sectors, government agencies, countries, and years; thus a given amount of R&D spending can represent varying amounts of technical activity.[1] Furthermore, whereas R&D dollars can be spent only once, the technical knowledge that they produce can be used many times over—or not at all.

Nevertheless, R&D spending figures provide an important starting

[1] R&D statistics typically do not include the production engineering activity of large firms, which can be comparable in magnitude to their development work. Many small firms are not in the habit of reporting their technical activity as R&D at all. Changing definitions over time confound analyses of historical trends, such as when the R&D portion of NASA's budget declined abruptly in 1984 because the Space Shuttle was declared no longer "developmental" but "operational." Similarly, merely by recategorizing its programs, the Environmental Protection Agency appeared to triple its basic research budget between 1981 and 1982.

point for analyzing the military's role in the nation's science and technology system. We begin this chapter by presenting the important R&D investment figures, giving a rough anatomy of the national R&D system, and tracing key changes over the past decades. Then we identify some of the factors that affect the contribution that defense investments make to the commercial technology base. The picture we derive is of a national science and technology system that is significantly, although decreasingly, affected by defense investments.

We go on to explore defense's role in supplying and drawing on the nation's pool of scientists and engineers, asking whether demand for technically trained personnel, or the technical "culture" engendered by military emphasis on pushing the state of the art, have adversely affected the nation's commercial competitiveness. We can guess at some possible effects, but quantitative substantiation—at least for recent years—is not available.

In fact, the problems facing quantitative evaluations of defense's influence on commercial technology generally prove intractable. In the last section of this chapter we show that useful conclusions can rarely be drawn concerning the economic impacts of defense-related R&D because the available methods for relating defense R&D investments to economic performance are inadequate. We are therefore limited to analyzing the investments themselves, rather than their consequences.

U.S. R&D INVESTMENTS

International Comparisons

The United States invests more in research and development than any other country in the world. In fact, from the standpoint of total national investment in R&D, the U.S. technological position should be impregnable, since it spends as much as the next nine largest R&D performers in the Western world put together: Japan, West Germany, France, the United Kingdom, Italy, Canada, the Netherlands, Sweden, and Switzerland (Table 4-1).[2]

But is this the appropriate measure? Comparing national R&D totals is equivalent to assuming that the fruits of every R&D project performed within a country are applicable throughout the country's economy, yet do not cross national borders. A picture that is somewhat more realistic is one in which the fruits of R&D "stick" to that sector of the economy

[2] Sources of data used in this chapter predate German unification.

TABLE 4-1:
National R&D Spending Comparisons, 1988
(billion U.S. $ according to OECD purchasing-power parities[a])

Country	R&D Spending	Defense R&D	R&D/GDP (Percent)	Nondefense R&D/GDP (%)
United States[b]	$133.7	$43.0	2.7	1.9
Japan[c]	47.0	0.4	2.6	2.6
Federal Republic of Germany	24.6	1.1	2.8	2.7
France	17.5	3.9	2.3	1.8
United Kingdom	17.0	3.3	2.2	1.8
Italy	9.1	0.6	1.2	1.1
Canada[d]	6.4	0.3	1.4	1.3
Netherlands	4.3	0.1	2.3	2.2
Sweden	3.6	0.4	2.9	2.6
Switzerland[e]	3.2	0.1	2.9	2.9
Non-U.S. Total[f]	$132.9	$10.2	2.3	2.1

Sources: Organization of Economic Cooperation and Development (OECD), *Main Science and Technology Indicators,* 1990 Number 2, except as noted. OECD defense R&D data are from government R&D budget appropriations and are not directly comparable to national totals, which are based upon surveys of R&D performers. Nevertheless, the difference is not likely to substantially affect the R&D/GDP. Unpublished National Science Foundation data drawing on additional data sources but not covering all the countries listed here give much the same numbers.

[a] Since currency exchange rates can fluctuate in ways that do not reflect the underlying costs of living (or, for example, the costs of performing R&D) in various countries, OECD has developed a set of "purchasing-power parities," based on the prices of a standardized basket of goods and services, to convert national currencies to dollars. These conversions are much more stable than exchange rates.

[b] Total 1988 U.S. R&D from National Science Foundation, Division of Science Resource Studies, *SRS Data Brief,* NSF 91-307, March 18, 1991. Defense R&D equals government outlays, from *AAAS Report XIV: Research and Development, FY1990* (Washington, DC: American Association for the Advancement of Science, 1989), p. 24, plus the authors' estimate for corporate-funded R&D devoted to defense in 1988 based on IR&D figures, as shown for 1990 in Table 4-5.

[c] Japanese data for 1988 are from Japan's National Institute of Science and Technology Policy, *Science and Technology Structure of Japan,* October 1990. They differ from Japanese R&D data reported to OECD, which treat researchers working "mainly" on R&D as if they were working entirely on R&D, thus somewhat overstating national R&D. After correcting for this effect, OECD's adjusted tables agree with the data quoted here.

[d] Canadian data used here, from telephone inquiry to the agency Statistics Canada, April 1990, include a program sponsoring R&D in the defense industry that is apparently not included in the government R&D data reported to OECD, increasing the defense R&D figure by about 50 percent over the figure reported by OECD.

[e] Swiss R&D data are scaled from 1986, the last year they are available from OECD, in proportion to GDP.

[f] Totals may not add up because of rounding.

in which the R&D is conducted. Under this assumption, every unit of the economy would need its own separate pool of R&D.[3] Consequently, a nation's technological inputs would be summarized not by its total R&D spending, but by its R&D investment per unit of economic activity—its R&D intensity. According to this index, the technological strength of the U.S. economy is comparable to that of its principal competitors, since the United States spends about the same fraction of its GDP on R&D (2.7 percent) as does Japan (2.6 percent) and West Germany (2.8 percent), and more than France, the United Kingdom, Italy, or Canada (Table 4-1).

The main difference between the United States and most of its economic rivals lies in the fraction of its R&D investment dedicated to defense. In 1988, the United States spent about $43 billion of its $134 billion investment in R&D on defense. The next nine performers of R&D in the OECD, which together spent the same amount on total R&D as the United States, spent only $10 billion of that total on defense (Table 4-1). The heavy defense orientation of American R&D is a major incentive for taking dual-use relationships seriously in U.S. technology policy.

The small ratio of nondefense R&D to GDP in the United States—1.9 percent—is commonly cited as indicative of underinvestment in commercial technology. While France and Great Britain are comparable with the United States by this measure, all lag substantially behind the West Germans (at 2.7 percent) and the Japanese (at 2.6 percent). Accordingly, it is argued, the "true" commercial investments of the United States are adequate relative to Germany or Japan only if defense technology makes an important contribution to civilian industry.

Yet this analysis, too, has its limitations. Counting only nondefense R&D in the numerator of this ratio implicitly assigns zero commercial value to defense R&D. Counting federal spending on R&D for nondefense purposes such as space and energy in the numerator implies that these activities have more commercial value than defense; this implication is difficult to defend in view of the similar technical and

---

[3] This assumption is also implausible, since it implies that the goods and services embodying R&D are not traded between industrial sectors. And like the first picture, it ignores flows of technology across national borders. But this picture is a more realistic one in that barriers to the diffusion of R&D benefits do exist. Innovations with one purpose can be incompatible with or irrelevant to other applications. Alternatively, it might take a lot of money or time to transfer them; or the work might be classified, proprietary, or otherwise subject to controls on information flow.

bureaucratic character of government contracting for defense and non-defense R&D. And counting all contributions to GDP, including government purchases, the denominator overestimates the value of the economic activity to which nondefense R&D contributes directly.

Further analysis requires a better picture of the relationship between national R&D systems and national economies. To make more sense of the national R&D accounts, we take a closer look at the R&D system—its defense and nondefense parts, and their interactions—in the following sections.

R&D Spending in the United States

The United States spent more than $145 billion on research and development in 1990. Just over half of this, 51 percent, came from private industry. The federal government provided the bulk of the remainder, 44 percent of the national total. Control of federal R&D spending is thus the most visible and easily understood source of policy leverage over the technology base. Including the government R&D funds that go to industry, private industry performs more than 70 percent of the nation's R&D (see Table 4-2).

The federally funded proportion of the nation's R&D has been declining for the past three decades, as industrially funded R&D has grown faster than federal R&D. Between 1960 and 1990, industrial R&D funding increased in real terms by a factor of almost 3.9—more than twice the real increase in federal R&D spending of just over 1.7 (see Figure 4-1).[4] Federal R&D spending did not keep pace with the growth of the economy over that period, and it even went through a period of real decline from 1967 to 1975. All told, the federal government's share of national R&D has dropped from 65 percent to 44 percent since 1960.

The U.S. share of the Western world's R&D has decreased as well. Between 1963 and 1988, real R&D spending among the non-U.S.

---

[4]R&D spending here is adjusted for inflation according to the GNP price deflator, which is appropriate for measuring the value of those goods and services that the country gave up in order to support R&D. Economists have developed deflators specifically for R&D, based on the real wages of the scientists and engineers who are performing it. These measures show that the cost of R&D has risen faster than goods and services in the economy as a whole. However, with no obvious way to measure R&D "output," these deflators cannot easily account for improvements in the productivity of technical effort made possible by improved instrumentation and computers; therefore they overestimate the rise in R&D costs per unit of "acquired knowledge."

TABLE 4-2:
1990 U.S. R&D Funding
(billion $)

| Funding Sources (read down): | Performers (read across): | | | | | |
	Federal Labs[a]	University-Run FFRDCs[b]	Industry[c]	University, Nonprofit, Other[d]	Total	Percent Funded
U.S. government	$16.1	$4.8	$31.3	$11.8	$64.0	44.0%
Industry	—	—	72.9	1.8	74.7	51.4%
University, nonprofit, other	—	—	—	6.8	6.8	4.6%
TOTAL (Billion $)	$16.1	$4.8	$104.2	$20.4	$145.5	100.0%
Percent performed	11.1%	3.3%	71.6%	14.0%	100.0%	

Sources: Row and column totals are from the table "National Funds for R&D Spending," *SRS Data Brief*, NSF 91-307, March 18, 1991. The entry shown here for R&D performed in universities, nonprofit, or other institutions and funded by industry is extrapolated (given adjusted row and column totals) from Table 2 in National Science Foundation (J.E. Jankowski), *National Patterns of R&D Resources: 1990*, Final Report, NSF 90-316 (Washington, DC: NSF, 1990), p. 18; the other entries above follow. Percentages calculated on unrounded numbers.

Note: Data in this table are based on surveys of performers. They do not necessarily correspond with federal budget data presented elsewhere in this chapter, in part because performers of R&D often expend federal funds in a year other than the one(s) in which various budgetary steps take place. Federal R&D budget totals are given separately for these different steps—authorization of funds, obligation, and outlay—corresponding to an agency receiving permission to award a contract, awarding it, and paying for it, respectively. As a result, the various totals differ among themselves typically by $1 or $2 billion, out of some $65 billion. Of the three, outlays should track the R&D performance data shown here most closely. Even so, the survey-based total shown here for federally funded R&D differs from the budget-based total for 1990 federal R&D outlays (Table 4-3) by $1.3 billion.

[a]Operated by the federal government.
[b]Federally Funded Research and Development Centers operated by universities under federal contract.
[c]Including FFRDCs operated by industry to account for less than 5 percent of the sector total.
[d]Including FFRDCs operated by nonprofit institutions, which are estimated in *National Patterns of R&D Resources: 1990* to account for less than 25 percent of the sector total.

FIGURE 4-1:
Components of U.S. R&D, 1953–1990

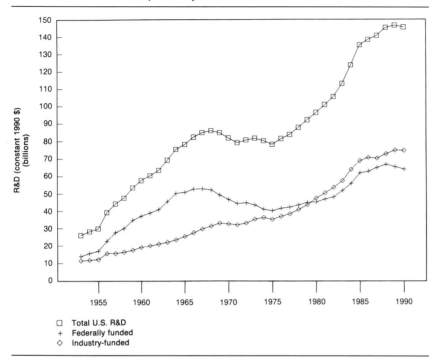

Sources: National Science Foundation (J.E. Jankowski), *National Patterns of R&D Resources: 1990,* NSF 90-316 (Washington, DC: U.S. GPO, May 1990), Table B-5, p. 47; and National Science Foundation, *SRS Data Brief,* NSF 91-307, March 18, 1991.

OECD countries increased by a factor of 4.0, nearly twice as much as the 2.1-fold increase in real U.S. R&D (Figure 4-2).[5] These two trends taken together make it clear that the relative importance of U.S. government spending—defense and nondefense—to the overall world technology base has declined significantly over the past 30 years.

Although federal R&D spending comprises a shrinking slice of the national R&D pie, it remains a large slice of the federal budget (Table 4-3). Few Americans realize that a significant fraction of the annual funding decisions made by the administration and Congress take the form of decisions about R&D. More than half of the federal budget consists of *nondiscretionary* or automatic expenditures such as interest

[5] National R&D figures are converted to dollars using OECD purchasing power parities. See note in Table 4-1.

FIGURE 4-2:
OECD R&D Breakdown, 1963–1988

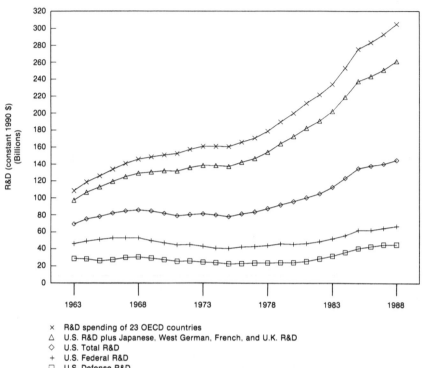

× R&D spending of 23 OECD countries
△ U.S. R&D plus Japanese, West German, French, and U.K. R&D
◇ U.S. Total R&D
+ U.S. Federal R&D
□ U.S. Defense R&D

Sources: U.S. Defense R&D and Total R&D: National Science Foundation (J.E. Jankowski), *National Patterns of R&D Resources: 1990,* NSF 90-316 (Washington, DC: U.S. GPO, May 1990), Table B-13, p. 55, and telephone update. R&D Spending of OECD Countries: Division of Scientific, Technological, and Industrial Indicators, Directorate for Science, Technology, and Industry, Organisation for Economic Co-operation and Development. OECD data do not include Luxembourg. Data for Yugoslavia were not used, since it was not affiliated with the OECD over the entire period; data for Turkey were not used since they are unreliable. Data for Japan were adjusted as described in Table 4-1. Missing data have been interpolated or extrapolated. Foreign currencies converted to U.S. dollars using OECD purchasing-power parities.

on the federal debt, veterans' benefits, farm supports, social security, and other entitlement payments. The remaining budget items, requiring annual appropriations, are considered *discretionary* in that the administration and Congress review and act on them every year. In practice, many discretionary expenses, too, are difficult to change in the short run. Only a relatively small fraction of the federal budget (certainly less than 40 percent, and perhaps quite a bit less) is subject to debate and possible redirection each year. R&D programs constitute at

TABLE 4-3:
R&D in the Federal Budget, FY1990
(outlays, in billion $)

Gross national product[a]		5,406
Federal budget[b]		1,252
Interest on debt	184	
Payments and grants to individuals, states, and localities	642	
National defense (DoD and DOE)	299	
National defense R&D[c] *Percentage of national defense budget that is R&D*	41.6 *14%*	
Other government activities	163	
Nondefense federal R&D *Percentage of other government activity that is R&D*	23.7 *15%*	
Total federal defense and nondefense R&D funding[d]		65.3

[a] Estimate for FY1990 derived from quarterly data given in *Economic Report of the President,* (Washington, DC: U.S. GPO, February 1991), p. 286.
[b] Subtotals do not add up because of $37 billion undistributed offsetting receipts. Data are from the FY1991 Federal Budget, as given in American Association for the Advancement of Science, Intersociety Working Group, *AAAS Report XVI: Research and Development, FY 1992* (Washington, DC: AAAS, 1991), Table I-1, p. 45.
[c] Does not include IR&D/B&P (see pp. 101 ff., below).
[d] Total based on budget data exceeds survey-based data in Table 4-2 by $1.3 billion.

least 14 percent of the discretionary budget (perhaps as much as 25 percent, depending on how the discretionary portion is defined). In this sense, R&D is a major preoccupation of the federal government—especially since overall caps on federal spending enacted by legislation such as Gramm-Rudman-Hollings and the Budget Enforcement Act of 1990 have the effect of pitting R&D increases directly against other proposed growth in federal spending.

U.S. Defense R&D: A Declining Share of the World Technology Base

Defense R&D is the dominant mission-related R&D expenditure in the United States, with health a distant second (Table 4-4).[6] This level of DoD support dates back to World War II, which was a watershed in

[6] This statement is true for research plus development. For basic and applied research alone, DoD funds a much smaller share of the federal total.

TABLE 4-4:
Federal R&D Funding, FY1990 (budget authority, in billion $)

Department of Defense[a]	37.9	
Department of Energy—defense[b]	3.1	
Total defense		*41.0*
National Institutes of Health (NIH)	7.2	
NASA	7.1	
Department of Energy—nondefense	3.9	
National Science Foundation	1.7	
Department of Health and Human Services (other than NIH)[c]	1.3	
Department of Agriculture	1.2	
Environmental Protection Agency	0.4	
National Geological Survey	0.3	
National Oceanic and Atmospheric Administration	0.3	
Department of Education	0.2	
National Institutes of Standards and Technology	0.1	
Bureau of Mines	0.1	
All other	1.5	
Total nondefense		*25.5*
Total R&D		*66.4*

Source: AAAS Report XVI: Research and Development, FY1992 (Washington, DC: AAAS, 1991), Table I-7, p. 52. Budget authority (the ability to spend money) differs in any given year from outlays (actual spending) such as those given in Table 4-3. Totals do not add because of rounding.

[a]Includes $1.2 billion for R&D support in DoD budget categories other than the RDT&E account such as military construction and military personnel. Excludes Independent Research and Development (IR&D) and Bid and Proposal (B&P) reimbursements, which appear as overhead in procurement accounts.

[b]Includes R&D for nuclear weapons, naval nuclear reactors, and some nuclear safeguards and arms control monitoring work.

[c]Mostly mental health.

respect to both the role of R&D in the national economy and the importance of government support of R&D in defense.

Before the war, the federal government was a minor factor in the national research picture. In 1938, for example, federal and state R&D together comprised only 19 percent of the national R&D effort, which in turn came to only 0.4 percent of national income.[7] Moreover, most federal R&D was nonmilitary, with the Department of Agriculture providing about 40 percent of federal R&D that year.[8]

During the war, the defense R&D effort under the civilian Office of Scientific Research and Development (OSRD) grew dramatically with the creation of many new laboratories staffed and managed largely by academic scientists on leave from their partially shut-down home universities. At the war's end, OSRD Director Vannevar Bush successfully argued for continuation of the new wartime mechanisms for supporting publicly funded research and development in private firms and universities.[9] As it happened, Congress did not pass the act authorizing the National Science Foundation until 1950. But by then Director Bush's central ideas had taken root in such agencies as the Office of Naval Research and the Atomic Energy Commission, as well as the National Institutes of Health. Through these agencies, themselves new, and the fledgling NSF, Washington created an entirely new relationship to science and technology.[10]

From the explosion of the first Soviet atomic bomb in 1949 until the mid-1960s, the driving force for science policy remained the military-technological competition with the Soviet Union. This race challenged the science and technology system of the United States across a wide spectrum of technical activity, and it maintained defense budgets and military R&D at levels unprecedented in peacetime.

The federal share of the nation's R&D reached 66 percent in the early 1960s, with defense- and space-related work accounting for about 85 percent of the federal total. Defense R&D dropped as a share of the federal total in the early 1960s as the space race intensified; later in the decade it recovered somewhat, but shrank again in the 1970s.[11]

---

[7] Vannevar Bush, *Science: The Endless Frontier*, Report to the President (Washington, DC: U.S. GPO, July 1945), p. 81. The government figure does not include grants to agricultural experiment stations.

[8] Harvey Brooks, *The Government of Science* (Cambridge, MA: MIT Press, 1968), p. 24.

[9] Bush, *Science: The Endless Frontier.*

[10] Harvey Brooks, "National Science Policy and Technological Innovation," in Ralph E. Landau and Nathan Rosenberg, eds., *The Positive-Sum Strategy: Harnessing Technology for Economic Growth* (Washington, DC: National Academy Press, 1986), pp. 119–168.

[11] National Science Foundation, *National Patterns of R&D Resources: 1990*, NSF 90-316 (Washington, DC: National Science Foundation, 1990), Table B-13, p. 55. The defense share fell from 80 percent in 1960 to 51 percent in 1965 as NASA R&D grew.

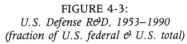

FIGURE 4-3:
U.S. Defense R&D, 1953–1990
(fraction of U.S. federal & U.S. total)

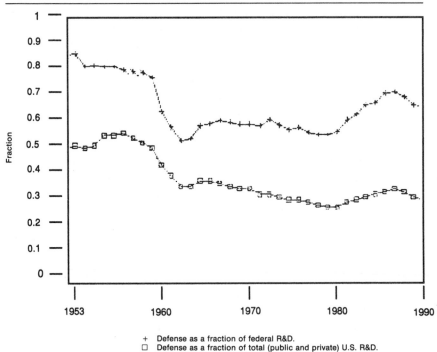

+   Defense as a fraction of federal R&D.
□   Defense as a fraction of total (public and private) U.S. R&D.

Sources: National Science Foundation (J.E. Jankowski), *National Patterns of R&D Resources: 1990,* Final Report, NSF 90-316 (Washington, DC: U.S. GPO, May 1990), Table B-13, p. 55, and telephone update. Defense R&D does not include IR&D/B&P.

Not until the Carter-Reagan buildup in the 1980s did defense R&D again constitute more than two-thirds of the government R&D pie (Figure 4-3).

The increase in defense R&D in the 1980s has been much remarked upon. But in broader context and viewed over a longer time, defense R&D spending has declined in relation to the world investment in science and technology. In 1960, half of U.S. R&D was defense-related; today a third is (Figure 4-3). Steady growth of foreign R&D spending over the same period means that the importance of the U.S. defense component to the overall stock of technology in the Western industrialized world has substantially decreased. In dollar terms, U.S. defense R&D accounted for one-third of the West's technological resources in the 1960s; by the 1990s—despite the Carter-Reagan buildup—it accounted for less than one-seventh.[12]

Components of U.S. Defense-Related R&D

Not all defense-related R&D is found in the Department of Defense budget. Although DoD's RDT&E (research, development, test, and evaluation) budget category funds more than 80 percent of the nation's defense-related R&D, two other elements must be added to it: military R&D in the Department of Energy, and corporate defense-related R&D partially reimbursed through the Defense Department's Independent Research and Development/Bid and Proposal (IR&D/B&P) program (see below).[13] Table 4-5 shows that when DoD RDT&E, DOE defense-related R&D, and IR&D/B&P spending are added together, they account for over 31 percent of national R&D investments. This total does not include R&D funds invested by private companies to develop defense products, if those companies do not receive IR&D reimbursements. There are no reliable estimates for how much defense-related R&D is miscategorized for this reason, but it is likely to be small compared to the IR&D total simply because most Pentagon purchases—especially of high-tech products—are from companies that participate in the IR&D program (see Chapter 6).[14]

R&D Spending within the Department of Defense. Defense R&D consumes a significant fraction of the defense budget (Table 4-6) as well as a large fraction of the federal R&D budget. Although DoD maintains an extensive system of laboratories, research and development centers,

---

[12]Figure 4-2 shows the U.S. share of OECD R&D from 1963 to 1988, the period over which OECD data are available. In 1963, the U.S. spent fully 64 percent of total OECD R&D; by 1988, only 48 percent. The drop in U.S. share of total OECD R&D from 1960 to 1990 can be conservatively extrapolated by assuming that the U.S. share was no higher in 1960 than in 1963, and no lower in 1990 than in 1988. In 1960, 52 percent of all U.S. R&D was defense-related (not including IR&D, for which data earlier than 1964 are difficult to obtain). U.S. defense R&D therefore comprised an estimated 33 percent of the OECD total in 1960. In 1990, the defense share of U.S. R&D was 28 percent (not counting IR&D), comprising 13 percent of the OECD total. *Sources:* OECD data as described in Figure 4-2; National Science Foundation, *National Patterns of R&D Resources: 1990*, Table B-13, p. 55; and other sources cited in this chapter.

[13]The testing and evaluation (T&E) portion of RDT&E fits within the normal definition of "R&D" and will not be distinguished here.

[14]NASA programs like the Space Shuttle plainly have a rationale that is partly military. But the agency's investments in R&D that is primarily defense related are small. The Department of Defense retains funding and control over the nation's military activities in space, paying NASA for launch services. (Since 1983, DoD's military space budget has exceeded NASA's total budget.)

TABLE 4-5:
U.S Defense and Nondefense R&D Spending
(Conduct of R&D and Facilities, FY1990)

	Outlays (billion $)		Percentage of Total
Defense-Related	46.0		*31.3%*
DoD total R&D		38.5	
DOE defense-related R&D		3.1	
Portion of IR&D ceiling + ½ B&P ceiling[a] not already contained in the DoD R&D budget		4.4	
Nondefense-Related	100.8		*68.7%*
Federal government civil R&D		23.7	
Industry self-financed R&D excluding IR&D and B&P ceiling amounts given above		70.3	
University and nonprofit-financed R&D		6.8	
Total	146.8		(100%)

Sources: Industry, university, and nonprofit data from Table 4-2. Government data are outlays from agency budget justifications for the FY1992 federal budget. IR&D and B&P data are from *Independent Research and Development and Bid and Proposal Cost Incurred by Major Defense Contractors in the Years 1989 and 1990*, Defense Contract Audit Agency, 1991.

[a]The IR&D ceiling is the amount of corporate-funded R&D accepted by DoD as having potential military relevance and eligible for partial reimbursement (see text). The B&P ceiling is similarly the amount of potentially reimbursable corporate expenses for preparing bids and proposals that have military relevance. B&P funds can and do fund technical activity but also support nontechnical administrative expenses.

and test sites, in which it conducts much of the exploratory development of technologies that may someday be incorporated into weapon systems, the greatest share of its RDT&E budget is spent in industry (see Table 4-7).

Defense-related R&D within the Department of Energy.   DOE funds

[15] Universities receive 22 percent of DoD's basic and applied research funding. However, since research is such a small fraction of DoD R&D, this amounts to only 3 percent of DoD R&D. National Science Foundation, *Federal Funds for Research and Development: Fiscal Years 1988, 1989, and 1990*, NSF 90-306.

TABLE 4-6:
Defense Department Budget (Budget Authority, FY1990[a]*)*

Budget Category	Billion $	Percentage
Operations and Maintenance	88	30%
Procurement	81	28%
Personnel (active duty and retired)	$79	27%
Research, Development, Test, and Evaluation	*36*	*12%*
Military Construction	5	2%
Other	3	1%
Total	$293	100%

Source: *Report of the Secretary of Defense to the President and the Congress,* January 1991, p. 109. Totals do not add because of rounding.

[a] Budget authority data do not correspond exactly with outlay data given in other tables. See note in Table 4-1.

R&D to support the military tasks it inherited from the Atomic Energy Commission. DOE R&D covers nuclear weapons, production of special nuclear materials, naval nuclear reactors, management of waste from defense reactors, security for nuclear weapons, and certain aspects of arms control verification. In 1990, DOE's defense-related R&D budget totaled $3.1 billion, the fifth largest federal science and technology investment (see Table 4-4).

IR&D/B&P. Corporate R&D activities partially reimbursed by the Pentagon's IR&D and B&P programs are important to any analysis of defense R&D, both because of their dollar amounts and because their technical content is determined at least in part by industry rather than by the Defense Department. Although IR&D and B&P activities must be relevant to defense needs to be eligible for government support, they need not be exclusively defense-oriented. In fact, the government assumes that if a firm or corporate division doing IR&D has any commercial business, that business benefits from the IR&D as well. As a result, the government pays for only that fraction of a firm's IR&D that corresponds to the firm's government sales, so as not to subsidize the firm's commercial operations (see Box 4-A). Even so, IR&D could in principle provide a useful mechanism for encouraging firms to conduct R&D relevant to both commercial and defense needs. The 1991 Defense Authorization bill recognized this potential, encouraging defense contractors to pursue IR&D that will "enhance the industrial competitive-

TABLE 4-7:
Breakdown of DoD RDT&E Funding

By Character of Work/Budget Category (1990)[a]			By Performer (1989)[b]	
6.1:	Research	2.6%	Intramural[c]	26.1%
6.2:	Exploratory Development	6.6%	FFRDCs[d]	4.6%
6.3A:	Advanced Technology Development	15.9%	Industry	65.4%
	SDI:	9.7%		Universities, Colleges, Nonprofits	3.2%
	Non-SDI:	6.2%		
6.3B:	Advanced Development	12.1%	Foreign and Other	0.6%
6.4:	Engineering Development	30.1%		
6.5:	Management and Support	7.6%		
6.6:	Operational Systems Development	25.2%		
Total		100%	Total	100%

[a] *Sources:* U.S. Department of Defense, Office of the Comptroller, *RDT&E Programs (R-1): Department of Defense Budget for FY1992 and FY1993,* February 4, 1991. Categories correspond to DoD R&D budget ("6") categories; see Box 4-B. Data are 1990 budget authority for DoD's RDT&E account. This account does not include the cost of buildings devoted to R&D nor the salaries of military personnel performing R&D. In tabulations of total DoD R&D such as that in Table 4-4, such items are included with "development," but they are not assigned to a "6" category.
[b] *Source:* National Science Foundation, *Federal Funds for Research and Development: Fiscal Years 1988, 1989, and 1990,* NSF 90-306. Percentages are based on 1989 estimated federal obligations for DoD R&D, the latest year that estimates including the effect of congressional actions on the budget were available.
[c] In-house R&D at government laboratories, plus the administration of both internally performed and externally performed R&D.
[d] Federally Funded Research and Development Centers, typically owned by the government but managed by universities, industrial firms, or nonprofit institutions. Just over half of this funding is spent in university-managed FFRDCs.

ness of the United States," including "development of technologies useful for both the private commercial sector and the public sector."[16]

In practice, most IR&D and B&P funds go to the 100 or so largest defense contractors, and in most cases the company division participating in the IR&D program does all or most of its business with DoD. In

[16] National Defense Authorization Act for Fiscal Year 1991, P.L. 101-510, Section 824, November 5, 1990. The act also encourages IR&D that supports "critical technologies" or furthers environmental goals.

BOX 4-A: INDEPENDENT RESEARCH AND DEVELOPMENT COST RECOVERY

IR&D and B&P expenses are treated by government procurement regulations as indirect or overhead costs. Like administrative salaries, rent, and utility bills, they increase a firm's total costs, but cannot be attributed to specific contracts. DoD by law can only fund IR&D which is "of potential interest to the Department of Defense."[1] As long as that test is met, however, it doesn't matter whether the project has commercial relevance as well. Moreover, all intellectual property resulting from IR&D belongs to the firm, which should make it not only possible but attractive for companies to take advantage of IR&D reimbursements to exploit technologies having dual-use potential. In practice, however, the segregation into defense and nondefense components within the firm would tend to discourage much of this sharing.

Determining the amount of IR&D/B&P that a firm can collect has been a complex and ritualized process, although pending legislation for fiscal year 1992 may simplify it considerably. For each participating firm or division, DoD annually negotiates separate ceilings on the amount of IR&D and B&P it will consider eligible for cost recovery.[2] These ceilings depend on the firm's past performance on IR&D and B&P, on previous reimbursements and trends, and on projected government sales. IR&D negotiations also involve technical assessments by government evaluators of the objectives, viability, and progress of the firm's IR&D portfolio.

Within a single division, IR&D and B&P costs—like other indirect costs—are assumed to pertain to all business activities, both government and commercial. So as not to subsidize the division's commercial business, the government reimburses only the fraction of the firm's allowable indirect costs equal to the government share of the division's total business. Reimbursements are made by increasing the overhead rate the firm is allowed to charge the government on each of its contracts and subcontracts. Correspondingly, the costs of IR&D and B&P appear to the government as incremental additions to the cost of each contract, and there is no single budget item or account for IR&D/B&P.

This IR&D administrative procedure has several implications. First, although the government has no direct role in managing a firm's IR&D

---

[1] Section 824, National Defense Authorization Act for Fiscal Year 1991, P.L. 101-510, November 5, 1990. This language superseded the previous statutory requirement that IR&D have a potential relationship to a military function or operation.

[2] Firms receiving payments are required to engage in negotiations if they receive $7 million or more in government IR&D/B&P payments, a threshold most recently raised in the 1991 Defense Authorization bill. (Individual corporate divisions receiving at least $700,000 may choose to negotiate IR&D as well.) Firms below the threshold are reimbursed according to a formula based on sales and IR&D expenses. However, legislation is pending to eliminate the ceilings and negotiation requirements for fiscal year 1992.

program, the process of ceiling negotiation and technical review subjects the firm's IR&D program to government scrutiny and some measure of indirect government influence. However, since a firm's IR&D program is considered proprietary, there is no public disclosure of its nature or even of its magnitude. Congress has little direct say in the appropriation of billions of dollars of public funds, and there is little opportunity for independent assessment of the program's value. Both industry officials and technical personnel within DoD, however, believe IR&D to be valuable.

Second, by paying for only a fraction of IR&D expenses, the government assumes that the firm's commercial business—which is paying for the rest—also benefits. If IR&D were assumed to be irrelevant to the rest of the firm's business, the government would have no basis for not paying for all of it. On the other hand, since all of the eligible IR&D done by a division that does only government business can be reimbursed, even when that division is part of a company dominated by commercial business, the government implicitly assumes that the R&D does not cross division boundaries to benefit commercial work elsewhere in the firm. Third, since IR&D is added as an overhead to existing government contracts, only those firms presently doing government business can benefit. Neither IR&D nor B&P can help a firm break into government contracting.

---

effect, the program allows a relatively small stable of large firms that are under more or less constant service to the Defense Department to look beyond current work to the future needs of DoD. Since the firms make their own decisions (subject to government review) about what new technologies to explore, their IR&D constitutes a hedge against gaps and fads in DoD's own planning process.

Defense contractors told DoD they spent about $4.9 billion on IR&D in 1990 and another $2.4 billion in the preparation of bids and proposals (B&P), an activity that also frequently involves technical work.[17] Through negotiations with participating firms, DoD recognized $3.8 billion of the IR&D and $2.0 billion of the B&P expenses as eligible for federal cost sharing. After attributing a proportionate share of these amounts to the participating firms' commercial business activities, the government paid the remainder—$2.2 billion for IR&D and $1.4 billion for B&P—in the form of increased overhead rates on existing contracts. Assuming that all the IR&D has some defense relevance,

---

[17]"Independent Research and Development and Bid and Proposal Cost Incurred by Major Defense Contractors in the Years 1989 and 1990," transmitted by the Assistant Director, Policy and Plans, Defense Contract Audit Agency, to the Director, Cost, Pricing, and Finance, Office of the Deputy Assistant Secretary of Defense for Procurement, July 5, 1991.

and that half of the B&P effort is technical in nature, IR&D/B&P was responsible for some $4.8 billion of defense-related R&D in 1990. Some of this IR&D—an amount we estimate as about $0.4 billion—is already included in the DoD RDT&E budget, since it appears as overhead on R&D contracts.[18] The remaining $4.4 billion, however, must be added to the DoD and DOE contributions to estimate total national spending on defense-related R&D.

The Dual-Use Contribution of Defense R&D

The impacts of defense R&D on the economy depend not only on the amount of that R&D, but also on its character and distribution. The breadth of applicability of these investments, the institutions in which they are spent, and the subject areas they address are all important in determining their influence on the commercial technology base.

Breadth of applicability. By definition, technical activity becomes more directed toward the solution of immediate problems as one proceeds from basic research to applied research to development. Basic research should be much more likely than development to lead to innovations of general applicability. By this measure, most defense R&D should have limited significance for nondefense purposes, since 91 percent of it is devoted to development, 7 percent to applied research, and only 2 percent to basic research.[19] In other words, most

---

[18] We assume that the IR&D/B&P expenses constitute the same fraction of R&D contracts as they do of all contracts in aggregate. Since in 1988, R&D contracts constituted about 15 percent of total DoD contract dollars, we estimate that they contain 15 percent of the $2.2 billion IR&D costs (or about $0.3 billion), and the same fraction of the $1.4 billion B&P costs (or $0.2 billion). Assuming all of the IR&D and half the B&P funds support technical activity, we find $0.4 billion worth of technical effort in the IR&D/B&P expenses are included as overhead in the DoD RDT&E budget. *Sources:* RDT&E contract dollars (for contracts over $25,000, the vast majority of the funding) from Department of Defense, Washington Headquarters Services, Directorate for Information Operations and Reports, *500 Contractors Receiving the Largest Dollar Volume of Prime Contract Awards for RDT&E, Fiscal Year 1988.* Total DoD contract dollars from U.S. General Services Administration, *Federal Procurement Data System Standard Report, Fiscal Year 1988 Through Fourth Quarter.* The factor of 15 percent is assumed to apply for 1990 as well.

[19] Self-financed industrial R&D, which is overwhelmingly commercial in nature, is 69 percent development, 25 percent applied research, and 6 percent basic research. For government nondefense R&D, the ordering is reversed, at 26 percent development, 31 percent applied research, and 43 percent basic research, reflecting the emphasis placed on general support of the national science and technology base. Government data from Intersociety Working Group, American Association for the Advancement of Science, *AAAS Report XVI: Research and Development in the Fiscal Year 1992 Budget,* Table I-6, p. 50; industrial data from National Science Foundation, *National Patterns of R&D Funding, 1990,* Tables B-5, 7, 9, 11; pp. 45, 47, 49, 51.

defense R&D is "D," not "R." See Box 4-B for a description of the categories into which DoD divides its R&D investments; see Table 4-7 for the distribution of funding over these categories for 1990.

BOX 4-B: R&D CATEGORIES

The National Science Foundation classifies R&D into three categories. *Basic research* is defined as research "directed towards increases in knowledge . . . without specific application toward processes or products in mind." *Applied research* is defined as "knowledge or understanding for determining the means by which a recognized and specific need may be met," and *development* is "the systematic use of the knowledge or understanding gained from research directed towards the production of useful materials, devices, systems, or methods, including design and development of prototypes and processes."[1] DoD uses a more finely graduated system, designated by numbers ranging from 6.1 (Research) to 6.6 (Operational Systems Development), tied to DoD's procedures for approving and funding weapons programs.[2] When DoD compiles R&D spending statistics for National Science Foundation surveys, it reports category 6.1 as basic research, 6.2 as applied research, and the remaining categories taken together as development.

6.1 Research. Research, in DoD parlance, emphasizes "fundamental knowledge for long-term national security needs." Most 6.1 work is organized according to scientific and engineering discipline.

6.2 Exploratory Development. At the 6.2 stage, funding is organized into broad areas of military applications such as "aerospace propulsion" and "surface ship technology." 6.2 funding supports applied work up to and including construction of breadboard hardware for exploring feasibility. But 6.2 stops short of supporting large experiments involving special-purpose hardware.

6.3A Advanced Technology Development. Category 6.3 is divided into two stages. In category 6.3A, funding starts to be organized by military system (e.g., "National Aerospace Plane") or by particular mission (e.g., "Strategic Defense Initiative"). Expensive hardware and full-scale testing are supported under 6.3A, but test articles do not necessarily resemble anything that would eventually end up in the field.

6.3B Advanced Development. Category 6.3B typically funds technology demonstrators or prototypes developed in response to identified mission needs. The hardware begins to resemble a weapons system that might actually go into the field, and effort turns to working

[1] National Science Foundation, *National Patterns of R&D Resources: 1990,* NSF 90-316, (Washington, DC: U.S. GPO, 1990), p. 34.

[2] Office of Technology Assessment, U.S. Congress, *The Defense Technology Base: Introduction and Overview* (Washington, DC: U.S. GPO, March 1988), pp. 54–55.

out the details necessary to move from a laboratory demonstration to something that will work under operational conditions. Contractors may explore competing designs.

6.4   Engineering Development. Funds in this category pay for designing, building, and testing production-model prototypes and for working out the details of manufacturing, operations, and maintenance. Final products are clearly identified (e.g., "Midgetman missile" or "C-17 transport aircraft").

6.5   Management and Support. Category 6.5 is really an overhead account that supports test ranges, test vehicles, laboratory maintenance, and management studies of the R&D complex. Costs of laboratory personnel and of special-purpose test facilities come from program budgets.

6.6   Operational Systems Development. Once a system has been approved for production, R&D funding shifts out of 6.4 into 6.6, which supports testing of operational hardware and any modifications to the design approved during manufacturing or after the system enters the inventory. Much work of this sort in the commercial world would probably not be considered R&D at all.

As in all taxonomic schemes, there is a certain fuzziness around the edges of these categories. The Defense Science Board has noted, for example, that some work in the development categories is erroneously categorized as 6.2 or even 6.1.[3] Conversely, programs in categories 6.3 or 6.4 may pay for the solution of problems that, if pursued for their own sake, would instead be categorized as 6.2 or 6.1; the more an R&D program challenges the state of the art, the more it may encounter problems late in development that require substantial basic and applied research to solve. The military services and agencies exhibit different mixes among R&D budget categories, a fact that probably represents different management conventions at least as much as the somewhat different roles played by technology in the accomplishment of their missions.

[3] *Report of the Defense Science Board 1987 Summer Study on Technology Base Management* (Washington, DC: DoD, Office of the Undersecretary of Defense for Acquisition, December 1987), p. 7.

On a dollar-for-dollar basis, it is plausible that defense basic and applied research investments are more dual use than development expenditures. But research constitutes less than one-tenth of the department's total R&D. Due to its sheer size, the downstream portion of DoD R&D (development, production engineering, and so forth) may well have a greater aggregate effect on the overall national technology base. Likewise, technical activity funded through the DoD procurement account, which outweighs the RDT&E account by a factor of two, also

TABLE 4-8:
Estimated Sources of Funds for R&D in Industry, 1990
(billions $)

Industrial Sector	Federal R&D Funds	Sector's Share of Federal R&D Funds	Industry's Own R&D Funds	Total R&D Funds	Federal Share of Sector R&D
Aerospace	$17.4	51%	$6.9	$24.3	72%
Electrical machinery and communications	6.1	18%	10.1	16.2	38%
Machinery	1.9	6%	13.1	15.0	13%
Chemicals	0.2	1%	10.5	10.7	2%
Autos, trucks and parts, and other transportation equipment	1.5	4%	9.4	10.9	14%
Professional and scientific instruments	0.7	2%	3.6	4.2	16%
Petroleum	0.0	0%	1.9	1.9	1%
Food and beverage	0.0	0%	1.5	1.5	0%

Rubber products	0.3	1%	1.1	1.4	23%
Fabricated metals	0.1	0%	1.1	1.2	10%
Stone, clay, and glass	0.1	0%	1.1	1.1	4%
Paper and pulp	0.0	0%	0.7	0.7	0%
Iron and steel	0.0	0%	0.6	0.6	0%
Nonferrous metals	0.0	0%	0.3	0.3	6%
Textiles	0.0	0%	0.3	0.3	0%
Other manufacturing	0.0	0%	1.0	1.0	2%
Total manufacturing	$28.3	84%	$63.0	$91.4	31%
Nonmanufacturing	$5.5	16%	3.2	8.7	64%
Total	$33.9	100%	66.2	100.1	34%

Source: Battelle, *Probable Levels of R&D Expenditures in 1990: Forecast and Analysis* (Columbus, OH: Battelle Memorial Institute, December 1989). These figures differ somewhat from others in this chapter because (1) they are pre-1990 estimates, and (2) they are based in part on different sources.

has considerable potential to affect the commercial technology base, particularly where manufacturing technology is concerned (see Chapters 3 and 10). Technology policies must therefore inquire into DoD's development and procurement practices, not just its research programs.

The difference in breadth of applicability between research and development funding is particularly relevant to the effects of the Carter-Reagan defense buildup. From 1980 to 1990, DoD RDT&E increased by 80 percent in real terms, growing even faster than total DoD spending. However, this growth heavily favored downstream over upstream activities. While DoD's development expenditures grew by 94 percent in real terms during this period, basic and applied research funding remained nearly flat, growing by less than 5 percent. Indeed, DoD's basic research budget grew the least of any major R&D agency during the 1980s. Since 1965, it has fallen almost one-third in real terms, and applied research has been cut in half. The surge in defense R&D spending over the past decade has not been in budget categories that would be expected to contribute the most to the commercial technology base.

Distribution over industry. The bulk of DoD's RDT&E dollars are spent in industry (Table 4-7), which might be thought better able to reap commercial benefits from those investments than government laboratories or universities would be. However, two important factors constrain potential dual-use synergies. The first is the limited extent to which companies or corporate divisions conducting defense R&D also do commercial work and have the incentives or the ability to relate the two. We return to this subject in Chapter 6. The second caveat is that these investments are distributed very unevenly across industrial sectors. Figures are not available giving the distribution of DoD funds alone, but since DoD accounts for 85 percent of all government-funded R&D performed in industry, the data in Table 4-8 reflect the pattern of DoD investments.[20] About half of all federal R&D spending in industry goes to aerospace, with another fifth to electrical machinery and communications. Indeed, federal R&D funding constitutes over 70 percent of all R&D in the former industry, and almost 40 percent in the latter. But many other important economic sectors get little direct benefit from federal R&D funding.

U.S. R&D Investments: Summary

Although defense R&D has constituted a declining share of national and world technology investments for decades, it still dominates tech-

---

[20] National Science Foundation, *Federal Funds for Research and Development: Fiscal Years 1988, 1989, and 1990,* NSF 90-306.

nology investments by the federal government, and it still comprises a substantial fraction of total national R&D spending, public and private. On a dollar-per-dollar basis, defense R&D's impact on commercial technology is limited for a number of reasons: most of it is spent on development of specific military systems having limited commercial relevance; most of it is spent in companies or corporate divisions with limited opportunities to turn it to commercial advantage (see Chapter 6); and most of it is concentrated in a few industries of special importance to defense, notably aerospace and electronics. Nevertheless, the sheer magnitude of the military R&D effort and the even larger budgets devoted to the procurement of high-technology military systems compensate in part for the limited dual-use potential of each DoD R&D dollar. The influence of military R&D on commercial technology goes well beyond what would be indicated by DoD's research budget alone.

HUMAN RESOURCES—THE ULTIMATE DUAL-USE RESOURCE?

Patterns of investments in human resources are as important as the patterns of R&D dollar investments in discerning relationships between defense and commercial technical activities. From a strictly numerical standpoint, about 15 percent of technical professionals in the labor force receive some support from DoD.[21] But this says nothing about mobility between defense and nondefense work. To what extent do scientists and engineers (S&Es) move readily from one activity to the other? Or are the training and experience acquired on defense projects irrelevant to—or worse, harmful to—effective performance in civilian industry?

Mobility

Since the technical knowledge used in both military and commercial work is much the same (Chapter 2), we might expect that the S&Es in commercial and military industry should be quite interchangeable. Indeed, scientists and engineers destined for defense and for commercial employment receive essentially the same undergraduate and graduate education. However, Chapter 2 also showed that the defense and commercial innovation processes (at least within large companies) are

---

[21] David K. Henry and Richard P. Oliver, "The Defense Buildup, 1977–85: Effects on Production and Employment," *Monthly Labor Review* (August 1987), Table 7, p. 10.

quite distinct. It could therefore also be that after gaining substantial working experience, S&Es would have trouble moving—e.g., because habits of mind, tacit know-how, and other experience-based learning predispose them to one type of work or the other.

There is ample evidence, however, that S&Es move both ways in large numbers—fully one-quarter of the defense S&E workforce crossing the boundary to or from civilian work in a four-year period, according to Lerner's research (see Box 4-C). Yet interviews with high-level technical managers in industry testify to the difficulties that large firms experience in getting effective performance from technical people transferred between defense and commercial divisions. How can these apparently contradictory findings be reconciled? The explanation probably lies mainly in the design, management, and marketing aspects of the work rather than in deficiencies or gaps in purely technical training or knowledge. The climate in which design choices are made differs greatly between defense and commercial sectors (Table 2-3), so that the "craft" experience or tacit know-how acquired in military projects is a poor guide for making design choices in commercial projects even when the purely technical knowledge involved is similar. This difference, which extends to management and organizational culture, may be most important for those S&Es in project management positions; engineers at lower levels should have an easier time adjusting. Hence the difficulties reported in interviews may not apply to all of the technical workforce, but only to the experienced project leaders who are more likely to come to the attention of top managers.

BOX 4-C: MOBILITY OF SCIENTISTS AND ENGINEERS

The most informative data on the mobility of S&Es between defense and civilian work comes from the analysis by Lerner of the National Science Foundation's Survey of Scientists and Engineers (SSE).[1] The SSE database comes from a survey of 138,000 individuals in eleven "cohorts," defined according to occupation as reported in the 1980 Census. The same population was surveyed in 1982, 1984, 1986, and 1989. Response rates were rather poor, varying from 71 percent of the original population in 1982 to 44 percent in 1986, and there was no follow-up on nonrespondents, which could bias the results. The data-

[1] Joshua Lerner, "The Mobility of Corporate Scientists and Engineers Between Civil and Defense Activities: Evidence from the SSE Database," Science, Technology, and Public Policy Program Discussion Paper 90-02 (Cambridge, MA: John F. Kennedy School of Government, August 1990).

base provides information on changes in industry or occupation between 1982 and 1986, but does not include job switching between firms
if the individual stayed in the same occupation and industry class.

The most important result is that 24.2 percent of respondents with
defense-related positions in 1982 had switched to civilian-oriented employment by 1986, despite the fact that the period 1982–1986 was one
of maximum defense buildup.[2] Moving in the other direction, 26.5 percent of the S&Es in defense-oriented jobs in 1986 had been in nondefense jobs four years earlier. This is a switch rate from defense to
nondefense or vice versa of about 6 percent a year, compared to a job
change rate for individuals at least 25 years old with four or more years
of college of 4.3 percent.[3] From indirect evidence it appears that most
of the movement took place *within* dual-use firms—i.e., firms with both
defense and nondefense divisions.

With respect to specific industries, of the S&Es working in defense
in 1982, 38 percent of those in the nonelectrical equipment industry,
18 percent in aircraft, and 15 percent in ordnance moved out of defense
within the next four years. Of those primarily engaged in defense in
1986, 53 percent in nonelectrical equipment, 21 percent in aircraft, and
17 percent in ordnance had been in nondefense positions four years
earlier. The fraction who shifted into defense is fairly close to the fraction who left; the fact that it is slightly higher presumably reflects the
expansion of defense activity between 1982 and 1986. What is most
striking is the size of the two-way flow.

The impression that emerges from this analysis is one of rapid circu-
lation between defense and nondefense positions in industry. From
other elements of the SSE survey data, Lerner also shows that the
defense and nondefense S&E populations are fairly similar in characteristics such as demographic, economic, and educational variables.
Those in defense appear to be somewhat older, less likely to be female,
more likely to be native born, more likely to have a Ph.D., and more
likely to be at a "plateau" in their careers, but the differences are quite
small.

[2]The defense-intensity of all employment went from 5.3 percent in 1980 to 6 percent in
1985, while the defense-intensity of manufacturing employment went from 6 to 9 percent
during the same period. Military R&D expenditures nearly doubled, in real terms, from 1980
to 1987.

[3]James P. Markey and William Parks II, "Occupational Change: Pursuing a Different
Kind of Work," *Monthly Labor Review* (September 1989), pp. 3–12; cited in Lerner, "Mobility
of Corporate Scientists and Engineers," p. 3.

This mobility suggests that the difficulties of converting from military to civilian production (or vice versa) may not be as great as perceived by high-level managers, but that if conversion is to take place,
it will have to be by movement of technical people into new organizations with different leadership, hierarchies, and culture, rather than by

attempting to convert an intact organization to a new function. In other words, the problem of conversion in either direction applies more to teams and organizations than to individuals.

Supply

The huge buildup of U.S. military R&D in the 1950s and 1960s had the effect of creating and sustaining a much larger stock of S&Es than in any other Western economy. Other countries did not attain similar levels of S&Es (as a proportion of the labor force) until 20 years later. The Servicemen's Readjustment Act of 1944 (the G.I. Bill of Rights), which provided funds for returning World War II veterans to continue their education, greatly assisted in this buildup.[22]

Many of these veterans gained technical training and experience with sophisticated equipment during the war. The G.I. Bill gave them an unprecedented opportunity to pursue their aspirations toward technical careers.[23] At the same time, massive military R&D investments provided extraordinarily challenging jobs where these people could exercise and enhance their skills. As civilian opportunities appeared, they then helped the United States bring to fruition commercial developments in semiconductors, computers, communications, and jet aircraft. Potential competitors with more limited high-level technical labor could not keep up. In 1965, the percentage of the U.S. workforce consisting of graduate scientists and engineers was at least three times greater than in the economies of our principal industrial rivals. Only in the early 1980s did this gap largely disappear. By 1988, the fraction of the Japanese labor force consisting of S&Es engaged in research and development equaled that of the United States and was increasing about twice as rapidly.[24] Furthermore, most scientists and engineers abroad, unlike those in the United States, were socialized in civilian

[22] Eli Ginzberg and associates, *Patterns of Performance* (New York: Columbia University Press, 1959), pp. 167ff. The G.I. Bill enabled more than 2.2 million ex-servicemen to attend colleges and universities; see Eli Ginzberg, "Scientific and Engineering Personnel: Lessons and Policy Directions," in Panel on Engineering Labor Markets, Office of Scientific and Engineering Personnel, National Research Council, *The Impact of Defense Spending on Nondefense Engineering Labor Markets* (Washington, DC: National Academy Press, 1986), p. 26.

[23] This policy has been described as "a massive investment in human capital, an investment of a size and scope never before contemplated . . . the single most important element in the postwar recovery . . . the domestic counterpart of the Marshall Plan." David T. Kearns and Denis P. Doyle, *Winning the Brain Race: A Bold Plan to Make Our Schools Competitive* (San Francisco: ICS Press, 1988), p. 101.

[24] National Science Foundation, *International Science and Technology Update: 1988*, NSF Report 89-307 (Washington, DC: National Science Foundation, 1988), p. 39.

industry from the beginning. The conditions imposed on the Germans and the Japanese after World War II, in particular, prevented them from engaging in military or even dual-use technologies. In Japan, therefore, many of the most experienced and creative aeronautical engineers applied their skills in the emerging automobile industry.

Impact of DoD Requirements on S&E Labor Markets

The potential for diversion of scarce talent into defense and away from civilian industry and commercial innovation became a matter of serious political concern in the early 1960s with the buildup of federal expenditures for the Apollo project and strategic missile programs such as Polaris and Minuteman.[25] The notion developed that the demands of defense and space projects were attracting the most talented scientists away from civilian industry and thereby were eroding U.S. competitiveness in world markets.[26] This argument lost currency with the cutback in space and defense expenditures in the 1970s and the appearance of a surplus of engineers, but it was revived again in the 1980s with the buildup of military expenditures under the Reagan administration, particularly in connection with the massive new investments in exotic advanced technologies that constituted the new Strategic Defense Initiative (SDI). Although the argument may have had some validity in the 1960s, it is much less applicable to the 1980s buildup.

Two rather different mechanisms have been alleged, both resulting in adverse impacts on commercial industry. The first is that federal projects, by increasing demand for S&Es, bid up their salaries. The second is that defense and space projects differentially attract the most talented and best-trained S&Es, not so much for higher salaries, but because of their technical challenge.

The first mechanism would raise the cost of commercial innovation directly. Since R&D is very labor-intensive, with salaries constituting

[25] See, for example, the report of the so-called Gilliland Panel of the President's Science Advisory Committee, *Meeting Manpower Needs in Science and Technology*, The White House, December 12, 1962. President Kennedy's science advisor, Jerome Wiesner, formed a joint panel the same year with the Council of Economic Advisors (CEA) to consider the possible negative impact of massive federal R&D projects on innovation in more traditional U.S. industries.

[26] An articulate early proponent of this view inside government was Michael Boretsky, an economist in the Department of Commerce. See Michael Boretsky, "Concerns About the Present American Position in International Trade," *Technology and International Trade* (Washington, DC: National Academy of Engineering, 1971), pp. 18–36. For a good summary of the main arguments in the 1950s and 1960s, see J. Herbert Hollomon and Alan E. Harger, "America's Technological Dilemma," *Technology Review* (July/August 1971), pp. 31–41.

about 70 percent of total costs, it was argued that foreign competitors would be able to undertake projects that would look unattractive to U.S. firms paying higher salaries. Indeed, between 1950 and 1965 the salaries of both entry-level and experienced physical scientists and engineers rose some 45 percent relative to median per capita income in the country as a whole.[27] There is also evidence, although it is more ambiguous, that relative salaries of engineers in the United States were higher and rising faster than in Europe and Japan.[28]

According to employment records, nearly 100,000 more engineers appeared in the labor force between 1950 and 1965 than graduated from higher education. Thus it seemed plausible (though unproven) that industry was upgrading technicians to take the place of graduate engineers who were transferred to rapidly growing defense and space projects.[29] In the 1960s especially, such moves would have been encouraged by DoD's emphasis, when selecting contractors, on breadth and depth of technical capability as judged by the stock of engineers and scientists with high educational qualifications.

According to the second of these postulated diversion mechanisms, the best and the brightest engineers would be attracted by the technical challenge of defense and space. Such projects, less constrained by economic and social considerations or by market uncertainty, pushed harder on the technical state of the art, and thus tended to have higher status in the professional value system of engineers. Commercial work, especially in "smokestack" and consumer product industries, was seen as less attractive, unable to compete for the best people. In the 1950s and 1960s, this value system was undoubtedly reinforced by the wartime experience of many industrial leaders who had been professionally socialized in a climate of urgent weapons developments with unlimited priority and resources, exemplified most strikingly by radar and the Manhattan Project. Indeed, the high proportion of industrial R&D funded by defense in the 1950s and 1960s tended to extend and reinforce the technically oriented (or technocratic) value systems absorbed during World War II.[30] Moreover, the professional experience of the 1950s and early 1960s had been a major testing ground for engineers,

[27] Hollomon and Harger, "America's Technological Dilemma," Figure 8, p. 34.

[28] Richard B. Freeman, *The Market for College-Trained Manpower: A Study in the Economics of Career Choice* (Cambridge, MA: Harvard University Press, 1971).

[29] Hollomon and Harger, "America's Technological Dilemma," especially Figure 9, p. 34.

[30] In 1962, nearly 70 percent of all U.S. R&D was supported by the federal government; of this, 93 percent was in defense, atomic energy, and space technology. See Harvey Brooks, "Government Support of Science," *McGraw-Hill Yearbook Science and Technology* (New York: McGraw-Hill, 1963), pp. 11–21.

scientists, and managers who subsequently rose to high positions in American corporations in later years and who may have shifted corporate cultures away from the requirements of commercial innovation.

Although both effects postulated here probably did occur in the 1950s and 1960s, the defense buildup of the 1980s (starting in 1979 after the Soviet invasion of Afghanistan) showed little or no escalation of engineering salaries relative to median personal income. Nor did starting salaries of engineering graduates show evidence of a differential between commercial and defense-oriented employment.[31] In fact, there is considerable evidence that starting salaries of S&Es have been declining relative to those of MBAs, lawyers, and accountants since 1985, a trend likely to accelerate with declining military R&D spending in the 1990s even if commercial R&D grows.[32] For one thing, the Carter-Reagan buildup of military R&D was considerably slower, in relative terms, than buildups immediately after the Korean War (1950–1953) and in the early 1960s.[33] At the same time, the supply of new graduates was considerably greater during the 1980s, owing to the entry of the baby-boom generation into the labor force. Furthermore, the 1980s military buildup began in a period of considerable slack in the economy owing to the severest recession since the end of World War II.[34]

On balance, defense cannot be said to have significantly tightened technical labor markets during the recent buildup, except in a few specialized fields such as computer science. Moreover, the immediate outlook appears to be for a surplus of S&Es rather than a shortage, although there has been a vigorous debate on this subject.[35] A period

---

[31] Robert K. Weatherall, "Defense and Non-Defense Employment: The View from Engineering School Placement Offices," in Panel on Engineering Labor Markets, *The Impact of Defense Spending,* pp. 83–91.

[32] Craig Stetson, "Immigration and U.S. Technical Personnel," Master's Thesis, John F. Kennedy School of Government, May 1991, based on data from Bureau of Labor Statistics, *Handbook of Labor Statistics,* Bulletin 2340, U.S. Department of Labor, August, 1989, and from R. Keith Wilkinson, *Science and Engineering Personnel: A National Overview,* NSF Surveys of Science Resources Series NSF 90-310; see also NSF, Division of Science Resources Studies, "Two-Year Decline Expected in Real R&D Spending," *SRS Data Brief,* March 18, 1991.

[33] Panel on Engineering Labor Markets, *The Impact of Defense Spending,* p. 5.

[34] Ibid., p. 34.

[35] Jeffrey Mervis, "Analysts Debunk Idea of Scientist Shortage, Citing Defects in Current Economic Models," *The Scientist,* May 13, 1991, p. 1; Alan Fechter, "Engineering Shortages and Shortfalls: Myths and Realities," *The Bridge,* vol. 20, no. 2 (Fall 1990), pp. 16–20; Peter W. House, "NSF Projections in Science and Engineering," pp. 21–22. For a balanced discussion of the intermediate to long-term outlook, see Eli Ginzberg, "Summary of Presentation: Implications of Changing National Priorities," presentation to the American Association for the Advancement of Science, February 15, 1991.

of rapidly declining demand today could lay the groundwork for future shortages, since graduation rates of scientists and engineers respond only after some years to fluctuations in the labor market. Defense and space cutbacks in the late 1960s and early 1970s did indeed generate a decline in the numbers of engineering graduates, which reached a minimum in 1976.[36]

One conclusion seems abundantly clear. Current econometric models of the supply and demand for technical people are simply not adequate to the task of predicting surpluses and shortages for more than a few years ahead.[37] This leaves room for wide differences of opinion about the longer-term (five- to ten-year) outlook.

Effect of Military Spending on Engineering Education

Has the U.S. investment in military technology driven engineering education in directions that were counterproductive to the capacity to compete in civilian markets or to maintain a high-quality civil infrastructure? This issue is difficult to resolve, in no small part because of the difficulty in separating the indirect effects of the external S&E labor market from the direct effects of DoD support for R&D and facilities on the campuses.

The impacts could come about in several ways:

1. through the effect of DoD research support on the orientation and expectations of faculty and students in engineering schools;
2. through the effects of the employment market on the career plans of students and on their choice of program, together with the influence of these student preferences on faculty selection and on the content of the engineering curriculum; and

[36] Richard B. Freeman, "A Cobweb Model of the Supply and Starting Salary of New Engineers," *Industrial and Labor Relations Review*, vol. 29, no. 2 (January 1976), pp. 236–248. Subsequently, the numbers of engineering bachelors degrees nearly doubled from 1976 to 1985 (apparently stimulated by a turnaround in industrial demand for engineers from 1976 to 1979), with the increased supply then contributing to the salary slide of engineers after 1985.

[37] There are many difficulties in making projections, especially in the near to intermediate term. For one, the number of students who enroll in engineering appears to be very sensitive to the current demand for graduates. However, the number of graduating engineers depends on enrollments several years earlier. The supply of engineers is also very sensitive to the modes of utilization of engineers in industry, and the rate of dropout of employed engineers (and to a lesser extent, scientists) from engineering activity. For these reasons, there is a high degree of flexibility in engineering labor markets.

3. through the consulting opportunities available to engineering faculty, which could indirectly influence their teaching and their choice of research problems for graduate and even undergraduate students.

1. Academic research and teaching.   The most dramatic development in engineering education following World War II was the growth of academic research and the expansion, relative to the natural sciences, of Ph.D. and, later, postdoctoral programs in engineering. Prior to the late 1940s, engineering education had typically terminated with the bachelors degree; the percentage of engineers going on to graduate training was much smaller than in the sciences, and restricted to a small number of elite schools. The new emphasis on graduate education was facilitated by the expansion of research support, mostly from agencies of the federal government. While federal support for academic research grew rapidly in all S&E disciplines from 1946 to 1967 (at an average annual rate of close to 15 percent), it grew faster in engineering, especially after 1960. In the 1950s, a very high proportion of this support came from the three military services, especially the Office of Naval Research (ONR), and from the Atomic Energy Commission. In 1954, for example, defense provided more than 60 percent of all federally supported academic research. In the 1960s and 1970s, the fraction of academic research support coming from defense agencies declined until, by the mid-1980s, it accounted for only about 8 percent of the total. However, DoD remains a major patron for research in engineering schools, and supplied 32 percent of all funds for academic engineering even in the late 1980s: 50 percent in electronics and electrical engineering and 42 percent in aerospace engineering, though only 20 percent for mechanical, 6 percent for civil, and 4 percent for chemical engineering.[38] Most engineering research sponsored in universities by DoD, AEC, and NASA was "engineering science"—i.e., investigations of natural phenomena underlying engineering practice—rather than engineering design, manufacturing operations, or the construction and testing of prototype equipment. Only in chemical engineering was there continued emphasis on process design and manufacturing. Where design and prototype testing were involved in academic research in other engineering disciplines, they were likely to be supported

[38] Figures are for 1986–1987, from National Science Foundation, "Possible Areas of Impact of DoD Funding Cut," briefing package, 1990.

by DoD or NASA, these being the only agencies providing funding in sufficiently large packages to permit academic departments to engage in such downstream engineering activities. In consequence, such distinctively "engineering" activity (as opposed to engineering science or phenomenology) to which students and faculty were exposed firsthand tended to be DoD-supported and hence representative of the DoD culture rather than the commercial culture. Furthermore, DoD and NASA support concentrated in the elite engineering schools, whose Ph.D. graduates often went on to teach at other colleges and universities.[39] How much this actually influenced the orientation and attitudes of students can only be speculation, however.

The shift toward engineering science and the increased emphasis on R&D (as opposed to design, manufacturing, testing, and the management of the workforce) were in part a response to the opportunities available through federal funding, and in part a reaction to the World War II experience in which scientists rather than engineers had often been the most visible players in radical technological developments. This led to a widespread view that the prewar education of American engineers had been too closely tied to current industrial practice, leaving graduates with insufficient depth in science to cope with emerging new technologies based on such advances as quantum mechanics, solid-state physics, and nuclear physics.[40] This led in turn to what may have been, in retrospect, an overreaction in the direction of engineering science, encouraged and perpetuated by the availability of funds for research.

During the entire postwar period, and especially during the rapid expansion of graduate education in the 1960s, engineering faculties were heavily colonized by people whose original training had been in the natural sciences, particularly solid-state, atomic, and molecular physics. This especially affected electrical engineering (the largest discipline) and aerospace engineering, but also mechanical, and, to a much lesser extent, civil and chemical engineering. Thus the implicit mission of engineering education, especially in the elite research universities, became to train people for participation in radical breakthroughs in technology, rather than for the incremental innovation that constitutes the bulk of commercial innovative activity. Meanwhile, industrial engi-

[39] National Science Foundation, "Possible Areas of Impact of DoD Funding Cut."

[40] This view was very forcefully articulated by a special committee established in 1948 and chaired by Vannevar Bush to advise on the future use of the McKay bequest for engineering and applied science at Harvard. See "Report of the Panel on the McKay Bequest to the President and Fellows of Harvard College," 1950.

neering departments emphasized operations research and other mathematical techniques, many of them spawned by the war, which also eroded the exposure of students to design and manufacturing. The universities significantly changed the reward system in the engineering profession, elevating the status of rigorous science-based analysis and radical innovation at the expense of product and process development through meticulous execution and continuous incremental improvement.

A matter of some debate is how much of this reorientation of engineering education can be attributed to the influence of defense spending in universities and how much would have occurred anyway in response to the changing labor market for S&Es. It is certainly true that little external funding was available to support research in engineering schools relevant to design and manufacturing, compared to what was available to support upstream activities removed from commercial production.[41]

2. The labor market.   Table 4-9 shows the percentage of engineers receiving DoD support in three major disciplines.[42] It shows that engineering employment as a whole is more DoD-intensive than employment of researchers in academia. Admittedly, there are problems interpreting these data because the definitions are fuzzy. Moreover, the members of the academic populations in the first two columns of the table are much more likely to be Ph.D.s or prospective Ph.D.s than are members of the engineering workforce as a whole; therefore the figures in column 3 are not strictly comparable to columns 1 and 2. Nevertheless, the results do suggest that defense support of academia, even in the most defense-intensive engineering disciplines, is not by itself enough to bias people toward the defense culture. If anything, the defense presence in academia appears considerably less than it does in the engineering workforce as a whole. S&Es who go into defense-oriented work in industry and government are slightly more likely to have an advanced degree than those in nondefense work, but the

---

[41] For an articulate discussion of the corporate innovation culture of the postwar era, particularly the segregation of the R&D function on rural corporate "campuses" separated from operations, see Martin Kenney and Richard Florida, *The Breakthrough Illusion* (New York: Basic Books, 1990). The authors attribute this corporate culture to distortions generated by DoD support of industrial research. Although we agree with their diagnosis of the problem, we feel that their arguments for attributing it primarily to the influence of defense spending are not very persuasive. See also Harvey Brooks and Maryellen Kelley, "From Breakthrough to Follow-Through," *Issues in Science and Technology*, vol. 5, no. 3 (Spring 1989), pp. 42–47.

[42] National Science Foundation, "Possible Areas of Impact of DoD Funding Cut."

TABLE 4-9:
Percentage of U.S. Engineers Receiving DoD Support, by Discipline

Discipline	Academic Employment[a]	Graduate Students[b]	All Engineers
Electrical engineering	17%	13%	29%
Aerospace engineering	23%	34%	53%
Mechanical engineering	13%	8%	19%

Source: National Science Foundation, "Possible Areas of Impact of DoD Funding Cut."
[a]Includes both teaching and research positions.
[b]Supported graduate students whose support (e.g., research assistantship) comes from DoD.

difference is surprisingly small.[43] This does not support the argument that defense is disproportionately attracting people with advanced degrees to the detriment of commercial activity.

In the period of the greatest dominance of the federal R&D budget by defense and space in the late 1950s and early 1960s, a study of the initial employment of Ph.D. physicists in industry showed that about 85 percent of them entered the two industry sectors, aerospace and electronics, that received the largest share of R&D support from DoD (and NASA). During the same period (roughly 1958 to 1964), a much higher fraction of Ph.D. physicists were going into industry than was true in any subsequent period.[44] This was before the rapid take-off in the number of engineering Ph.D.s referred to earlier, and it provides an additional hint that employment markets might have been influencing the direction of engineering education, especially in its emphasis on engineering science. Graduate engineering education, in particular, was responding to the labor market as demonstrated by the recruitment of physicists in the late 1950s.

3. Faculty consulting. Heavy DoD funding of research in industry and government laboratories also created consulting opportunities for

[43]Of the defense-oriented S&Es, 14.8 percent have seven or more years of post-high school education, compared to 12.3 percent of the nondefense S&Es. Lerner, "The Mobility of Corporate Scientists and Engineers Between Civil and Defense Activities. Table III, p. 12.

[44]Harvey Brooks, "The Dynamics of Funding, Enrollments, Curriculum, and Employment," in Martin L. Perl, ed., *Physics Careers, Employment, and Education,* American Institute of Physics Conference Proceedings No. 39 (New York: American Institute of Physics, 1978), p. 97.

university faculty. Defense agencies in the 1950s and 1960s sought to reach out to the entire U.S. technical community for advice and new ideas. Senior faculty often brought their younger colleagues and even graduate students into this work, especially the "summer studies" sponsored by DoD in its intense search for new ideas. Defense contractors, like defense agencies, defined mission-related technical problems, broke them down, and farmed out parts suited for academic research to faculty members and their graduate students on a contract or a consulting basis. The ample funds available to DoD agencies, and the relative absence of cost discipline on R&D activities, suggest that opportunities for consulting were heavily concentrated in defense, although we have no statistical data to verify this. Although much defense-oriented consulting (unlike DoD-supported research within universities) was under security classification, there was a large interacting community inside the classified fence, whereas consulting for commercial firms tends to be more circumscribed by proprietary considerations. All this meant that faculties had readier access to the defense-oriented culture in industry and government than to commercial or civilian-oriented culture, especially until the end of the 1960s, when U.S. involvement in Vietnam made ties to DoD unfashionable in academia.

Defense Investment in Technical Human Resources: Summary

Defense demand after World War II greatly accelerated the nation's production of scientists and engineers, boosting their percentage of the labor force in the 1950s and 1960s to levels far above that of other industrialized countries. But to the extent that defense experience proved ill-suited to commercial innovation, this may have been a mixed blessing. At any rate, it was a temporary one, with Japan surpassing the United States in the S&E fraction of its labor force by 1988. Although defense demand for S&Es in the 1950s and 1960s drove up wages and may have preferentially absorbed the most qualified and best trained S&Es, it would be difficult to conclude that these effects outweighed the contributions defense made to the technical workforce. Moreover, there is little evidence that defense demand has squeezed the supply of scientists and engineers at any time since the mid-1960s.

The evidence is mixed as to how well scientists and engineers working on defense projects can adapt to the needs of the commercial sector. Survey data indicate a higher degree of interchange than the anecdotal evidence from corporate managers would imply, suggesting that technical skills transfer more easily than managerial ones. In addition to direct transfers of S&Es, defense influence on engineering education, whether directly through support of academic research or indirectly through faculty consulting or labor market signals to engineering students, could also affect the commercial technical workforce.

There is little question that a dramatic reorientation of the engineering profession occurred in the postwar years, coincident with and reinforced by the growth of defense technology investments. Furthermore, new technical jobs, particularly in the 1950s and early 1960s, were heavily concentrated in the industries and disciplines most closely connected with defense, and many of the leading engineering managers of the 1970s and 1980s were probably recruited from this cohort. Although the technical knowledge required for defense work does not differ very much from that required for commercial engineering, the economic and business aspects of the defense environment lead to very different managerial orientations. Defense influence almost certainly contributed to the overemphasis now placed on radical innovation, as compared to continuous, incremental product and process improvement, and also to the isolation of R&D from other activities within firms. While it is clear that defense has also contributed to the lack of prestige and the low status of manufacturing, both in industry and in engineering education, it is less clear to what extent defense has been a prime factor in the deterioration of the United States' relative performance in manufacturing. Other factors, including the long heritage of Taylorism from the early twentieth century, may have been equally important.[45]

Defense contributions to and utilization of the nation's base of technical human resources has had considerable influence, both positive and negative, on the commercial sector. As the defense share of the national technology effort has declined, these influences have probably declined as well. Measuring them quantitatively—much less coming up with a net assessment of their effects on commercial innovation—is not at all straightforward. In the following section we discuss the limitations of attempts to quantify dual-use interactions.

---

[45] Taylorism is a factory management approach first developed with the rise of mass production for continentwide markets. See Harvey Brooks and Michael Maccoby, "Corporations and the Work Force," in John R. Meyer and James M. Gustafson, eds., *The U.S. Business Corporation: An Institution in Transition* (Cambridge, MA: Ballinger, 1988), Chapter 6, pp. 113–131. See also the discussion of manufacturing in Chapter 10 of this book.

CAN WE MEASURE DUAL-USE INTERACTIONS QUANTITATIVELY?

The statistics detailing DoD's involvement in the national technology base that have been presented so far in this chapter—by way of R&D funding and the technical workforce—say nothing about DoD's effects on commercial activities. Even though military R&D must in the final analysis be justified on military grounds, reliable estimates of the net contribution or cost to society of each dollar spent on military R&D would be useful for policy purposes.

The effects of defense spending on the pace and direction of techno-logical change are typically analyzed through case studies and historical examples. But those cases singled out for analysis are usually selected precisely because in retrospect they have had a particularly strong influence—positive or negative—on commercial technology. They do not tell us much about the *average* impact of military spending upon the commercial sector, and they are not easily generalized.

Thus it would be useful to have a quantitative analysis of the military-commercial technology relationship that included both ordi-nary and extraordinary cases. The first requirement for any such analy-sis, as we pointed out in Chapter 3, is to identify an explicit standard of comparison against which the commercial impacts of defense R&D investments are to be evaluated. Without such a standard, any answer is possible. Even with such a standard, however, existing quantitative methods—typically econometric studies of defense R&D spending—are too crude and too limited to be useful guides to understanding the dual-use relationship, as we are about to explain.

This is not to say that econometric analysis cannot help in under-standing the economic impacts of R&D. It can.[46] But these methods work better for privately funded than for publicly funded R&D. More-over, they provide more reliable results at low levels of aggregation—that is, for a single firm or industry—because the inputs and (espe-cially) the outputs of R&D can be more precisely identified. In contrast,

---

[46] For a review, see Office of Technology Assessment, *Research Funding as an Investment: Can We Measure the Returns?* Technical Memorandum (Washington, DC: U.S. GPO, April 1986). See also Edwin Mansfield, "R&D and Innovation: Some Empirical Findings," in Zvi Griliches, ed., *R&D, Patents, and Productivity* (Chicago: University of Chicago Press, 1984), pp. 137–154; Zvi Griliches, "R&D and Productivity: Measurement Issues and Economic Results," *Science*, vol. 237 (July 3, 1987), pp. 31–35; Zvi Griliches, "Econometric Estimates of R&D Returns: Measurement Problems, Results, and Policy Implications," prepared for the conference *Technology and Investment: Critical Issues for the 90's*, sponsored by the Royal Swed-ish Academy of Sciences, Stockholm, January 21–24, 1990.

measuring the economic value of defense R&D requires addressing the direct and indirect effects of public R&D funds on a wide range of industries, for which conceptual and methodological problems limit the conclusions that can be drawn.

Returns to R&D

Essentially all attempts at quantitative study of dual use begin with econometric models that seek relationships between the inputs and outputs of R&D. The inputs are typically dollar values for R&D spending, the outputs some measure of technological innovation as reflected in the growth of output or productivity. Almost all these studies reach very negative conclusions about the dual-use relationship. They claim that defense R&D is, in aggregate, harmful to national economic performance.

Such econometric studies generally proceed in two steps. First, they try to estimate the economic rates of return on investments in defense R&D, almost invariably finding that the returns on defense (or government) R&D investments are less than the returns on private R&D. This fact, if indeed true, would hardly be surprising (see the Chapter 3 section, "Assessing the Benefits of Spinoff: Compared to What?"). Most private R&D is undertaken with the express aim of increasing economic output, whereas most defense (or government) R&D has commercial benefits as an incidental by-product.

The difficulty rises when these analyses take a second step and try to show that defense R&D displaces or "crowds out" private R&D. Since the private R&D is more valuable, according to this analysis, the net consequences to the economy can be adverse even if the total R&D pie is bigger as a result of the defense contribution.[47] If one could rely on both of these steps, then increasing defense R&D would automatically decrease economic performance. It is therefore worth paying close attention to the assumptions and techniques that underlie this argument.

Economywide Output and Input Measures

First, how does one estimate the fraction of total factor productivity (or whatever indicator of economic performance one is measuring)

---

[47] For example, see Frank B. Lichtenberg, "The Impact of the Strategic Defense Initiative on U.S. Civilian R&D Investment and Industrial Competitiveness," *Social Studies of Science*, vol. 19 (1989), pp. 265–282, for an argument that defense R&D is harmful to national economic competitiveness.

that is due to technological innovation as opposed to other inputs? Generally these studies proceed by a process of elimination. They model the contributions of capital, labor, and materials—and in fancier analyses intangible factors like worker skill and attitude, management skill, and economies of scale and scope—and subtract them. Whatever economic output is left over is attributed to innovation. (Note that any errors in the model used to estimate the contributions of other factors will lead to even larger uncertainties in the part attributed to innovation.) Next, innovation is attributed to technological effort, and technological effort to R&D spending. Before turning to the calculation of inputs, let us pause a moment longer on the calculation of outputs.

Those who collect and analyze statistics on the economy do best at capturing the value of technological innovation when it fits into existing economic niches and accounting categories. But it is harder to assign value to qualitative improvements brought about by new technology, particularly entirely new products (say, xerographic copiers) and new industries created over long time periods (say, computers). Moreover, the benefits of innovations that flow back to manufacturers in the form of profits are easier to measure than those accruing to consumers, who by purchasing the new products are able to do things they could previously not do at all, or only at a higher price. The latter benefits can only be estimated as a "consumer surplus" that reflects how much of a premium customers would on average have been willing to pay for the new product.

It follows that calculations of returns on investment in technology probably err systematically by giving more weight to process innovation (i.e., making existing products more efficiently) than to product innovation (i.e., creating new types of products or improving existing ones). Moreover, they will do a better job at estimating the benefits of incremental improvements in specific products, as compared with the far-reaching effects of radical innovations such as the computer, which can generate long-term structural change in the economy and give rise to whole families and successive generations of new products. If defense R&D contributes more strongly to product than to process innovation, as is frequently alleged, and to radical innovation more than to incremental progress, defense R&D may be undervalued by present methods of econometric analysis.

So much for the calculation of economic output. How does one handle technological input? The first thing to note is that the calculations generally ascribe technological progress to organized R&D only. But R&D is not the only source of technological contribution to economic growth, simply the one most easily quantified. R&D statistics exclude downstream activities such as "learning-by-doing" on the part of both producers and users of technological products. Neither do they capture technology embodied in purchases of manufacturing tools, professional services, or licenses.

Moreover, the value to be attributed to R&D spending depends on how long ago it was performed. Very recent R&D counts less since it has not yet had a chance to contribute to products and processes of commercial value. Similarly, R&D from many years ago must be discounted since knowledge becomes obsolete. It turns out that the estimated rates of return to R&D are very sensitive to the assumed depreciation rate of knowledge, which may be quite different in different industries.[48]

Most calculations tend to treat R&D as a homogeneous quantity and do not break it down into its components—not even the traditional categories of basic research, applied research, and development. Attempting such a breakdown would cause further difficulties, since government and industry do not account for R&D in fully consistent ways, and neither do different industries or different countries. A few studies that have made this distinction—sometimes questionable as to their interpretation—suggest that returns to basic research greatly exceed those to development.[49]

Relating Output to Input: Spillover

The analyst must live not only with the prospect of substantial estimation errors on both the input and the output sides, but with ambiguities in relating the two. To begin with, firms that spend money on R&D do not always capture the benefits. We have already mentioned the difficulty in measuring those benefits that accrue to their customers. In addition, the fruits of R&D also "spill over" to other firms that read the results in journals, hire away their employees, or "reverse-

---

[48] Jacques Mairesse and Mohamed Sassenou, "R&D and Productivity: A Survey of Econometric Studies at the Firm Level," *STI Review*, no. 8 (April 1991) (Paris: OECD, 1991), pp. 9–44, especially p. 30.

[49] Edwin Mansfield, "Technological Change and Economic Growth," *U.S. China Conference on Science Policy* (Washington, DC: National Academy Press, 1985), pp. 1–18, especially p. 13; Edwin Mansfield, "Basic Research and Productivity Increase in Manufacturing," *American Economic Review*, vol. 70, no. 5 (December 1980), pp. 863–873; Griliches, "Econometric Estimates of R&D Returns: Measurement Problems, Results, and Policy Implications"; Robert E. Evenson, "Comparative Evidence on Returns to Investment in National and International Research Institutions," in Thomas M. Arndt, Dana G. Dalrymple, and Vernon W. Ruttan, eds., *Resource Allocation and Productivity in National and International Research* (Minneapolis: University of Minnesota Press, 1977), p. 237.

engineer" new products or processes at a fraction of their original development cost.[50] Such spillovers, added to the consumer surplus, will tend to make the so-called social return on R&D investments—that is, the benefit to society as a whole—larger than the "private" return to the firms that funded the R&D. Spillover across industries is particularly important in the case of a radical innovation that results in the development of a whole new industry over the course of many years. In such a case, the benefits become so diffused through the economy that it is difficult to determine which long-past R&D investments should get the credit.

Industrywide or economywide studies capture social returns more effectively than studies of particular firms or narrow sectors, leaving analysts with a choice: either measure returns within a sector and thus miss the spillover, or aggregate many sectors (mixing computers, airplanes, pharmaceuticals, and whatever), capturing the spillover at the price of muddying the results. Either alternative limits the usefulness of the analysis. A second problem in relating economic outputs to technological inputs comes with assuming that dynamic industries are dynamic because they do lots of R&D. The reverse can also be true. Companies with forward-looking management that deal with fast-moving and profitable technologies tend to do R&D. But the reason they are profiting may not be that they are doing R&D, simply that their technology is promising and their management is forward-looking. Or, even more simply, perhaps only the dynamic and successful firms can afford high levels of R&D spending.[51] Third, most analyses measure average rates of return, not marginal rates of return. Since one would expect the marginal returns to be less than the average returns, the quantitative implications of these analyses are likely to overstate the shifts at the margin, which is precisely where policy changes have their effects.

"Crowding Out"

Even if publicly funded R&D is less productive economically than privately funded R&D, defense (or other public) R&D can be shown to

---

[50] Two attempts to grapple with the impact of spillovers are Adam Jaffe, "Technological Opportunity and Spillovers of R&D," *American Economic Review*, vol. 76 (1986), pp. 984–1001; and F.M. Scherer, "Inter-industry Technology Flows and Productivity Growth," *Review of Economics and Statistics*, vol. 64 (1982), pp. 627–634.

[51] It should be possible to sort out cause from effect by correlating output measures with R&D investments from earlier years—as indeed is done in the weighting procedure described above by which very recent and very old R&D are both discounted—but this process can be meaningful only when there is sufficient time variation of R&D spending.

be a drag on the economy only if it displaces, or "crowds out," private activity that would be economically more valuable. Can such a conclusion be justified? One form of crowding out supposedly occurs within firms that do both defense and nondefense business: growth in defense business is said to bleed scientists and engineers away from the nondefense side of the house. But in view of the segregation of defense and nondefense activities, even within single companies, this type of crowding out seems unlikely. A less direct form of crowding out, in which government demand for technical manpower drives up S&E wages, may well have occurred during the rapid defense/space buildup in the early 1960s, as discussed earlier in this chapter. However, such effects have not been observed more recently.

Arrayed against the possible crowding-out effects would be defense's contribution to what might be described as economies of scale and scope in R&D. Defense R&D has increased the long-term demand for highly trained people, paid for a great deal of widely applicable technical knowledge, and supported the research infrastructure of instrumentation, materials, and supporting services. In this manner, defense support raises the productivity and efficiency of the whole R&D "industry," with benefits shared by both defense and commercial sectors. These benefits of scale, however, are very difficult to quantify. There are also possible offsetting negatives—for example, the lack of cost discipline on defense R&D, which may lead to bad habits that spill over into commercial practice.

The Limitations on Econometric Studies of Dual-Use Relationships

From this review of techniques used to estimate the net social cost or benefit of defense R&D spending, we conclude that existing methods shed little light. It is not surprising that the economic output attributed to a dollar of government R&D should fall short of its privately funded counterpart. Most government R&D goes for national defense, space exploration, and health (mainly the cure of disease), the value of which cannot be captured in market prices. Although defense and space R&D lead to tangible end products, those products are sold to a single customer—the government—at "prices" that do not reflect their value to society, but rather are based on costs. Therefore the only possible way that government R&D can contribute to economic growth is through spillover. For privately funded R&D, on the other hand, econometric analysis can in principle capture both the direct returns to the investor and the spillover effects in other parts of the economy.

We have also seen that the economic benefits resulting from government R&D are likely to be undervalued, that the crowding out of private R&D by government R&D has not been demonstrated (at least for recent decades), and that we have no way to know how the commercial technology base would have fared in the absence of the research infrastructure supported by defense.

Not all analyses have all the difficulties sketched here. With continued effort they should improve, although some of the limitations are fundamental. For the time being, guidance for policy can come only by assembling a picture of the impacts of defense R&D piece by piece—the goal of this book.

WHAT THE NUMBERS SHOW

- The United States invests more by far on R&D than does any other nation ($145 billion in 1990), spending as much as the next nine largest R&D performing nations in the OECD combined. However, more than 30 percent of U.S. R&D is devoted to defense, a share that is four times higher on average than that of other large R&D spenders in the OECD.
- Growth in commercial R&D has outpaced government-funded R&D in the United States for decades, as has growth in world R&D compared with the U.S. total. U.S. defense R&D funding has therefore been steadily decreasing as a share of the world's R&D output, dropping from one-third of all OECD R&D in 1960 to less than one-seventh by 1990.
- Defense R&D places overwhelming emphasis on development rather than research, a distribution that was further skewed during the Carter-Reagan defense buildup. The commercial relevance of much defense R&D is limited, since defense development relates primarily to specific weapon systems and has less general applicability than research. However, the sheer size of DoD's development and procurement budgets means that their aggregate effect on commercial technology may nonetheless be substantial.
- Defense R&D is unevenly distributed across industrial sectors; aerospace and electronics receive the largest amounts. Among academic disciplines, only in engineering is DoD funding a substantial fraction of federal support.
- The rate of interchange of scientists and engineers between defense and nondefense work seems higher than would be pre-

dicted from interviews with senior technical managers. Perhaps this is because "bench-level" technical skills are interchangeable between the two sectors, while management training and organizational experience are less so.

- Defense funding since World War II has substantially expanded the technical human resource base of the United States. In the 1960s, there was evidence that government demand drove up wages to the detriment of commercial industry, but such crowding out has not been evident recently.

- Defense and space technical challenges helped give rise to a technocratic value system that put lesser value on skills important to commercial innovation, such as manufacturing. But defense is not the only cause of such a devaluation. Other factors are likely at least as responsible as defense for U.S. industry's lack of attention to manufacturing.

- Econometric methods provide little guidance for measuring the net social cost or benefit per dollar of military R&D, and we must therefore analyze case studies or qualitative trends to assess the impact of defense R&D funding on the economy as a whole.

5

Military-Commercial Segregation: Defense Policies and Their Effects on Dual Use

Private industry designs and manufactures practically all of the tens of billions of dollars worth of weapons systems and matériel the United States buys each year. Except for a small proportion of export sales, the defense industry that has grown up to meet DoD's needs has no other customers.[1] In an attempt to preserve some of the strengths of free enterprise and competitive markets in this largely single-customer, DoD-controlled activity, while maintaining standards of accountability and oversight associated with public expenditures, the government has put an elaborate structure of regulations and policies in place. These regulations and policies, which have grown up incrementally over the years, have had the effect of segregating defense contractors and their

Portions of this chapter draw on Gerald Epstein's previous work in Office of Technology Assessment, *Holding the Edge: Maintaining the Defense Technology Base*, Chapter 8 ("Lab to Field: Why So Long?") (Washington, DC: U.S. GPO, April 1989) and Appendix A ("The DoD Acquisition System") of Volume 2: Appendices (Washington, DC: U.S. GPO, January 1990).

[1] Although the United States is now the world's largest arms exporter, arms exports nevertheless constitute only 10 to 15 percent of annual production. Moreover, more than half of the exports are actually purchased by the U.S. government, through the Foreign Military Sales program, and transferred at cost (plus an administrative fee) to the recipient government. Therefore the U.S. government purchases more than 90 percent of U.S. weapons production. See Office of Technology Assessment, *Global Arms Trade: Commerce in Advanced Military Technology and Weapons* (Washington, DC: U.S. GPO, June 1991), pp. 6, 11.

suppliers from commercial industry, even though both may draw on the same suppliers, labor market, and knowledge base, and even though individual companies may do business in both sectors.

This segregation is of fundamental importance to the subject of this book, since it indicates why the defense and commercial technology bases reinforce one another much less than might be expected from an examination of the many technologies in which the two sectors share an interest. This institutional segregation, like the different engineering cultures of the two sectors, limits dual-use synergy.

This chapter describes how the Departments of Defense and Energy develop technology and apply it to fielded weapons; how the distinctive aspects of military production and government policy divide the economy into defense and commercial sectors; and how that segregation impedes defense access to state-of-the-art technology in the commercial sector. The divisions run deep, arising from strongly held views in Congress and the electorate concerning the need to maintain accountability, oversight, and fairness in government actions. Reducing the divisions will require fundamental changes in political attitudes.

Basic changes in future U.S. defense needs, which will no longer be driven primarily by military competition with the Soviet Union, complicate the situation. Defense budgets will fall from their cold war levels. Notwithstanding public admiration for the performance of U.S. high-tech weaponry in the Gulf War, defense technology investments are likely to fall as well, making access to commercial technology increasingly important for maintaining military effectiveness.

This chapter begins by describing how R&D fits within the defense acquisition system. It then describes how some of the "barriers" that impede the flows of technology between defense and commercial sectors result from fundamental conflicts in the goals of the two sectors, while others have to do with differing technical requirements and still others stem from institutional and procedural practices. The consequences of the ensuing segregation are described, along with the compromises that will be necessary to reduce the problem. Recognizing both the changing world situation and the existing difficulties in relating military and commercial technologies, the chapter concludes with an assessment of the challenge facing defense acquisition and technology development in the years ahead.

GENERATING AND FIELDING DEFENSE TECHNOLOGY

For the United States, technology has been the cold war's principal weapon. The requirement to maintain and extend its technological advantage, along with the need to hedge against all possible technical surprises, has driven the United States to invest heavily in defense technologies across the board.

Who Develops Defense Technologies?

This investment has spawned a massive technical infrastructure within the Department of Defense. But it is DoD's industrial contractors that have primary responsibility for the detailed design, development, and production of weapons systems. The rapid evolution of aerospace and electronics technology during the cold war—like that of military aviation between the world wars—demanded a flexibility that government-operated arsenals and shipyards were ill-equipped to provide.[2] Thus defense production became more or less privatized.

Industry does not act alone. Its designers draw from a menu of defense technologies that have been developed through DoD funding. As the buyer of military systems, the Defense Department specifies the requirements that those systems must meet; its proposal solicitations (often developed with substantial input from its in-house laboratories) and its contract awards signal which technical approaches should be pursued. Ultimately, the technological state of the art of deployed weapons systems depends on the competence of defense firms, on DoD's ability to fund and promote technical advances, and perhaps most important, on the time it takes a weapon to proceed from the drawing board to the field.

In contrast to the DoD approach, the design and production of nuclear weapons remains almost entirely in government hands. Since its formation after World War II, the Atomic Energy Commission and its successors have maintained exclusive ownership of nuclear laboratories and production facilities, with total control over the fissionable material used to make nuclear weapons.[3] Through the AEC's transformation first into the Energy Research and Development Administration (1974) and then the Department of Energy (1978), the structure of the nuclear production complex remained unchanged.

The Lawrence Livermore and Los Alamos National Laboratories,

---

[2] For discussion of how the government's inability to deal with rapid technological advance engendered its dependence on private industry, see Thomas McNaugher, *New Weapons, Old Politics: America's Military Procurement Muddle* (Washington, DC: Brookings Institution, 1989).

[3] Public Law 70-585: The Atomic Energy Act of 1946; August 1, 1946.

both managed by the University of California, compete with each other in designing new weapons. Sandia National Laboratory, managed by AT&T for the government on a nonprofit basis, handles all non-nuclear aspects of weapon design and engineering. The weapons themselves, together with the special nuclear materials they require, are manufactured by what is in effect a vertically integrated government arsenal structure, even though the individual facilities constituting this complex are contractor operated. The DOE design and production complex maintains responsibility for nuclear weapons throughout their entire life cycle. Almost all the technical expertise available to the Department of Energy resides in these contractor-operated facilities, which have been able to avoid some of the bureaucratic constraints imposed on other government facilities and civil service personnel.

DoD Technology Acquisition

The apparently straightforward sequence implied by DoD's R&D budget categories (6.1, 6.2, and so forth—see Chapter 4, Box 4-B) masks the fundamental division between the *technology base* activities at the beginning of this pipeline and the *systems acquisition programs* that follow. Technology base activities explore the potential of new technologies and extend the capabilities of existing ones.[4] Although conducted with an eye toward potential military applications, they are not necessarily expected to lead to specific weapons. Systems acquisition programs, on the other hand, are intended to meet specific military needs.[5] Even when R&D results promise new capabilities, the military services must balance that promise against the risks of relying on an unproven approach. Technological advances must "earn their way" into new systems; they are not automatically incorporated.

DoD technology base activities are largely managed, budgeted, and performed separately from systems acquisition programs, although the same institutions (national laboratories, industrial firms, and military commands) may be involved in both. In the DOE nuclear weapons laboratories, the two are more tightly coupled, with technology base and systems development combined under a single "level-of-effort"

---

[4]DoD defines its technology base programs as its basic research programs (category 6.1) plus those in exploratory development (6.2). However, the term "technology base" is increasingly used to refer to categories 6.1 through 6.3A, which are formally referred to by DoD as the "science and technology base."

[5]Systems development programs typically fall in categories 6.3B and 6.4, with category 6.6 representing R&D for operational systems in the field. See Box 4-B.

budget that remains relatively stable from year to year.[6] When many systems are being designed for DOE, less technology base work gets done, and vice versa.

The DoD technology base.   Both the individual services and defensewide agencies, primarily the Defense Advanced Research Projects Agency (DARPA) and the Strategic Defense Initiative Organization (SDIO), manage technology base programs. All are overseen and to some extent coordinated by the Office of the Secretary of Defense. Each service has a basic research agency (the Army Research Office, the Office of Naval Research, the Air Force Office of Scientific Research) that primarily funds work in universities. Exploratory development dollars, on the other hand, flow primarily to DoD's own laboratories and industry.[7]

The services generally fund and conduct research and technology development in areas relevant to their missions—underwater acoustics for the Navy, for example, or cruise missile guidance technology for the Air Force. Most of their effort seeks to improve capabilities that already exist. As a result, the services are not necessarily motivated to explore radical innovations that appear irrelevant or even threatening to their present roles and missions.

DARPA and SDIO were established, independent of the services, specifically to foster such revolutionary advances and develop them to the point where the services can take over. DARPA was created as the Advanced Research Projects Agency (ARPA) in 1958, following the Sputnik launch, with the mission of pursuing "advanced projects essential to the Defense Department's responsibilities in the field of basic and applied research and development which pertain to weapons systems and military requirements."[8] SDIO, formed in 1984, bundled to-

---

[6] The DOE nuclear weapons laboratories also conduct a significant amount of military and nonmilitary research, unrelated to nuclear weapons, that is separately funded by various agencies.

[7] Fifty-seven percent of DoD's basic research dollars are spent in universities, with government labs (including federally funded research and development centers) getting 31 percent and industry only 9 percent (in fiscal year 1989). The distribution is inverted for exploratory development: industry receives 47 percent, government labs 42 percent, and universities 10 percent. Data are from National Science Foundation, *Federal Funds for Research and Development: Fiscal Years 1988, 1989, and 1990,* NSF 90-306 (Washington, DC: U.S. GPO, 1990), Tables 13 and 20.

[8] Public Law 85-325 (February 12, 1958) authorizing the creation of ARPA, quoted in Burton I. Edelson and Robert L. Stern, *The Operations of DARPA and Its Utility as a Model for a Civilian ARPA* (Washington, DC: The Johns Hopkins Foreign Policy Institute, The Paul H. Nitze School for Advanced International Studies, November 1989), p. 4.

gether a number of ongoing research programs relevant to ballistic missile defense, soon adding several new ones.

DARPA is unique within the Defense Department in that it has a minimum of administrative layering and gives its program managers wide discretion to support technologies they consider promising. It operates no laboratories of its own, and until 1987 did not even have the ability to execute its own contracts, relying instead on the services to act as its contracting agents. It funds research in universities, industrial, and service laboratories, and has been granted the authority to enter into venture-capital-like cooperative arrangements with private companies.[9]

The breadth of its mandate has permitted DARPA to fund the development of pathbreaking technologies in areas such as computing that have had major importance for both the defense and the commercial sectors.[10] Its R&D budget for fiscal year 1990 was $1.2 billion. DARPA's farsighted view of long-term, high-risk technologies having high potential payoff has attracted the interest of policymakers who seek to extend this approach to commercially relevant technologies that are not necessarily important to the military. Some would like to broaden DARPA's mission to include development of commercial technologies, others to create an analog to DARPA with a purely civil mission within the Department of Commerce, or as an independent agency. However, it is not at all clear that DARPA's approach would work outside the military domain. DoD's need for military hardware requires and justifies significant front-end investments in technological exploration; its mission of preventing technical surprise requires investigating and assessing a wide range of high-risk technologies that don't necessarily end up in weapon systems. With only 3 percent of the Pentagon's RDT&E budget and less than one-half of 1 percent of total DoD funds, DARPA has enjoyed a great deal of autonomy. In any other setting, such an agency would attract much more attention, controversy, and "help."

[9] Granted in the National Defense Authorization Act for Fiscal Years 1990 and 1991 (Section 251, P.L. 101-189, November 29, 1989), this authority was exercised for the first time in April 1990, when DARPA invested in Gazelle Microcircuits Inc., a semiconductor firm specializing in gallium arsenide electronics. Controversy over this arrangement reportedly cost Craig Fields his job as DARPA director; he was transferred to another position later that month and resigned from government service shortly afterward.

[10] See Kenneth Flamm, *Targeting the Computer: Government Support and International Competition* (Washington, DC: Brookings Institution, 1987).

SDIO has even more autonomy than DARPA. As originally constituted, its funding was walled off from all other Pentagon operations, and its director reported only to the secretary of defense. The personal interest of President Reagan and his defense secretary propelled the growth in SDIO's budget, but it also generated a level of controversy and political visibility unprecedented for a technology base program.[11] SDIO survived the end of President Reagan's term in office, but it has been partially integrated back into DoD's organizational machinery. Moreover, its funding growth has not only slowed but turned around, with the agency taking a 1.5 percent cut in 1990 from its 1989 peak of $3.6 billion, and another 20 percent cut in 1991.

Systems acquisition.   Turning the fruits of technology base developments into weapons is the responsibility of each service's "buying command." These commands fund and oversee the design, development, and production of military systems. Nominally, acquisition programs begin when a service identifies a mission need that can plausibly be addressed by a new system. This may arise from the desire to exploit new technology, but it can also result from changes in defense strategy, obsolescence of current equipment, or changes in the perceived threat. If the proposed new system wins a share of the service's anticipated acquisition budget in competition against other service needs, and if it survives review in the Office of the Secretary of Defense, the Office of Management and Budget, and (for the largest items) the White House, the program formally begins as a line item—typically in budget category 6.3A or 6.3B—in DoD's overall budget submission to Congress.

The new program first enters an Exploration/Definition phase, lasting perhaps up to two years, that identifies alternative concepts to be demonstrated and establishes broad cost, schedule, and performance goals.[12] In the next phase, Demonstration/Validation, DoD solicits alternative concepts from industry—possibly in the form of prototypes—to be demonstrated and compared to one another. It is at this

---

[11] Much of the controversy hinged on the direct conflict between SDIO's goals and U.S. treaty obligations, together with what critics of the program perceived as its implicit commitment to demonstration, production, and deployment of ballistic missile defenses. When the same R&D programs had been conducted as technology base activities (largely in DARPA), unrelated to specific weapon systems, they did not attract much attention.

[12] The durations of various phases of the acquisition process are highly variable. These estimates are from Office of Technology Assessment, *Holding the Edge: Maintaining the Defense Technology Base:* Volume 2: Appendices (Washington, DC: U.S. GPO, January 1990), Appendix C, pp. 55–57.

stage—typically lasting two to three years—that new concepts from the technology base make the transition into proposed designs for next-generation military systems. At the end of the Demonstration/ Validation phase, the Pentagon selects one of the concepts for Full-Scale Development (FSD), budget category 6.4. After successful completion of FSD, which typically lasts three to six years, the system enters production.

In the past, the production contract would almost always go to the system's developer, which encouraged contractors to bid low for development contracts with the expectation of recouping their losses on production. Severe cost overruns were the predictable result. To curb this practice, DoD now often awards development contracts to two or more competing firms, with only one proceeding to production. Moreover, DoD on occasion recompetes production contracts, transferring technical data and design/development information to a second source and dividing production between the two in proportions that are readjusted in successive recompetitions. The approach was intended to control overall costs and improve quality, at the price of the initial investment needed to establish the second source.[13] However, savings depend on the size of the ultimate production run; DoD has concluded that dual-sourcing is not always cost-effective, and it will likely prove even less so in the future as the size of procurements drops because of declining defense budgets.[14]

Still, development contractors run the risk not only of losing the production contract, but also of having their design work turned over to a rival for production.[15] Consequently, they are reluctant to incorporate proprietary technologies or production processes in their propos-

---

[13] Jacques Gansler, *Affording Defense* (Cambridge, MA: MIT Press, 1989), pp. 186–188. A comparison of 12 empirical studies on second sourcing finds that no general conclusion can be drawn as to whether money is indeed saved; see James J. Anton and Dennis Yao, "Measuring the Effectiveness of Competition in Defense Procurement: A Survey of the Empirical Literature," *Journal of Policy Analysis and Management*, vol. 9, no. 1 (1990), pp. 60–79.

[14] Center for Strategic and International Studies (CSIS), *Integrating Commercial and Military Technologies for National Strength: An Agenda for Change*, Report of the CSIS Steering Committee on Security and Technology (Washington, DC: CSIS, March 1991).

[15] During the late 1980s, DoD pressured contractors still further, forcing them to share the cost of new systems developments. A team consisting of Lockheed, Boeing, and General Dynamics, winners of the Air Force's Advanced Tactical Fighter competition, invested more than $1 billion of their own funds in developing two prototype planes; the losing team of Northrup and McDonnell Douglas spent nearly as much. Richard Stevenson, "Air Force Chooses Lockheed's Design for Fighter Plane," *New York Times*, April 24, 1991, p. A1.

als. Divorcing design from production by awarding a separate contract for each stage also discourages contractors from investing in the additional development effort needed to minimize downstream manufacturing costs.[16] Moreover, it makes it difficult for contractors to adopt simultaneous engineering, an approach gaining currency in commercial production in which a product and its production process are designed at the same time so that trade-offs between performance and manufacturability can more readily be made.

DoD's difficulty in making effective trade-offs—between design and manufacturability, between cost and performance, and sometimes even between conflicting performance requirements—lies at the heart of its acquisition difficulties. Decisions made early in the program's existence, when it must compete against other military programs and nonmilitary needs to secure its place in the budget, drive up acquisition times and life-cycle costs (which include the costs of maintenance, repair, and logistical support). According to the Defense Science Board,

> There are great pressures to overpromise in order to survive the [funding] competition. Since the decisions are made by political processes among a large and diverse group of people, there is little pressure to discipline the process and to enforce realism. Clear-cut designs to meet the requirements are not allowed because they would interfere with the next step—competitive source selection. The result is a firm over-stated requirement which too frequently can neither be met nor changed.[17]

Partly as a result of these institutional pressures—but also as a consequence of the technological competition characterizing the cold war—DoD has been willing to pay a high premium for technological sophistication and performance. Commercial firms, on the other hand, must be much more cost-sensitive because their customers routinely weigh cost against function. The contrast between these two attitudes underlies many of the differences between the military and the commercial sectors and influences the degree of technological synergy achievable between the two sectors under existing policies. Meaningful change in DoD acquisition will require that it begin to weigh relationships between cost and performance at early stages and permit trade-

---

[16] Indeed, if price is a significant element in the award of development contracts, such additional development expenses could put a contractor at a competitive disadvantage.

[17] *Report of the Defense Science Board 1985 Summer Study on Practical Functional Performance Requirements* (Washington, DC: Department of Defense, Office of the Under Secretary of Defense for Research and Engineering, March 1986), p. 91.

offs between the two. Such trade-offs have always been central to product development in civilian industry.

FACTORS IMPEDING DUAL-USE RELATIONSHIPS

A number of factors frustrate technological interaction between military and commercial activities in industry: technological divergence, conflicts in fundamental goals, differences in functional requirements between military products and their civilian counterparts, distinctive aspects of military production, and oversight requirements imposed on government procurement in general. These factors encourage firms to conduct military business in divisions that are managed separately from commercial operations (see Chapter 6), often with separate workforces, production and research facilities, accounting practices, engineering design philosophies, and corporate cultures. They also make it difficult for the military to use technology that is available off-the-shelf from commercial suppliers.

Technological Divergence

Mention of barriers between commercial and defense industries implies that the two are doing fundamentally similar things. This overstates the case in many instances. Although military and commercial items share many underlying technologies, some military technologies—such as stealth and nuclear weapons—are irrelevant for commercial purposes. DoD's technical focus is concentrated mainly on aerospace and information technology, and it devotes much less attention to many other industries that are also important to the economy. This does not mean that potential synergies between military and commercial activities should not be exploited. However, it does mean that there are areas where synergies are not likely to be found.

Conflicts in Goals

The mission of defending the country differs fundamentally from that of promoting the competitiveness of U.S. industry. In some cases, the two sets of goals come into direct conflict. The very point of export controls, for example, is to prevent the sale of U.S. products that might convey certain technological capabilities to potential adversaries. These two objectives cannot be harmonized; rather, a deliberate choice must be made to promote one at the expense of the other.

Export controls.   The United States is a member of the Coordinating Committee for Multilateral Export Controls (CoCom), an informal, voluntary association of Western nations that agreed in 1949 to restrict exports of militarily relevant technology to the Soviet bloc. The United States has also imposed unilateral controls going far beyond the multilateral CoCom regime.[18] These policies—particularly their unilateral aspects—have hurt U.S. companies. Although the forgone sales to Soviet or Soviet bloc customers are minor, and accepted as a cost of national security, the damage results when U.S. companies must deal with the lengthy, convoluted, and unpredictable process of acquiring export licenses to sell to friendly nations. Exporters in other CoCom countries, often with equivalent levels of technology but far less cumbersome export procedures, have an advantage over U.S. firms. Compounding the problem, foreign-produced goods containing more than 25 percent U.S.-origin parts or components also fall under U.S. export controls if they are to be reexported to a third country. Consequently, foreign producers have an incentive to avoid U.S.-supplied components in order to avoid the concomitant U.S. restrictions.

Export control policy, like other aspects of defense policy, is now being reassessed in light of the overwhelming changes in what formerly was the Soviet bloc. But even as the East-West confrontation abates, controls—given new urgency by the Gulf War—will remain an important component of U.S. policy to restrict the proliferation of ballistic missiles and weapons of mass destruction. However they evolve, controls that limit the export of high-technology products or components will inevitably conflict with attempts to promote this country's export sales and improve its balance of trade.

Classification.   In cases where mere restrictions on exports are not sufficient to safeguard information or technology that is considered particularly sensitive, the military can prohibit its disclosure to any unauthorized individual, foreign or domestic, through the use of security classifications. Classification constraints may range all the way to keeping secret the very existence—let alone the technical details—of

---

[18]Two reports of the National Academy of Sciences provide further detail on export controls, their implications for the competitiveness of U.S. industry, and modifications that are taking or that should take place as a consequence of the changing world situation. Most recent is National Academy of Sciences, *Finding Common Ground: U.S. Export Controls in a Changed Global Environment* (Washington, DC: National Academy Press, 1991). This report extended the work done a few years previously by a committee chaired by Lew Allen, published in National Academy of Sciences, *Balancing the National Interest: U.S. National Security Export Controls and Global Economic Competition* (Washington, DC: National Academy Press, 1987).

multibillion-dollar programs. While classified products cannot be sold to any commercial customer, foreign or domestic, such products rarely have much commercial relevance. More significant for competitiveness are security restrictions that prevent a company from using technology developed for the military in its commercial business.

Differences in Requirements

Requirements for military procurements are established through a different process than those for commercial goods, and to meet different needs. Differences in design requirements for military and commercial products, of course, need not imply that the materials, components, and subsystems from which these products are made must also differ. But often they do, particularly since until recently the incentives to use commercial or "nondevelopmental" items in defense systems have been weak.

Cost/performance. Perhaps the most significant difference is that defense customers will pay a high premium for functional performance, whereas commercial users are much more cost-sensitive. For this reason, even when military and commercial products are based on a common technology and share the same components, they can differ markedly at the systems level. While market competition in the commercial sector rewards suppliers that offer the best performance for the price, no real market exists in defense. The U.S. government is practically the only buyer, a handful of companies are often the only vendors, and DoD sets the requirements in great detail. Many factors bias DoD performance requirements upward in such an environment, beginning with the emphasis placed on technological superiority as a cornerstone of national security policy. Operating at the edge of the state of the art, in the face of enemy systems explicitly designed to counter them, military systems must meet requirements that have no counterparts in the commercial world. A small edge in performance can be a matter of survival.

The structure of the acquisition process also serves to ratchet up performance requirements. The initial political bargaining contributes. So does the length of the acquisition cycle, which drives up performance requirements in two ways. First, systems must be designed to counter not the current threat, but rather the projected one a decade or more in the future when the system will first be deployed. Conservative planners err on the side of overestimating the future threat. Second, because DoD routinely underestimates development costs and overestimates future defense budgets, the inevitable funding shortages lead to delays, raising program costs still further. Moreover, when a new system is delayed, the system it was to replace must remain in service longer than intended. Not knowing when a successor system will get into the field or how long it will have to stay there, systems designers seek to build in additional margins of performance.

DoD acquisition managers lack adequate tools, incentives, and power to enforce trade-offs between performance and cost. Such trade-offs require realistic cost estimates. Artificially low estimates, intended only to hold a place in the budget in the expectation that supplemental funds will be found, defeat the purpose. At the same time, however, making trade-offs requires getting something in return. Unless at least some of the funds saved through relaxed performance requirements can be used to buy more systems—or otherwise reinvested to compensate for the lost performance—reform will never begin. If Congress insists on keeping all the savings, there won't be any.

Environmental and logistical requirements.   Since the location, nature, and starting date of the next war cannot be known in advance, military systems must be designed for very long shelf life and for a broad range of possible conditions (desert, jungle, arctic, or the European battleground that preoccupied cold war planners). These constraints impose development, testing, and production costs beyond those needed for commercial products, for which operating conditions are more predictable, even though they may be harsh.[19] On the other hand, many in industry argue that DoD requirements—such as for temperature range and shock resistance—are unrealistic and not based on actual operating experience.

Logistics considerations impose additional constraints on military systems. Even when commercial products exist that could satisfy military requirements, DoD is wary of depending upon suppliers whose future business plans are not under military control. Purchasing commercial items that may be superseded or discontinued by their manufacturers puts DoD at risk of having to maintain outdated products on its own (e.g., by paying premium prices to specialized contractors producing obsolete parts).

---

[19] However, increasingly strict, retroactively applicable product and environmental liability laws and court decisions are eroding predictability for some commercial products as well. George L. Priest, "The New Legal Structure of Risk Control," *Daedalus* (Fall 1990), pp. 207–227.

Distinctive Aspects of Military Production

Since the Department of Defense is the sole buyer for most military goods, including exports through the Foreign Military Sales program, it must by conscious direction accomplish much of the market regulation that classical economic theory assumes will happen automatically.[20] DoD cannot merely buy the products it wants today, but must also worry about preserving industry's capability to produce things it might want to buy tomorrow. It must take an overall systems view, buying all the complementary items and services needed to use a new piece of military hardware effectively. And the type, quantity, and production rate of military goods differ in major ways from those for commercial products.

Patterns of production and use.   Weapons may take 10 to 15 years to develop, and they may remain in service for decades more. Commercial products, on the other hand, can proceed from design through development and marketing to obsolescence in less than five years, particularly consumer products or high-technology items such as DRAMS or microprocessors. Whereas annual purchases of high-volume consumer goods can run into the millions of items, many military procurements are one-of-a-kind or are purchased in small lots.

It is possible to make too much of such differences, as we noted in the Chapter 2 discussion of the different technical cultures of military and commercial innovation (see Table 2-3). Military systems can be more appropriately compared to high-cost, low-volume, long-lived capital goods such as power plants, refineries, or telephone switching networks, with which they may also share design and manufacturing technologies (see Chapter 10). With the help of flexible automation, consumer products are being made in smaller batches to better meet customer demand. Even so, DoD does not have the same incentives as commercial producers to adopt modern manufacturing equipment and practices, nor do the sequential phases of its acquisition process make it easy to utilize these practices.

Systems.   Most DoD procurement dollars go, not for products or components, but for systems. These include hardware components, software, maintenance equipment, spare parts, and often training and simulation support. Equivalent systems purchases do exist in the commercial sector, especially in capital goods. However, the proliferation

---

[20] Gansler sets out 30 assumptions of the market economy that are violated in defense sales in *Affording Defense*, pp. 159–160.

of competing vendors that can rapidly exploit niche markets, adapt to changing demand, and offer a wide range of alternatives provides the commercial world with a flexibility and resilience lacking in the defense industry.

The defense industrial base.   DoD's responsibility for the defense industrial base introduces factors into its procurement decisions that have no commercial counterparts. These factors cut DoD off from many potential suppliers:

- *Competition:* If DoD wishes to preserve competition in the defense industry, maintaining several suppliers for each item it buys, it must purchase enough from each to keep that firm in business even if the products are not DoD's first choice.
- *Surge and mobilization:* The Defense Department must see that the nation maintains the capability to increase production in a crisis if it expects to fight wars that last longer than its munitions stockpile.
- *Nationality of companies and employees:* DoD generally does not offer contracts for classified research to companies that are not U.S. owned; foreign nationals cannot obtain U.S. security clearances.[21] DoD also has the authority to specify domestic sources of supply for items that might not be available from abroad in a crisis.

Requirements on Government Procurement

Defense acquisition policies and procedures are strongly influenced by DoD's institutional environment as a part of the federal government, under close supervision of Congress and many executive-branch agencies. Many of the rules and regulations imposed on all government procurement (including DoD) have deep roots in the American political process. These separate defense contractors from the mainstream economy.

---

[21] Several arrangements exist to permit U.S. firms that have come under foreign ownership to continue working on classified contracts. They require the foreign owners either to turn over management control of their U.S. subsidiary to U.S. citizen trustees or add a majority of U.S. citizen outside directors to the subsidiary's board. The president can block the sale of a U.S. firm to foreign interests on national security grounds even if the firm is not conducting classified research. The only time this authority has been exercised has been to order a trading company owned by the People's Republic of China to divest a recently acquired U.S. aircraft parts manufacturer.

Administrative and legal requirements.   Taxpayers and voters expect a level of accountability and oversight over government activities quite different from that expected of purely commercial transactions. Accounting standards, audit and disclosure requirements, and cost allocation rules for federal contracts are not only different from standard commercial practice, but also more restrictive and burdensome.[22]

Moreover, government seeks many goals in addition to efficient and economic procurement. Federal contracts contain provisions intended to protect the environment, promote socioeconomic development, prescribe particular sources of supply in order to preserve the industrial base, ensure access to all potential bidders, and promote the interests of particular congressional constituents. None of these clauses necessarily serves the immediate interest of procuring the best product for the military at a reasonable price. They can pose a particular obstacle to the procurement of off-the-shelf commercial products, since goods already in production may not comply. Commercial suppliers at lower tiers of the government contracting chain are also affected, since they are deemed subcontractors and must also comply.[23]

Cost-based contracting.   In a competitive market, customers typically do not know or care how much it costs to design, develop, and manufacture the goods they are considering purchasing. All that matters to them is whether the product seems worth its price. Profits are the reward for supplying products that have more value to their purchasers than they cost to produce and distribute. Strict price competition among bidders for major military systems, however, is rarely feasible. When military procurements pose substantial risk, contractors will not gamble on fixed-price bids. Instead, the government must negotiate prices based on actual contractor costs.

The government first used cost-based contracting during World War I, when uncertainties in labor and material costs made fixed-price contracting impossible, and a cost accounting section was set up within the office of the Army Chief of Ordnance.[24] Since then, requirements have grown in complexity and scope. Today they cover cost accounting standards (defining bookkeeping formats for government contracts),

---

[22] See CSIS, *Integrating Commercial and Military Technologies* for further discussion of federal accounting principles and contracting requirements and their influence on the government's ability to buy products from commercial suppliers.

[23] Ibid., p. 67.

[24] William Crozier, *Ordnance and the World War* (New York: Charles Scribner's Sons, 1920), pp. 18–19.

cost accounting principles (specifying which charges are acceptable), and cost or pricing data (ensuring that the price charged the government is reasonable). To permit the government to verify costs, contractors are required by law to disclose "all the facts at the time of price agreement that prudent buyers and sellers would reasonably expect to affect price negotiations significantly."[25] When the government is able to ensure a fair price by entertaining competitive fixed-price bids from a number of suppliers, it need not obtain cost or pricing data. However, as we explained earlier, fixed-price contracts are not feasible for projects with high technical risk.

Demands for cost or pricing data complicate government attempts to purchase commercial products or to deal with suppliers that operate primarily in commercial markets. Many companies prefer to forgo government sales rather than disclose the required information or simply to avoid the paperwork. According to the Semiconductor Industry Association, five of the top ten U.S. semiconductor companies will not accept DoD business if the contract requires certified cost or pricing data.[26] Although current policy now exempts contractors from having to supply cost or pricing data for items that are sold to the general public at established catalog or market prices, industry believes that this exemption is too narrowly interpreted. Moreover, qualifying for the exemption also requires paperwork and record keeping that firms find onerous. According to the deputy assistant secretary of defense for procurement, industry representatives claim that statutory demands for certified cost or pricing data are the most significant impediment "by far" to selling commercial products to the government.[27]

Rights to technical data.   The government may seek the right to use, duplicate, or disclose to others technical information concerning products it purchases, particularly in cases where the government has funded the product's development. Such data may be needed to maintain and operate equipment, to obtain competition among suppliers for future purchases, or to foster technological improvements.[28]

---

[25] Federal Acquisition Regulation, Section 15.801, originating in the Truth in Negotiations Act of 1962.

[26] Cited in CSIS, *Integrating Commercial and Military Technologies*, p. 20.

[27] Eleanor R. Spector, Deputy Assistant Secretary of Defense for Procurement. Speech before the 28th Annual National Educational Conference, National Contract Management Association, July 13–14, 1989, Los Angeles, CA.

[28] "Technical data" are here distinguished from title to patents deriving from government-funded work. Policy toward the latter has been changing as outlined in Chapter 3.

Conflicts surrounding rights to technical data have simmered for 35 years. At issue is the assertion of rights to a company's "background" technology that is incorporated in a defense product along with technology derived from government support; a small amount of Pentagon funding can compromise the proprietary rights of a company that develops technology important to its business with its own money. To prevent this, suppliers may segregate DoD development from internally funded efforts or refuse to accept government R&D contracts altogether.

Under present law, the government acquires rights to technical data in one of three categories: "unlimited" rights; "government-purpose license" rights, which provide for unlimited government usage (including the right to provide the data to third parties so they can produce goods for the government) but prevent the government from letting third parties use the data for commercial purposes; and "limited" rights, which prevent the government from disclosing the data without approval and prohibit its use for competitive reprocurement.[29] DoD cannot require that contractors relinquish rights to privately funded technology as a condition for contract award. Nevertheless, bids offering only limited rights can still be penalized if, for example, DoD determines that they will as a result lead to higher life-cycle costs.[30] Contractors are therefore greatly concerned that the government not make indiscriminate demands for technical data. These concerns extend to lower-tier suppliers as well, since prime contractors are generally required to pass these provisions through to their subcontractors.

Specifications and standards.   A complicated system of military specifications and standards governs how military purchases are made, what components may be used, and even how those systems are assembled and tested. This system is a consequence more of the procedural requirements of government contracting than of the technical and environmental requirements that weapons systems must meet.

DoD purchasing officers—like their counterparts across government—cannot exercise the discretion that private companies delegate to their purchasing departments, even for products that are obviously commercial in nature.[31] Members of Congress are particularly concerned that their respective constituents have as good a chance as any-

[29] CSIS, *Integrating Commercial and Military Technologies*, pp. 53–64.

[30] Ibid., p. 60.

[31] Steven Kelman, *Procurement and Public Management: The Fear of Discretion and the Quality of Government Performance* (Washington, DC: AEI Press, 1990).

one of winning government business, and the requirement for "full and open competition" as written into the Competition in Contracting Act of 1984 has been interpreted as requiring that all vendors capable of fulfilling a contract be allowed to bid. Government purchasers therefore seek objective criteria that will protect them from charges of favoritism and unfair competition by losing bidders and their congressional representatives.

Since specifying brand names would be seen as restricting competition, specifications are developed in what might seem to be absurd detail to ensure that anything meeting them will be acceptable. An Army spokesman explained the rationale behind a 10-page specification for sugar cookies: "Why do we do things like this, because we're stupid? No, because we have paid for something called a sugar cookie in the past and what we got was trash."[32]

The government need not award all its contracts to the lowest bidder. One of the major accomplishments of the Competition in Contracting Act was to recognize that contracts could be awarded on the basis of nonprice factors such as technical merit or quality and still be considered "competitive." Nevertheless, awards based on these more subjective criteria are more difficult to defend.

DoD studies of military specifications and standards have found that the existing structure "essentially meets defense acquisition needs," although it certainly has much room for improvement.[33] Other studies emphasize the costs of unique requirements:

> In general, where the government product or process specifications differ from commercial techniques, a separate facility will be established to produce technologies the DoD way, with no assurance of higher quality but a virtual guarantee of higher costs. This serves both to limit the ability of the defense contractors to expand into commercial arenas and to curtail the interest of commercial producers in performing defense business.[34]

Government processing requirements and testing procedures not only prevent manufacturers from integrating commercial and government production, but also require them to undergo lengthy and expensive

---

[32] Ralph Vartebedian, "Paper Rules Swell Cost of Defense," *Los Angeles Times*, May 21, 1988, p. 1.

[33] Report to the Secretary of Defense by the Under Secretary of Defense for Acquisition, *Enhancing Defense Standardization—Specifications and Standards: Cornerstones of Quality*, November 1988 (quotation on p. v).

[34] CSIS, *Integrating Commercial and Military Technologies*, p. 22.

requalifications if they wish to update or improve their defense production lines. Like other procurement regulations, specifications and standards typically flow down to subcontractors, giving rise to the same problems.

To the extent that specifications and standards remain necessary, reviewing outdated ones would help.[35] So, too, would making greater use of standards that have already been adopted in the commercial sector. But nongovernment standards predominantly address test methods, processes, recommended practices, and safety; relatively few (about 8,000 out of 35,000) pertain to products. Although DoD is working with industrial standards organizations to develop common product standards, manufacturers who are represented in commercial standards-setting bodies often oppose standards that might exclude their own products or harm their ability to obtain competitive advantage via unique product characteristics.[36]

There are two other ways to reduce the barriers posed by military specifications and standards: change the process of developing specifications so that existing commercial products are not precluded, or change the acquisition system to obviate the need for the specifications. The first approach requires that those developing the specifications know what is available commercially but do not tilt the selection toward any particular product. The second requires that government purchasers be given the latitude to select products based on market acceptance, best value, or past contractor performance, rather than adherence to detailed specifications at the lowest price.

The Office of Federal Procurement Policy within the Office of Management and Budget believes it has found a solution in the first approach. It has proposed a process that would bring market research into the specification development process in an open, competitive manner. First, any interested manufacturer would be invited to respond to a preliminary announcement of the government's anticipated requirements. Then the contracting officer would refine the initial requirements on the basis of information submitted, and the resultant procurement would be limited to those manufacturers that had re-

---

[35] Overdue for review were 7,200 documents as of September 1988. Report to the Secretary of Defense by the Under Secretary of Defense for Acquisition, *Enhancing Defense Standardization—Specifications and Standards: Cornerstones of Quality,* November 1988, p. vi.

[36] More than 100 nongovernment standards address sampling, analyzing, application, and testing of paint, but for more than ten years producers have prevented a major standards organization from issuing a product standard for paint. Consequently, the government must buy paint according to government specifications. Ibid., pp. vi, 9–10.

sponded.[37] It is not clear at this stage whether such a process would actually work, nor whether it would be seen as truly maintaining full and open competition. The next step should probably be a pilot program.

The second approach is more drastic, requiring a fundamental change in political values. Its implications and prospects are discussed later in this chapter.

CONSEQUENCES OF MILITARY-COMMERCIAL SEGREGATION

Thirty-five years ago, the principal effect of DoD's inability to access commercial suppliers efficiently was to waste money and time. While these concerns remain, there is a new and strikingly different factor today—the degree to which commercial products can contribute to DoD's needs in technologically sophisticated applications as well as mundane ones. When the President's Blue Ribbon Commission on Defense Management (the Packard Commission) recommended that DoD "make greater use of components, systems, and services available 'off-the-shelf,' " its arguments were not limited to cost savings. It also noted that "the process of procuring microchips made to military specifications involves substantial delay. As a consequence, military microchips typically lag a generation (three to five years) behind commercial microchips."[38] With the state of the art rapidly advancing in electronics, delays translate directly into performance penalties. Inability to tap commercial products thus impedes DoD's efforts to attain the performance edge it values so highly.[39]

Advances in commercial technologies are not, of course, limited to parts and components. Entire products and systems offered commercially off-the-shelf (COTS) can provide DoD with advanced capabilities otherwise obtainable only by initiating a new system development program. At the same time, however, such products can complicate logistical support, or lock DoD into depending on particular suppliers if the

---

[37] CSIS, *Integrating Commercial and Military Technologies*, p. 50.

[38] The President's Blue Ribbon Commission on Defense Management, *A Formula for Action* (Washington, DC: President's Blue Ribbon Commission on Defense Management, April 1986), p. 23.

[39] That performance advantages are obtainable with commercial components has been reiterated in more recent studies such as Office of the Under Secretary of Defense for Acquisition, *Report of the Defense Science Board on Use of Commercial Components in Military Equipment*, June 1989, p. 2.

rights to use or transfer proprietary technologies are not obtained. COTS products are no bargain if they fail to meet essential performance requirements, or if "minor" modifications to meet military needs turn out to be major.[40]

Even though COTS products might be able to replace some items developed for DoD, many military requirements will still demand specialized components and systems. Yet as commercial demand continues to drive technologies critical to defense, the needed expertise may no longer be found within DoD's traditional contractor base. According to the Office of Technology Assessment, "the ability of the military to achieve and maintain leading-edge technology in the future will depend in many cases on the health of the corresponding industry in the commercial sector of the economy."[41] Even a strong commercial sector may not suffice if firms are not interested in the Pentagon's business. OTA warned that "government practices have made it increasingly difficult for DoD to obtain state-of-the-art technology in areas where civilian industries are leading, making defense business unattractive to innovative companies and contributing to traditional suppliers leaving the defense business."[42]

To improve DoD's access to commercial products and manufacturers, many of the policies described earlier in this chapter will have to be changed. Although conflicts such as those imposed by export control regulations and security classification cannot be avoided, their consequences are probably not as severe as those of other policies that can, in principle, be mitigated: exaggerated and inflexible requirements, insufficient emphasis on cost, and administrative procedures that discourage commercial suppliers from seeking government business.

REDUCING MILITARY-COMMERCIAL SEGREGATION

The defense acquisition process has developed in response to forces originating not only within DoD but also from other executive branch agencies, Congress, and the public at large. Policy changes should not stop with the Department of Defense. Nor can they be imposed without taking account of the constraints that shaped the system in the first place. For example, much of the bureaucratic overhead embodied in defense acquisition supports policy objectives, such as ensuring ac-

---

[40] U.S. General Accounting Office (GAO), *Procurement: DoD Efforts Relating to Nondevelopmental Items*, GAO/NSIAD-89-51 (Washington, DC: GAO, February 1989), pp. 10–11.

[41] Office of Technology Assessment, *Holding the Edge: Maintaining the Defense Technology Base*, p. 176.

[42] Ibid., p. 34.

countability for the use of public funds, that the nation has decided—explicitly or otherwise—are at least as important as efficiency in procurement. Those decisions have to be recognized as choices among legitimate policy objectives. They can be reversed, but they cannot be evaded.

Political Obstacles

Whether the political system will permit fundamental change depends on public and political perceptions of crisis. In wartime, for example, peacetime constraints fall away.[43] Conversely, as the perceived national security threat facing the United States lessens, it may be harder to mobilize a political consensus to restructure acquisition policy. While there is nearly universal agreement that acquisition should be reformed, there is no consensus on what to do.

Box 5-A illustrates that what some view as a problem—in this case, the requirement that defense procurement be conducted using "full and open" competition—is seen by others as a proper exercise of governmental responsibility. The problems in reaching agreement on this point reflect the difficulties that must be confronted and resolved if acquisition policy is to change.

The story in Box 5-A illustrates the politically charged atmosphere in which acquisition policy is made. Many members of Congress, along with a substantial fraction of the American public, insist on constant and vigilant oversight of the defense industry and its relationship with the military. This attitude is reinforced by well-publicized incidents concerning exorbitant prices for spare parts, improper activities of consultants, and massive cost and schedule overruns on major weapons contracts. It also reflects long-standing suspicions harbored by many Americans and their congressional representatives with respect to the defense industry.[44] Reforms intended to reduce the bureaucratic bur-

---

[43] When the military ran short of ground receivers for the Navstar satellite navigation system during Operation Desert Shield, it purchased over 10,000 commercial receivers on short notice, even though they did not meet military standards for ruggedness. Vincent Kiernan, "Guidance from Above in the Gulf War," *Science*, vol. 251 (March 1, 1991), p. 1012.

[44] American attitudes toward the military-industrial complex are discussed in Benjamin Franklin Cooling, ed., *War, Business, and American Society: Historical Perspectives on the Military-Industrial Complex* (Port Washington, NY: National University Publications/Kennikat Press, 1977); Benjamin Franklin Cooling, *Gray Steel and Blue Water Navy: The Formative Years of America's Military-Industrial Complex 1881–1917* (Hamden, CT: Archon Books, 1979); and Paul Koistinen, *The Military-Industrial Complex: A Historical Perspective* (New York: Praeger, 1980).

BOX 5-A: JACK BROOKS VERSUS THE DEFENSE SCIENCE BOARD

The Defense Science Board's 1986 study *Use of Commercial Components in Military Equipment* decried the government's inability to use "commercial-style" procurement practices, for which it blamed the Competition in Contracting Act's requirement to seek "full and open competition" in government procurement.[1] The preface to the DSB report states that "really substantial change in our buying practices will require changes in our laws. A move to 'effective' competition with an appropriate level of 'common sense subjectiveness' and a move to limit or prohibit protests by losing bidders are needed before we can realize the benefits inherent in good commercial practices."[2]

This conclusion enraged Congressman Jack Brooks, then the chairman of the House Committee on Government Operations and a champion of the Competition in Contracting Act (CICA). Brooks asked the General Accounting Office (GAO) to examine the Defense Science Board panel's balance, the evidence it used, and the process by which it arrived at its conclusions.

Noting that the DSB panel's recommendations conflicted with positions that GAO had previously taken, GAO perhaps not surprisingly found that the panel's conclusions and recommendations "should be viewed with skepticism." According to GAO, the DSB panel based its findings not on factual evidence but on professional and legal opinion; it did not show that CICA had impeded off-the-shelf purchases; nor did the panel address the long-term consequences of changing the requirement for full and open competition. Perhaps most significantly, said the GAO, the DSB panel "did not address whether there are not fundamental differences between commercial and government operations and accountability which give rise to the need for different procedures."[3]

Chairman Brooks himself was not so circumspect. In a statement accompanying the release of the GAO review, he agreed that "increased procurement of commercial products is a goal we all share and would no doubt save American taxpayers billions of dollars." However, he blasted the DSB panel's recommendation to adopt commercial-style competition, which he said "attempts to legitimatize restrictive competition to a few selected companies. It ignores the fundamental differences between commercial and government operations and is just another way of sole-sourcing to your favorite vendor. . . . The Competition Act was passed in 1984 over strong objections from

[1] Final Report of the Defense Science Board 1986 Summer Study, *Use of Commercial Components in Military Equipment*, Office of the Under Secretary of Defense for Acquisition, January 1987.

[2] Ibid., p. xi.

[3] U.S. General Accounting Office, *Competition Act: Defense Science Board Recommended Changes to the Act*, GAO-NSIAD-89-48 (Washington, DC: GAO, November 1988), pp. 2–5.

the Defense Department. It is obvious the DOD and parts of the defense industry haven't given up and will go to any length, including the use of bogus studies, to get their way."[4]

[4]Quotations by Chairman Brooks are from "GAO Criticizes Defense Science Board Attack on Competition," News Release, House Committee on Government Operations, November 10, 1988.

den of contracting with the government cannot appear to compromise oversight of public expenditures.

Similar disagreements swirl around the role of federal procurement in promoting socioeconomic goals, such as greater access by minority-owned firms to government business. The Packard Commission, for example, decried the multitude of advocates and interests that buffet acquisition program managers, making demands that threaten cost and schedule. Although these objectives each might be desirable in itself, the commission said, taken collectively they severely complicate procurement.[45]

The current chairman of the House Government Operations Committee brings a very different perspective. Representative John Conyers has made clear his view that government procurement is an important tool of social policy. "[A] question that I find intriguing is, how do we go beyond the objectives of efficiency and better management to ask to what degree the promotion of social goals fits into more efficient procurement processes?" Referring to goals such as ameliorating regional unemployment (an objective that has a particularly high priority in Swedish defense procurement policy—see Chapter 7) or ensuring that opportunities exist for qualified minority entrepreneurs, Conyers argued that government must pursue social objectives in its own activities to lend credibility to social legislation that it imposes on the rest of the country. "How can we ask the private sector to do more than we, in the government, are doing ourselves?"[46]

Chairman Conyers stated he would not want to sacrifice efficiency and good management to pursue social goals. However, when efficient acquisition comes into conflict with social policy—as eventually it must—it seems clear that Conyers (or Chairman Brooks before him)

[45]President's Blue Ribbon Commission on Defense Management, *A Formula for Action*, p. 8.

[46]Sandra Sugawara, " 'Finding Dirt Isn't the End of Our Job': Conyers' New Agenda for Operations Panel," *Washington Post*, December 18, 1988, pp. H1, H4.

would balance the two differently than would, say, the Packard Commission or the Defense Science Board. So too would they likely disagree on appropriate trade-offs between efficiency and accountability. While officials within the defense industry and many within the Department of Defense would give industry a freer hand, reducing the procedural and administrative "overheads" associated with government contracting, many in Congress and others in the Defense Department who are responsible for seeing that congressional mandates are carried out will be leery of giving too much discretion to acquisition managers. A further question is whether letting a firm link its military and commercial activities confers an "inappropriate" subsidy, or whether it serves public policy. (This tension is a running theme of Washington debates over federal technology transfer policies—Chapter 3.)

Sweetening the Pot

In principle, doing business with DoD could be made attractive enough for many more firms to put up with the difficulties. The most straightforward step would be to raise the profit margins allowed on defense contracts. But an electorate that believes fraud and overpricing are major contributors to the defense budget will not be likely to support such a policy.[47] Many voters and government officials share the suspicions voiced by an Army spokesman: "there is an attitude in this country that having a government contract is a license to print money."[48]

A more palatable mechanism for broadening defense access to commercial suppliers might be to liberalize cost recovery for Independent Research and Development (IR&D). This could be done by relaxing the requirement that projects must be of potential interest to the Department of Defense, by otherwise making more of the contractor's research program eligible for cost sharing, or by increasing the percentage of eligible costs actually reimbursed. Since the performance of R&D

---

[47] Seventy-five percent of survey respondents identified fraud and overpricing by defense contractors to be either the greatest (50 percent of the responses) or the second-greatest (another 25 percent) cause of waste in defense spending, over such other possible responses as incompetent management by the Defense Department, pork-barrel defense projects, and unnecessary new weapons systems. *A Quest for Excellence: Final Report to the President by the President's Blue Ribbon Commission on Defense Management* (Packard Commission), Appendix L: "U.S. National Survey: Public Attitudes on Defense Management," prepared by Market Opinion Research, June 1986, Appendix to Final Report, p. 215.

[48] Quoted in Vartebedian, "Paper Rules Swell Cost of Defense," p. 1.

generally has positive externalities—benefits that extend outside the firm—the public might view more generous IR&D procedures as preferable to other rewards.[49] On the other hand, IR&D may be controversial enough as it is. Many observers already consider it an improper subsidy for large defense contractors. Although closely overseen by DoD, most information concerning IR&D projects and budgets is kept proprietary, and Congress has little say in the expenditure of these billions of dollars of public funds. The program has often become a target of those seeking "waste, fraud, and abuse" in DoD or trying to cut defense expenditures.

More to the point, IR&D and B&P as presently structured have little attraction for firms not already doing business with DoD. In fact, they serve as barriers to entry. Companies without contracts cannot collect IR&D or B&P payments, yet they must compete against established contractors who do receive IR&D/B&P support. The program can serve a useful purpose, provided the political issues are resolved, in encouraging defense contractors to adapt their commercial technology for military use, or to pursue a new technology that may have either military or commercial applications. By liberalizing IR&D cost recovery for technical activity in certain high priority areas such as manufacturing, government could also use the program to signal its interest in these areas.

Consequences of Dependence on the Commercial Sector

High-technology markets span national boundaries, and this is likely to increase. Increased DoD dependence on commercial products inevitably means increasing dependence on foreign products. However, provided that DoD manages this dependence so that it does not lead to vulnerability—reliance on a single source for critical goods, or on multiple sources that could all be interrupted as a result of a single geopolitical event—foreign technology should be seen as an asset, not a threat. Where practical, DoD will have to maintain a diversity of suppliers, avoiding dependence on single firms or possibly even single nations as

---

[49] IR&D funds in any given year are indistinguishable from earnings and in the short run could be converted into uses that the public might choose not to subsidize: higher profits, executive bonuses, or even political contributions. However, under the present system of IR&D negotiations, DoD does not look kindly on firms that do not spend their previous year's IR&D funds appropriately.

it seeks to create "managed interdependence" among the major allied economic powers.[50]

In general, reversing the trend toward increased foreign dependence is neither feasible nor desirable. Restricting DoD purchases to domestic producers closes off access to superior technology that might be available to potential adversaries. And given the increasing investments needed for industry to remain at the technological frontier in high technology, DoD funds alone will not be enough to sustain the domestic technological infrastructure in more than a few areas. Bolstering the technological competence of domestic firms, even foreign-owned ones, offers better prospects for both defense and commercial industry than trade protection—a limited tool that can become counterproductive. It will also be important to improve DoD's ability to draw more from, and contribute more to, civilian industry. Other options to redress potential vulnerabilities include stockpiling parts or assemblies and maintaining backup domestic production capabilities, even if these are inefficient.

Recommendations

Fundamental changes in acquisition policy ultimately depend on reassessing the relative weight given to efficient acquisition compared with other policy goals. Some—but not all—of the following recommendations imply such a shift in political values.

- DoD should encourage trade-offs between cost and performance, avoiding unrealistic performance objectives and keeping military requirements (for parts and components, if not entire systems), where possible, in line with the attributes of existing commercial products. However, DoD will not make cost/performance trade-offs unless it gets to keep at least some of the savings, instead of seeing them diverted to other national needs. Those responsible for establishing requirements will also need to learn what is available in the commercial marketplace.
- DoD (with new legislative authority, if necessary) should exempt commercial manufacturers from some of the administrative, legal, and accounting requirements of the federal procurement code, and eliminate the flowdown of such requirements from prime contractors to suppliers that make commercial products

[50] See in particular Theodore H. Moran, "The Globalization of America's Defense Industries: Managing the Threat of Foreign Dependence," *International Security*, vol. 15, no. 1 (Summer 1990), pp. 57–99.

DoD could otherwise use. This change would require a clear test of "commercialness"—one that itself did not require burdensome reporting—and would also imply accepting possible dilution of the public policy goals that those clauses were intended to further.

- DoD should substantially relax its demands for technical data, particularly the right to release this data to third parties.
- DoD and Congress should allow government purchasing agents and contracting officers to exercise professional judgment, basing their decisions on market acceptance, best value, and past contractor performance, instead of adherence to voluminous, government-specific standards. This would require giving DoD personnel the training and support needed to take on this new responsibility.
- DoD and Congress should foster experimentation and pilot programs to provide experience with new procurement practices. Pilot programs are needed not only to see what the consequences of new policies might be, but also to see how hard those policies might be to implement administratively.[51]

THE FUTURE OF THE DEFENSE TECHNOLOGY BASE

Even without the end of the cold war, DoD would have found it difficult to continue with business as usual because of its increasing dependence on technologies whose progress is driven by the commercial sector. DoD can no longer afford requirements and practices that preclude it from using commercially developed parts and systems, but these are likely to persist absent a crisis serious enough to motivate change. DoD will either have to learn how to draw on the commercial sector effectively or be satisfied with whatever level of technological competence it can instill in those firms still willing to do business with it.

Technological superiority is likely to remain vital to DoD's mission—not, as before, to compensate for an opponent's numerical superiority, but to counter potential adversaries who may have purchased technologically sophisticated arms on the international market.[52] But the overall scope and scale of DoD's R&D effort, budgeted for 1992 at

---

[51] Some of these recommendations echo those in Center for Strategic and International Studies, *Integrating Commercial and Military Technologies.*

[52] See, for example, Office of Technology Assessment, *Global Arms Trade.*

more than 50 percent of a shrinking procurement budget, will almost certainly lessen. DoD will have to establish the capability to handle a wide variety of contingencies without building systems to address each one. To accomplish this, the technology base will have to become a "technology reserve" that "would not consist of complete 'on-the-shelf' engineering designs that would quickly grow obsolete, but instead would consist of a dynamic exploratory program that would pursue a wide range of technical possibilities and, importantly, that would keep together groups of scientists and engineers with knowledge of important military problems."[53] With the "warm start" provided by such a reserve, DoD would be able to regenerate military capabilities should the international security situation facing the United States demand it.

This is a challenging task. DoD will have to find ways to remain technologically aware without doing a great deal of full-scale development and production. It must maintain links to commercial suppliers without the ability to offer them lucrative productive contracts. Declining procurement budgets will lessen whatever incentive firms now have to invest their own R&D in defense-related areas in the hope of future defense sales. These conditions suggest what is needed:

- Flexible technology base programs that keep a wide range of options open, that keep skilled teams together, and that are financed at a higher level than today at the upstream end of the acquisition process—basic research, exploratory development, and prototyping.
- R&D programs with the primary objective of technology assessment rather than ultimate production.
- Cooperative R&D agreements involving commercial firms to develop dual-use technologies.
- Strengthened ties with commercial industry, including the ability to purchase commercial parts, components, and products.
- Increased emphasis on upgrading, rather than replacing, existing equipment.[54]

Whether funding for technology base activities does or does not increase, total R&D and procurement budgets will drop. Political de-

---

[53] Report of the Carnegie Commission on Science, Technology, and Government, *New Thinking and American Defense Technology,* August 1990 (New York: Carnegie Commission), p. 23.

[54] This list draws on Institute for Defense Analysis, *The Future of Military R&D: Towards a Flexible Acquisition Strategy,* July 1990, p. 5.

mands for cumbersome accounting procedures and oversight may lessen as well, permitting reforms. Without a change in the political climate, it will be difficult to institute streamlined contracting procedures that will help a more flexible defense technology base program maintain ties with the commercial sector.

REINTEGRATING THE TECHNOLOGY BASE

Since World War II, U.S. military strength has relied on technological supremacy built on a massive, sustained investment in R&D. Much of this R&D, and essentially all defense production (except for nuclear weapons), has taken place in private industry. However, the defense industry has become isolated from commercial industry, which limits potential synergies.

This separation has many causes. Some military technologies are simply irrelevant to commercial needs. In other cases, operating conditions or logistical needs impose unique requirements. Institutional barriers—notably DoD's stress on performance over cost and its difficulty in managing trade-offs—stand between the military and commercial sectors, even where they use the same technology. But the most important cause may well be the requirements placed on government procurements in the pursuit of accountability, oversight, fairness, and other social goals.

Because the barriers between DoD (and its contractors) and the commercial sector are deeply rooted in the American political system, closer integration will require a fundamental rethinking of competing policy objectives. Greater integration would mean relaxed procurement requirements and greater discretionary authority for procurement officials—both of which are likely to be politically contentious. If such changes occur, they will reflect the recognition that DoD's isolation now threatens its technological edge, as it did not when defense technology was superior to anything the commercial sector could provide.

6

Dual-Use Industry: Structure and and Strategy

President Eisenhower's famous departure speech, warning against the political influence of the "military-industrial complex," reinforced an image of the defense industry that is held by many Americans: a collection of large, monolithic contractors that have few commercial interests and enjoy too close a relationship with their government customers. As the defense procurement scandals of recent years demonstrate, there is more than a little truth to this image. But it is flawed in several respects.

In this chapter we shall see that defense firms, more often than not, are parts of larger corporations whose majority business is commercial (and often international). And while defense dollars move in a highly concentrated flow through a relatively small number of prime contractors, subcontracts touch a very large number of firms, located in every state. It is natural, therefore, to look for management opportunities to increase synergies between military and commercial operations within firms, if not among them. But if firms with both commercial and military business (dual-use firms) are structured in ways that impede technical exchange between the two—as often happens, for reasons described in Chapter 5—then expectations for spinoff and other forms of technology sharing are likely to go unrealized. Technology transfer is difficult within a single firm; it is much more so between firms in a competitive industry.

This chapter explores the structure of the defense industry—

defense/commercial relationships within individual firms, the degree and significance of subcontracting, and the extent to which industrial sectors serve both military and commercial markets. Do firms balance their commercial and government work to seek mutual leverage? How might managers look for dual-use technology opportunities?

These issues are important to the economy, for they illuminate the extent to which defense procurement is a boon or a burden for competitiveness. They may help define public policies for strengthening the industrial base on which defense and economic success depend. These questions are also important to the defense industry, for they illuminate its alternatives in responding to post-cold war restructuring, as well as to international competition.

THE STRUCTURE OF THE DEFENSE INDUSTRY

Concentration

Defense purchases of goods and services, including research and development, are concentrated in a relatively small number of prime contractors. The top 10 defense contractors in 1988 won contracts worth $46.7 billion, 34 percent of the $137.0 billion worth of defense contracts that exceeded $25,000 issued that year; the top 100 took two-thirds of the total.[1] McDonnell Douglas, the largest contractor in 1988, alone received $8.0 billion or 5.8 percent, which accounted for 53 percent of its total revenues. Even so, many commercial firms are much larger; McDonnell Douglas ranked 35th in sales in the *Fortune* 500 that year.

The concentration of military contracts in a relatively small number of firms does not imply that these firms specialize in defense; quite the contrary. Among the U.S. corporations that are consistently among the top 100 defense contractors, DoD prime contracts average less than one-tenth of their total business. Defense contracts comprised more than 75 percent of 1988 total sales in only three of these companies,

---

[1] U.S. Department of Defense, *100 Companies Receiving the Largest Dollar Volume of Prime Contract Awards, Fiscal Year 1988*, DIOR/P01-88 (Washington, DC: Department of Defense, undated [probably 1989]). This document covers contract actions over $25,000. Smaller contracts constituted only 2 percent of DoD contract actions for FY1988 but accounted for 89 percent of the contract dollars, according to "Standard Report: Fiscal Year 1988 through Fourth Quarter," U.S. General Services Administration (GSA), Federal Procurement Data Center (Washington, DC: GSA, 1989), p. 3.

and more than 50 percent only in another six. Despite being the largest defense contractors, these firms are overwhelmingly civilian-oriented.[2]

Segregation of Defense Business

Even though we have just seen that large defense contractors are predominantly commercial from a corporationwide perspective, this does not mean that the divisions or business units doing defense work integrate military with commercial production. In fact, the opposite is generally true; most defense business is done by divisions that work mostly for DoD.

This can be seen by analyzing data from DoD's Independent Research and Development/Bid and Proposal (IR&D/B&P) program, where cost reimbursements for a given division depend on its ratio of defense sales to total sales (see Chapter 4). In fiscal year 1988, 416 companies or corporate divisions—all those receiving more than a certain threshold of IR&D/B&P cost recovery—were required to negotiate their IR&D with DoD. These divisions had $118 billion worth of DoD business, two-thirds of their total revenues of $178 billion.[3] For the same fiscal year, DoD contract actions for all purposes totaled $149 billion.[4] Using this $149 billion to approximate all direct sales to DoD (even though it is not exactly the same thing, since contract actions represent commitments to future expenditures, not current payments), we find that purchases from those 416 divisions constitute about 80

---

[2]These conclusions are based on data from 67 firms listed in Appendix 6-A. The figure of one-tenth, however, must be interpreted with caution. Only prime contracts (i.e., sales directly to DoD) are considered "defense business" in this analysis. All defense subcontracts and intermediate goods shipped by defense contractors are included under the "civil sales" figure, which is obtained by subtracting prime contracts from total sales. We have no reliable data for the share of DoD prime contract dollars that are subcontracted out or spent on intermediate goods, but we assume later in this chapter that about half is subcontracted, and about half of this represents one prime contractor subcontracting to another. On this basis—i.e., that large defense contractors on average perform subcontracts for other prime contractors equal in value to one-quarter of their own defense prime contracts, and that these subcontracts show up in the data as civil sales, we estimate the true military sales (including subcontracts) for these firms to be 25 percent greater than the amount of their defense prime contracts, and their true civil sales to be smaller by the same amount. Making this adjustment, we find that the ratio of civil to military sales of the companies listed in Appendix 6-A is still greater than 8 to 1—down from 10 to 1, but substantial nevertheless.

[3]Defense Contract Audit Agency, *Summary: Independent Research and Development and Bid and Proposal Cost Incurred by Major Defense Contractors in the Years 1988 and 1989*, P-7730.15 (Washington, DC: Defense Contract Audit Agency, March 1990).

[4]Federal Procurement Data Center, U.S. General Services Administration, "Standard Report: Fiscal Year 1988 through Fourth Quarter," p. 2.

percent of all DoD purchases. Therefore we find that most DoD purchases are from firms or divisions of firms that do mostly DoD business.

Research and Development

Military R&D is even more concentrated than total military sales, with the top 10 recipients of 1988 RDT&E contracts garnering $10.9 billion, nearly half of the $22.5 billion in RDT&E contracts over $25,000. The top 100 recipients won 86 percent of the total.[5] Not surprisingly, most of the top prime contractors also perform much of the RDT&E: the top 15 RDT&E contractors and the top 15 prime contractors have 11 firms in common.

The company-by-company data listed in Appendix 6-A show a striking difference between military and commercial R&D intensities—ratios of R&D to total sales. Aggregating data for all the firms in Appendix 6-A, the ratio of defense RDT&E contracts to all defense contracts is 17 percent, more than four times larger than the ratio of civil R&D to total civil sales for the same companies. Recognizing that the "civil R&D" figures include IR&D, which is partially reimbursed by DoD, this disparity becomes even more pronounced. When the firms are grouped according to their primary line of business, every group shares this disparity except oil companies, which do essentially no military R&D. The very high R&D intensity of defense reflects the emphasis placed on technology by DoD. It also indicates a significant "cultural" distinction between military and civil business, even in firms that do both.

This discrepancy between military and commercial R&D intensities is also apparent from an examination of the DoD budget. In 1990, DoD's RDT&E budget of $36.6 billion was fully 45 percent as large as the $81.4 billion spent for non-R&D procurement, or 31 percent of the $118 billion combination of the two. By contrast, among U.S. manufacturing firms that perform R&D, including defense firms, R&D averages 4.8 percent of sales. Even the highest of high-technology companies rarely report R&D at more than 10 or 12 percent of civil sales. Many firms do not report doing any R&D at all. Some of the disparity

---

[5] Department of Defense, *500 Contractors Receiving the Largest Dollar Volume of Prime Contract Awards for RDT&E, Fiscal Year 1988* (Washington, DC: OSD Washington Headquarters Services, Directorate for Information Operations and Reports, undated [probably 1989]). This report, like the prime contract award listing, generally includes contracts for all of a firm's divisions and subsidiaries within the parent company total. However, some subsidiaries are still listed separately. The total given here for the top 15 firms includes those of their subsidiaries that are listed separately in the top 500.

between military and civil R&D intensities is due to accounting inconsistencies, particularly since many of DoD's purchases of goods and services are not included in the procurement and R&D accounts that we have taken to be the analog of commercial revenues.[6] But most of the discrepancy is real, a striking illustration of DoD's penchant for technical virtuosity, its investment to eliminate the possibility of technical surprise, and its tolerance for expensive R&D programs that fail to proceed to procurement, or are never intended to.

Subcontracting

The size of the prime contracts shown in Appendix 6-A, and the relatively small number of firms that win them, tell us that control over defense production is quite concentrated. But the prime contractors do not perform all of this work themselves. To the extent that prime contractors specialize in defense marketing and systems integration, subcontracting development and production to others, they have less of a role in mediating dual-use interactions. Second-tier subcontractors may be more important dual-use institutions.

"Second-tier" refers to firms that supply components, subassemblies, specialty materials, and other items—including services—to prime contractors (usually larger firms) that are responsible for integration and assembly of end products or systems. "Third-tier" manufacturers produce standardized parts and materials that are priced in competitive commodity markets. (The distinction between tiers is fuzzy, since one firm can easily span two or sometimes all three.) The share of prime contract funds that are spent on subcontracts has been estimated at about 50 percent, we further assume that half of these subcontracts are with other prime contractors.[7] Accordingly, perhaps only

[6]The FY1990 defense RDT&E appropriation, in budget authority, is from Department of Defense, *RDT&E Programs (R-1)*(Washington, DC: Department of Defense, February 4, 1991). The procurement figure is from Department of Defense, *Report of the Secretary of Defense to the President and the Congress* (Washington, DC: Department of Defense, January 1991), p. 109. The average R&D to net sales ratio for U.S. manufacturing companies that perform R&D is a 1988 value, from National Science Foundation, *National Patterns of Science and Technology Resources: 1990*, NSF 90-316 (Washington, DC: National Science Foundation, May 1990), p. 68.

[7]The portion of prime contracts that are subcontracted varies from case to case, but it has remained fairly stable, at about 50 percent, over the past 25 years. See Kenneth R. Mayer, "Patterns of Congressional Influence in Defense Contracting," in Robert Higgs, ed., *Arms, Politics, and the Economy: Historical and Contemporary Perspectives* (New York: Holmes and Meier, 1990), p. 220.

one-quarter, or $34 billion, of the 1988 prime contract dollars find their way outside the prime contractor community to second- and third-tier vendors.

Whereas most third-tier firms employ standardized production technology, competing on cost and delivery within well-documented quality standards, second-tier manufacturers often develop strategies around a particular technological specialty that gives them competitive advantage. When they are able to apply their expertise to learn from and solve problems shared by commercial and military clients, these companies form a powerful linkage between the two sectors. An example typical of these companies is Lord Corporation (see Box 6-A), which defines its business as "vibration and noise isolation." Its products are found in the rotors of military helicopters, the landing gear of commercial airliners, and the shock mounts of outboard motors. While defense products constitute only 15–20 percent of Lord's business, their demanding technical requirements allow Lord engineers to try out new ideas before commercial markets are ready to accept them. In meeting defense needs, Lord uses the same team of engineers and the same production facilities that it applies to commercial aviation and related markets. In this way, defense requirements stimulate technical progress that is directly applicable to commercial opportunities without the need for complex technology transfer arrangements.

Although prime contract dollars are highly concentrated in a few companies, DoD contracts and subcontracts nevertheless touch a great many manufacturing firms. According to a 1988 Census Bureau survey of selected manufacturing industries, half the establishments that employed more than 20 people shipped products either directly or indirectly to the Department of Defense. Of an estimated 39,600 establishments, some 12,300 plants (31 percent) were primes, shipping directly to a defense agency; 17,000 (43 percent) were subcontractors. These two groups have considerable overlap, as 9,600 plants were both primes and subcontractors.[8] These data do not tell us how much of the defense dollar reaches subcontractors, but they suggest that participation is very broad.

---

[8] Todd Watkins, of Harvard's Kennedy School of Government, compiled this information using data from the *Current Industrial Reports: Manufacturing Technology 1988* (Washington, DC: Department of Commerce, Bureau of the Census, May 1989). The industries covered—metal fabrication, machinery, electrical and electronic equipment, transportation equipment, and instruments—are those categorized by two-digit Standard Industrial Classification (SIC) codes 34 through 38. See footnote 12 of this chapter and Appendix 6-B.

Box 6-A: The Lord Corporation—A Second-Tier Dual-Use Firm

Founded in 1924 by Hugh C. Lord, this privately held company was built on products made of rubber bonded to steel.[1] Today the company's "cash cow" is an adhesive called Chemlok, invented in the 1950s by the founder's son, Thomas Lord, and a gifted chemist, Donald M. Alstadt, the current chairman. Lord and Alstadt did not make the mistake so common to single-product, high-tech startups, like the buggy-whip maker who goes out of business because he did not realize he was in the vehicle accelerator market. Instead of defining the business as adhesives manufacturing or rubber-to-metal bonding, they saw the firm as a technology company solving industrial problems in vibration isolation and noise control.

Lord still has strong lines of business in elastomer bonding and industrial adhesives. But today the company is developing advanced materials and computer-controlled active systems for both military and commercial clients using signals from motion or acoustic sensors to anticipate vibration and nullify it. In many cases, no rubber bonding or adhesive is involved in the advanced products supplied to their traditional markets.

Thus Lord Corporation is very much a science and technology company. Not only does it define its business in functional technological terms, but it also recognizes the new and more important relationship between science and engineering. Lord lists its core technologies as "materials science, mechanical dynamics, surface science, electromechanical systems, and specialty polymers and chemicals."[2] Although it would be considered only middle-sized, Lord has established a sophisticated research laboratory, and assiduously cultivates relationships with university science departments and small high-tech startups with particularly interesting materials skills.

From its modest beginnings in a basement, Lord Corporation has grown to $250 million in annual revenues. About a third of the firm's sales come from the aerospace segment of its business, a mixture of defense and commercial projects. On the defense side, Lord has worked for Sikorsky on the Black Hawk and CH-53 helicopters and the Bell-Boeing V-22 Osprey tilt-wing transport; on the commercial side, the firm is a supplier for Boeing's 737, 757, and 767 aircraft. All this is done in one division; military work is not segregated. Thus a single engineering team learns lessons from both markets, meeting stringent performance demands on military projects and equally stringent reliability and cost demands on the commercial side.

[1] Lord Corporation internal documents and interview (LMB, August 1990) with J. Freeman, vice chairman, Lord Corporation.
[2] Lord Corporation *1989 Annual Review*.

Because the company's business is defined in technological terms, multiple market strategies are not only natural, but inevitable. Management does not need to make a deliberate effort to see how defense-driven technology might be used in a commercial product or how commercial ideas could be incorporated to meet a defense requirement. All of Lord's work is client driven, subject to the overall strategy of leveraging the company's technical strengths. Since Lord products must almost always be design-optimized for a particular application, the key technical people at Lord have traditionally been the applications engineers. Product research and development, as it is conventionally thought of, was until recently a relatively modest activity, but is now being encouraged to create new business opportunities. Nevertheless, applications engineering remains a key function, and when applications engineering teams develop a close working relationship with the customer's product design engineers, Lord not only can do a better (and quicker) job for the customer, but also can build itself into a customer's future product plans.

This kind of relationship, if established with both civil and military systems integrators, creates a powerful mechanism for technology sharing between government and private markets. For that reason, second-tier firms like Lord Corporation are perhaps more important agents of public-private synergy than are the large systems integrators and end-product manufacturers that receive most of the attention in discussions of competitiveness.

HOW IMPORTANT IS DEFENSE PRODUCTION IN U.S. MANUFACTURING?

Defense's overall economic impact can be measured in several ways. From budgetary outlays, we find that the fiscal year 1990 defense budget of $299 billion constituted 5.5 percent of the nation's GNP for the same period.[9] A similar value is derived from the Commerce Department's National Income and Product Accounts, which show that purchases of goods and services for national defense in fiscal year 1990 totaled $307 billion, or 5.7 percent of GNP. Disaggregating this expenditure one step further, we find that defense purchases of durable goods (very close in amount to DoD's procurement budget line item) came

[9]This figure includes DoD spending plus DOE nuclear weapons programs plus a small amount of defense-related activities outside DoD such as Selective Service and Civil Defense.

to $81.1 billion in fiscal year 1990, comprising 8.7 percent of the nation's total durable goods production—one and one-half times DoD's share of GNP. On the other hand, defense consumed only 0.9 percent ($11.1 billion) of the nation's nondurable goods production, mostly in the form of petroleum products and ammunition.[10] These figures have dropped since 1987, the high-water mark of the Carter-Reagan defense buildup, when defense absorbed 11.6 percent of durable goods production and 6.5 percent of GNP.[11] Defense's 31 percent share of the nation's R&D is disproportionately high compared to any of these measures.

The 50 Largest Defense Industrial Sectors

Although of modest aggregate impact in terms of dollars, DoD purchases are very unevenly distributed across industry. A Defense Department economic model has been developed to estimate the distribution of defense spending across states and across industries. Results from this analysis (Appendix 6-B) show that the communications "electronics," and aircraft industries dominate defense procurement, as they do defense R&D.

Based on this model and on Commerce Department data, Appendix 6-B presents data on the 50 industrial sectors that had the highest defense shipments in 1987.[12] Although uncertainties in the data make it difficult to interpret quantitatively, two points stand out. The first is that few of these sectors are dominated by defense business. For only 9 of the 50 does defense account for more than half of sector output, and the defense shipments of all 50 sectors constitute only 17 percent of their total output. As a fraction of sector output, defense shipments

[10] Department of Commerce, *Survey of Current Business,* vol. 711, no. 1 (January 1991), Tables 1.1, p. 6, and 3.7B, p. 11.

[11] Defense spending in 1986 was larger in real terms than it was in 1987, and defense goods and services were a larger share of GNP. But the defense share of durable goods during the 1980s peaked in 1987.

[12] "Industrial sector," as used here, refers to a single category in the Standard Industrial Classification, used by the Commerce Department to collect data on U.S. economic activity. This scheme categorizes every economic "establishment" (factory, store, hotel, and so forth) in the United States. At the degree of detail used here and in Appendix 6-B, manufacturing activity is subdivided into some 450-odd sectors such as "aircraft" (SIC 3721) or "machine tools, metal cutting type" (SIC 3541). These industrial sectors are called "four-digit SIC sectors," since each is denoted by a four-digit number. With some modification, the same categories are used by the Defense Department economic model. In this discussion and in Appendix 6-B, "shipments" from an industrial sector represent the total sales value of all products manufactured by factories classified in that sector. Defense shipments include shipments to DoD (direct sales) and its contractors (indirect sales).

range from practically all (e.g., ordnance, space vehicle equipment, ammunition, and shipbuilding) to hardly any (e.g., the petroleum, paper, motor vehicles, auto parts, and photographic equipment and supplies sectors).

The second point is the importance to defense of industrial sectors considered to be "high tech." Economists classify more than a fifth of the manufacturing sectors (100 out of some 450) as high tech, responsible for 36 percent of all manufacturing value-added (see Appendix 6-B).[13] However, a much higher fraction (36 of 50, or 72 percent) of the 50 sectors with the largest DoD shipments qualify as high tech. These 36 sectors account for about 65 percent of the total value-added of the 50 sectors. If we assume that the ratio of total value-added to total shipments for these sectors also applies to that portion of their output devoted to defense, fully 80 percent of the defense value-added from these 50 sectors comes from the 36 high-tech sectors.

The share of output devoted to defense affects the degree to which dual-use relationships might be found in an industrial sector. The defense-exclusive industries such as ordnance could hardly be considered dual use. Civil-dominated industries such as paper and auto parts, with appreciable defense business but much larger civil markets, might likewise have little incentive to seek dual-use synergies because of defense's small market leverage. However, there may be niche markets in these industries where defense applications could have important dual-use potential.

In between these extremes, 20 sectors listed in Appendix 6-B have somewhat balanced levels of defense and commercial production, with between one-fifth and four-fifths of their shipments destined for DOD. These are

- radio/TV communications equipment (SIC 3622);
- aircraft equipment, aircraft, and their engines (3728, 3721, 3724);
- industrial trucks and tractors (3537);
- guided missiles and space vehicles (3761); space propulsion (3764);
- instruments (3811, 3823, 3832); automatic controls (3822); measuring and controlling devices (3829); and process control instruments (3823);

[13] The classification criteria are those of Ann Markusen, Peter Hall, and Amy Glasmeier, *High-Tech America* (Boston: Allen and Unwin, 1986). See Appendix 6-B. "Value-added" is the difference between revenues received for goods and services and the cost of the materials, energy, and contract work that went into them.

- machine tools (3541);
- turbines and turbine generator sets (3511); motors and generators (3621);
- electronic components (3679); semiconductors (3674); electronic connectors (3678);
- primary aluminum (3334); and
- plating and polishing (3471).

With the exceptions of primary aluminum and plating and polishing, these sectors are all high tech, and they might be fruitful candidates for dual-use technological synergies. However, the degree to which firms can seek and attain such synergies will depend on their organization and management—the subject of the next section.

CHARACTERISTICS OF DUAL-USE FIRMS

Given our examination of the largest defense contractors (Appendix 6-A) and defense-intensive industry sectors (Appendix 6-B), how are we to distinguish firms whose managements might view dual-use technology as strategically important? We have seen that the 50 largest defense manufacturing sectors derive 83 percent of their revenue from civil markets, implying considerable overlap (at that level of disaggregation) between goods and services needed for defense and for the commercial economy. Overlap continues at the level of the firm, where we found that defense prime contracts constitute only 9 percent of the total business of the largest defense contractors. But these measures do not tell us to what extent military and civil work are integrated within the firm. Indeed, we have seen that most large contractors segregate defense and commercial business in separate divisions, linked only at the highest levels of corporate management. In lieu of systematic information that can provide details on the products these firms manufacture, the production processes involved, the technologies used, or the management methods employed, we must draw on case studies to address how firms relate their military and commercial work.

We have already presented one such case, describing the Lord Corporation's ability to meet both military and commercial demand with a single production facility and engineering workforce. Several more will be discussed here. The purpose is not to draw firm conclusions, but to ask the following:

- How many firms have sufficient balance between military and civil lines of business that they have the opportunity to seek competitive advantage by sharing or transferring technology, or by seeking RDT&E contracts from defense in the expectation of commercial reward? How does the incentive to exploit cross-sectoral opportunities compare with other incentives that drive managerial decisions at the corporate level?

- What mix of government and private business is most likely to be conducive to success in technology sharing and transfer, assuming that the motivation to do so is there? Under what circumstances does defense work threaten the success of a firm's commercial work?

Requirements for Success

Generalizations about the circumstances under which firms are motivated to focus on technology shared across public-private markets are difficult to draw, not only because the incentives are complex, but also because both the incentives to transfer technology and the ease of accomplishing the transfer must be addressed. Compatibility of technological skills is a requirement; so too is complementarity of technological assets. For technology to flow effectively from one sector, institution, activity, or team of people to another, there must be common experience, shared intent, and ability to cooperate, as well as a strong motivation to do so. This circumstance is more likely to prevail in firms with a degree of balance between military and civil activities, and where both depend on similar technologies. The most obvious circumstance in which this condition prevails is when, in a large or small firm, the same engineers support both market sectors. In smaller second-tier firms that define their business by their technology, this will likely be the case. Among larger companies, we might expect to find shared technology in firms such as Raytheon or Rockwell International, which possess strong and similar technological skills in both markets.

Not all such companies function in this way, however. There are many possible motives for firms to engage in both commercial and defense lines of business. We can identify five, although they are not mutually exclusive:

1. diversification (conglomerates);
2. balancing business cycles (diversification within related industry sectors);

3. technological integration (companies with businesses defined in terms of technology that seek economies of scale and scope);
4. acquiring defense technology to support a commercial business strategy; and
5. providing a public service (making technological capabilities available to meet national needs).

Diversification (conglomerates).   The most obvious motive is that of the successful holding company, which assembles a portfolio of businesses paying primary attention to return on capital. The nature of the market served and the technology needed to serve it are secondary concerns, and executives of conglomerates presumably have no particular interest in seeking advantage from technological synergy across lines of business.

Historically, the best example is ITT Corporation under Harold Geneen. Geneen had been the number-two executive at Raytheon Company, itself the antithesis of a conglomerate, notwithstanding its range of defense and civil businesses. Geneen left Raytheon after a dispute with the firm's chairman over Geneen's desire to make business acquisitions.[14] In the heyday of conglomerates in U.S. business, ITT was the fast-growth leader. Today this strategy looks like playing from weakness, given superior Japanese access to cheap capital and the importance of management focus on production, quality, and cost.

Balancing business cycles (diversification within related industry sectors).   Firms in highly cyclical businesses may believe—or hope—that swings in government and commercial demand will be at least uncorrelated, even if they do not run counter to one another. To the extent that is true, being in both markets helps hedge against downturns in either alone. From the standpoint of promoting stable revenues and earnings, this scenario does not require that the firm be able to shift resources rapidly between the two markets, only that the sum of both businesses be more stable than either separately. However, the ability to shift resources between sectors would enhance the effectiveness of this strategy; thus management may have an interest in technological flexibility across lines of business. Raytheon Company is a good example of such a firm (see Box 6-B).

Technological integration (companies with businesses defined in terms of technology that seek economies of scale and scope).   A firm may derive

---

[14] See David Warsh, "How Raytheon Flourished without Corporate Fads," *Boston Globe,* May 20, 1990, pp. A1, A3.

BOX 6-B: RAYTHEON—A BALANCED DEFENSE AND COMMERCIAL COMPANY

"What are the prospects for Raytheon?" asked then-CEO Tom Phillips in a company newsletter in December 1989.[1] "It is well that we have become and remain a diversified company." Raytheon's defense and civil sales have been roughly 50-50 for a good many years. Of its 1989 sales of $8.79 billion, 39 percent derived from nondefense businesses: Beech Aircraft, D.C. Heath publishing, and several energy and home-appliance units, including Amana Corporation. Defense earnings are even more prominent than defense sales at Raytheon, but they have slowly but steadily declined as a fraction of total earnings from 1985 (when defense profits were 88 percent of the total) to 78 percent in 1989. This is more a reflection of improvement in Raytheon's civil businesses than a decline in defense profitability. Beech Aircraft, for example, earned $44 million in pretax profit in 1989, up from $28 million in 1988.

Statistically, Raytheon looks very much like a dual-use firm—one that consciously seeks to relate its defense and its commercial activity. But Raytheon executives echo the views of most balanced defense and civil market companies when they say, as Phillips does, "There is nothing magic about 50-50." Joseph Shea, senior vice president for engineering, notes that except for Raytheon's corporate research laboratory there are few deliberate efforts by management to share technologies across the boundaries of civil and military divisions.

Nevertheless, Raytheon demonstrates a number of interesting examples of dual-use achievements besides the microwave oven and ship-borne radar described in Chapter 3. The Beechjet project began as a purely civil executive jet program—a joint venture with Mitsubishi called Diamond II. Mitsubishi built a manufacturing plant for the venture, but prior to first production Beech bought out its partner and moved the entire production facility—numerically controlled machines, program tapes, and all—to Wichita, Kansas. In 1986, the first aircraft, called Beechjet 400A, was produced; and in 1990, Beech won an Air Force contract with a potential of over $1 billion for up to 211 Beechjets.[2] The main civil/military transfer here was not of technology but of marketing know-how. "Beech was able to draw on the resources of Raytheon's government group—in particular the Missile Systems Division—to help with the [Air Force] proposal," Dennis Picard, now chief executive officer of Raytheon, told the stockholders. Picard has described a number of instances in which Raytheon used its govern-

---

[1] This discussion is drawn from Frederic M. Biddle, "Raytheon's Peace Plan," *Boston Globe*, January 23, 1990, and from interviews conducted by Harvey Brooks, Lewis Branscomb, and Craig Stetson with a number of Raytheon officials, including Thomas L. Phillips, then CEO, and Joseph Shea, senior vice president for engineering.

[2] *Raytheon News*, June–July 1990, pp. 1ff.

ment marketing know-how as a complementary asset in marketing an essentially civil technology to the public sector.

A more typical example of military technology supporting a commercial venture is Beech's Starship, a commercial aircraft constructed of advanced composite materials. Another example is the MILVAX computer mentioned in Chapter 3. At the 1990 annual meeting, Picard made a clear statement of Raytheon's intent. "By bringing together our government and commercial groups on a project-by-project basis, we can create synergy. We can make the company greater than the sum of its parts."

both military and commercial sales from a common technology base in which it specializes. This is characteristic of second-tier vendors such as Lord Corporation (Box 6-A), and may also be true of large prime contractors such as Boeing, Hughes (Box 6-C), or General Electric's jet engine business. Such firms presumably have an interest in keeping as much of their engineering capability as possible common to both government and commercial lines of business so it can be rapidly redeployed as opportunities come and go. In doing so, they also put themselves in a position to capture economies of scope—the benefits that accrue from applying a given set of tools and techniques to the different but related problems that the two markets offer.

Technological integration can also be advantageous in industries with high capital costs, such as jet engine and aircraft manufacturing, where it is desirable to have large production runs of both military and civil products. The internal organization of these companies is rarely as unified as in the Hughes example in Box 6-C, but such firms as Boeing have nevertheless managed the combination of military and civil production to good effect. So too have General Electric and the Pratt & Whitney division of United Technologies in jet engines.

Acquiring defense technology to support a commercial business strategy. Some firms have set out to acquire a military business for the purpose of acquiring technology that arose outside their industry. Rockwell Standard Corporation's acquisition of North American Aviation was an early, highly visible, and initially unsuccessful attempt (see Box 6-D).

Public service. Another motivation can be a sense of obligation to make the firm's capabilities available to government, perhaps coupled with the hope that the firm might more easily approach government for sympathetic attention in times of stress. One example is the IBM Corporation, where government-contracts work is segregated in the Federal Sector Division. While substantial by the scale of the defense

BOX 6-C: HUGHES AIRCRAFT—A DUAL-USE COMMUNICATIONS SATELLITE MANUFACTURER

Entering the satellite business in July 1963 when its Syncom II became the Earth's first geosynchronous artificial satellite, Hughes Aircraft Corporation has long been a leading supplier of commercial satellites. Hughes, as part of General Motors, is also the sixth largest U.S. defense contractor (Appendix 6-A).

At the heart of Hughes' dual-use strategy is an organizational structure designed to exploit the common technical ground among its satellite customers. This structure has two main components: program offices and divisions. Differentiation along customer lines occurs in the program offices, while the engineering and manufacturing divisions stress interchangeability and commonality among all projects, both military and civil.

Three program offices serve as direct links to customers—one each for DoD, NASA, and commercial programs. These offices are generally responsible for the task organization and completion of specific projects. Correspondingly, each program office has its own complement of systems engineers, who focus on the particular demands of that office's customers.[1]

The design and fabrication of spacecraft subsystems is centered in the engineering and manufacturing divisions. In order to capture the benefits of scale and retain the flexibility to interchange parts and manpower when needed, these two divisions serve all programs, regardless of the structure of the individual customer's contract. One implication of this organizational design is that technical manpower in the engineering and manufacturing divisions is entirely interchangeable among projects—a feature not found in the program offices.[2]

As a result of the commonality in design, manufacturing, and testing, most satellites built by Hughes are extensions of a "generic spacecraft bus." That is, they contain identical propulsion and power systems, bearings and mechanisms, altitude control sensors, digital computers, and structural members. As Albert D. Wheelon, former CEO of Hughes, has noted, "All the spacecraft were built to the same standards, independent of customer, contracting method, and level of in-plant inspection."[3]

Civil and military satellites are not identical, though. Because of different communications frequencies, substantially different radio subsystems must be designed. Further, because DoD missions tend to be more technologically sophisticated, its spacecraft are typically

---

[1] Albert D. Wheelon, *Space Policy: How Technology, Economics, and Public Policy Intersect*, Program in Science, Technology, and Society Working Paper No. 5, Massachusetts Institute of Technology, May 1989, pp. 76–77.

[2] Ibid., p. 77.

[3] Ibid., p. 74.

larger than the more cost-sensitive commercial designs. The combination of these different size and weight characteristics, along with contractual obligations, explains the different launch vehicles used for each: DoD on the larger Titan, and commercial satellites on a host of rockets, from the Chinese "Long March" to the U.S. Space Shuttle.[4]

Hughes' latest satellite design, the HS 601, illustrates its strategy. Power capability on a satellite is a function of solar panel size as well as the capacity of on-board batteries. In the HS 601 design, extra solar panels can easily be attached to the satellite structure, and battery capacity may be expanded by the insertion of more plates. Therefore the "HS 601 can be adapted to a wide spectrum of customized payloads without requiring bus redesign." The first two customers and their proposed missions demonstrate Hughes' dual-use strategy. The first HS 601 was sold to Australia to provide mobile satellite communications, direct TV broadcast service, and expanded coverage of New Zealand and the Australian outback. The second HS 601 was sold to the U.S. Navy as its UHF Follow-On satellite, providing tactical communications among air, sea, and land forces around the globe—a decidedly different mission.[5]

DoD has a great "fear of being accused of having given the successful contractor an edge by virtue of this inherited design."[6] As a result, Hughes has adopted a common bookkeeping system for both civil and military projects, as well as DoD procedures for cost and schedule control, contract management, and quality control. With consistent bookkeeping the government can track cost allocations more easily when facilities or services are shared. More important, Hughes builds its generic components to the standard of its most demanding client, which is not always the military.[7] Consequently, efforts to reduce the overhead for commercial programs serve to reduce the costs of military programs as well.

The dual-use strategy employed by Hughes may provide further benefits as defense expenditures decrease. Should the DoD component of Hughes' business fall, the fact that it has pursued a production process common to both military and civil customers will enable an easier transition. The lesson from Hughes is clear: it is possible for firms to succeed in both military and commercial markets by exploiting the common ground—in production and technical design—between these two types of customers, if functional requirements and management practices permit.

[4] Ibid., p. 78.
[5] "A World Without Boundaries," Hughes Space and Communications Group brochure No. 890333 (1989); and idem, "Dynamic HS 601," No. 890315 (1989).
[6] Wheelon, Space Policy, p. 80.
[7] Ibid., p. 78.

BOX 6-D: ROCKWELL INTERNATIONAL—AN AMBITIOUS DUAL-USE ACQUISITION

Prior to its merger with North American Aviation in 1967 to become North American Rockwell, the Rockwell Standard Company had been an automotive parts and related mechanical manufacturing company. It was flush with cash, and its president, Al Rockwell (son of founder Willard Rockwell), reportedly saw North American as a gold mine of commercializable technology. North American, for its part, was the prime contractor for the Apollo program and at the pinnacle of its fame in 1967; the first mission to the moon was only months away. But by then its NASA contracts were already coming to an end, and Lee Atwood, CEO, saw little follow-on business after Apollo. He was on the lookout for a diversification partner.[1]

Whatever Al Rockwell's expectations were before the merger, he was to be disappointed afterward. Teams of engineers sent to find and transfer technology had at best limited success. Peter Cannon, who was brought in from GE to head up a New Product Development Office in 1973 and later became vice president and chief scientist, described how Rockwell attempted to transfer North American technology to Rockwell products: "After the merger, the commercial businesses wrote product requirements and threw them over the wall to North American engineers, where defense planners evaluated them. Most failed." By 1973, John Moore, senior manager in the electronics-producing Autonetics division, is said to have remarked: "There is more technology in the wastebaskets of North American than the combined company could use in a century." Rather than reflecting Al Rockwell's original optimism, however, Moore was noting in discouragement that the capability to absorb technology is just as important as the capability to create it.

The biggest effort in technology transfer in the automotive field was antiskid truck brakes, which were mandated by government safety regulations in 1973. At the time North American had experience in antiskid aircraft brakes, while Rockwell had 100 percent of the front axle truck brake market. Bill Panny, a senior executive with Rockwell, moved to the West Coast for six months with the North American engineering team: "I dinked around but got nowhere." Finally, Al Rockwell said, "Move engineering to Detroit." They concluded that the only successful transfers were through the movement of people.

Even this move was unsuccessful, at least in the short run. Aircraft

---

[1] Interviews (LMB) with William Panny, a senior executive of Rockwell International until he moved to Bendix Corporation in 1977; R. Foxen, who came to be senior vice president for strategic management of Rockwell; and Peter Cannon, who was Rockwell's vice president and chief scientist before his retirement in 1988.

and truck brake requirements are quite different, and costs for the truck brakes developed with aircraft technology were too high and misestimated. In any case, the government canceled the safety requirement. But the electronics engineers who were moved to Detroit did, after a period of time, successfully inject the new ideas of digital electronics into the Rockwell auto parts mechanical engineering culture.

With a change in corporate management in 1978–1979, Rockwell (by now called Rockwell International) was revitalized. Today it is primarily in the applied electronics business, with revenue in recent years roughly balanced between commercial and government markets. The Autonetics division, in particular, has become more adept at applying technology to its two markets. Design teams for both defense and commercial products reside on the Anaheim campus, but they keep their books separate. Rockwell International was the prime contractor on the Minuteman missile electronics, for which it developed a four-bit microprocessor. Management saw the commercial opportunity in calculators, establishing a marketing "skunk works" under a commercial manager, while the microprocessor continued to be produced by the defense electronics organization. In the 1980s, Rockwell was the prime source of microelectronics for both Apple computers and Casio consumer electronics, and it now builds advanced processors for commercial airline cockpits, among many other devices and systems.

Cannon says, "RI prefers not to speak of 'technology transfer' but rather 'accommodating technology flow.'" Every exchange of technology is viewed as a transaction between cooperating parties, both gaining some advantage. People are trained to work in both military and commercial cultures, and chief and deputy chief engineers are actively moved between the military and the commercial product organizations. Thus, the dreams of Rockwell and Atwood were ultimately realized, though only after many years.

---

industry, with $1.3 billion in military contracts in 1990 (not including NASA and FAA), the Federal Sector Division is dwarfed by IBM's 1990 total sales of $69 billion.[15] IBM has sustained this business unit over many years, even though, with its profit margin so much less than that of IBM's commercial sales, it was frequently a drag on IBM's overall profitability. Throughout the thirteen years IBM was subject to a Justice Department antitrust suit, it was not lost on Washington that IBM sustained vital elements of U.S. submarine sonar, military avionics, and space system computer capabilities (see Box 6-E).

[15]IBM 1990 sales from *Forbes*, April 29, 1991, p. 164.

BOX 6-E: IBM'S FEDERAL SECTOR DIVISION—CORPORATE LINKAGE TO FEDERAL INTERESTS

In the late 1970s, the chief scientist of IBM asked Frank T. Cary, then chief executive officer, "Why does IBM operate the Federal Systems Division? Since the government limits profit margins on contracts to levels substantially below those enjoyed by IBM from its commercial computer business, the FSD lowers the profitability of the company and makes it harder to satisfy expectations of financial analysts." "Because IBM couldn't *not* have an FSD," was the answer.[1] Cary went on to explain that IBM could not deny its own government access to the company's massive R&D capabilities for meeting national security needs.[2] But he held the division's head-count to 10,000 people to limit negative profitability impact. FSD was then, and is still, the source of only 2–4 percent of IBM annual revenue, and accounted for a similar fraction of the workforce. When asked about his expectations for spin-off technology of use to the commercial divisions, Cary replied that his expectations were not great, but it was certainly appropriate for his chief scientist to look for such opportunities and try to realize them.

In that interview, Cary did not suggest that IBM ever expected a quid pro quo from government, other than to be paid for the contracts FSD might win. However, the Justice Department's antitrust suit against IBM was still grinding on, exacting a high price from IBM and the government alike. IBM technology, like that of Bell Telephone Laboratories, was a recognized national asset, even if that fact would not determine the outcome of an antitrust case.

The earlier history of IBM's involvement with the Department of Defense is, of course, quite relevant, since IBM was one of a half dozen firms to explore the digital computer in collaboration with government as computers emerged from World War II.[3] The critical government contributions were advanced applications: the SAGE early warning system and the HARVEST computer for the intelligence community were two examples. But the commercial computer industry was weaned very fast from government support in the late 1950s, when IBM's revenues grew at 24 percent annually.[4] With the introduction of the IBM System/

[1] Discussion with IBM Chairman Frank T. Cary, conducted by one of the authors (LMB), then IBM chief scientist, in 1978. IBM's Federal Sector Division had earlier been called the Federal Systems Division.

[2] IBM's R&D, $4.4 billion in 1988 (Appendix 6-A), accounted for fully 7.3 percent of all U.S. industrially funded R&D in that year.

[3] See Chapter 3, Box 3-B. Richard Nelson, *Government and Technical Progress* (New York: Basic Books, 1982); Kenneth Flamm, *Targeting the Computer: Government Support and International Competition* (Washington, DC: Brookings Institution, 1987).

[4] Emerson Pugh, Lyle R. Johnson, and John H. Palmer, *IBM's 360 and Early 370 Systems* (Cambridge, MA: MIT Press, 1991), p. 647.

360 in 1964, the computer market took off and the industry experienced annual growth rates of 15 percent and higher year after year. By 1966, the government share of the installed U.S. base of computers had fallen to only 10 percent.[5] In IBM's view, the company had a substantial historical debt to repay.

Finally, there is the question of ongoing technology relationships. During the 1980s, IBM's FSD participated in the Defense Department's VHSIC program, intended to accelerate the use of high-density VLSI chips in the military. This was initially a story of "spin-*on*," as IBM made available for its VHSIC work the integrated circuits designed for its commercial applications. Later, IBM's commercial chip factories benefited from early production learning on the FSD production line.

The most obvious example of product spinoff was the IBM 3838, a front-end array processor to help the oil exploration industry analyze seismic data. This product was a direct derivative of a sonar processor designed for Navy submarines. While the commercial version got off to a slow start (in part because the algorithms were hard for users to alter—a characteristic the Navy welcomed), demand kept rising even as IBM considered discontinuing production. The 3838 (like many spinoff products) was not a big financial success, but it served as the progenitor of the array processors offered as part of the System 3090 Vector system, IBM's successful offering for the commercial supercomputer market.

In the 1980s, IBM enjoyed an important spinoff of technological capability from an unexpected quarter. For many years, FSD had taken on very large and difficult software jobs for NASA and the military. In the 1980s, IBM's commercial customers—international banks and insurance companies—wanted to rationalize their multiple, incompatible international networks, interconnecting large numbers of computers from many different manufacturers. IBM first turned to FSD, which set up a special team of 200 experienced military systems engineers to work on a major project for Bank of America. In time the team expanded into a division; it trained many people who went back to their IBM subsidiaries around the world. Finally this group joined with FSD to form the IBM Systems Integration Division, with two parts: one serving government customers as before, the other supporting large international commercial customers. Thus after many years of looking for spinoff from defense hardware, IBM found the greatest value in the experience of highly trained people with a very special technical skill: systems integration of global computer and telecommunications networks.

---

[5] Flamm, *Targeting the Computer*, p. 107.

Organizing the Firm: Technology Sharing as a Design Criterion

However attractive a centrally managed, top-down implementation of corporate technology strategy might seem for such purposes as promoting military/commercial technology sharing, most firms organize instead to achieve accountability for meeting the business plan. Such an approach requires delegating enough freedom of action to the managers of profit centers to make the necessary tactical decisions; thus operating divisions are likely to have considerable autonomy.

Motivations for keeping military divisions at arm's length do not apply to most commercial divisions. Nevertheless, commercial divisions can also find themselves relatively isolated. The most common complaint of senior technical managers of large high-tech firms is the reluctance of the engineers in one division to accept advice or criticism from those in another, to pay attention to relevant work outside the division, or to share technical assets with other divisions of the firm. Even when engineers are eager to cooperate across organizational lines, technology transfer is slow and difficult. It is no surprise to find widespread skepticism in American industry that military technology can have a strongly stimulating effect on commercial competitiveness.

It is a cardinal principle of technical management that innovators must not only accept challenging goals, but be free to solve problems as they arise. Thus line managers are given freedom to select technical strategies and to change them on the fly. Technology is assumed to be too complicated and too situation-specific to be dictated from above. This view was clearly expressed by George Sarney, senior vice president for strategic planning at Raytheon.[16]

> Decentralized decision making at the lowest level is the best way to manage R&D, since it is most responsive to the customer. Centralized R&D is too slow; ideas are not readily accepted from above. A flat organizational structure creates a customer-driven, market-pull form of R&D. Technology transfer initiated from above becomes technology push, which is rarely effective in commercialization of products.

Technology sharing—or internal technology transfer—is a management challenge that must be taken seriously if much success is to be expected. But not all managers would agree with Sarney. John T.

---

[16] Interview with Craig Stetson, of Harvard University's John F. Kennedy School of Government, August 10, 1989.

Hartley, Jr., Harris Corporation's president in 1980, said,

> Technology transfer has to be a top-down process, as it is at Harris. Unless you tell the managers of your businesses that part of their promotions, salaries, and punishments will depend on technology transfer, nothing will happen. Technology transfer is an unnatural process. Once you have divisions, you have barriers—people who follow their own interests.[17]

Thomas Watson, Jr., retired chairman and CEO of IBM, notes in his memoirs that his father, IBM's founder, refused to permit organization charts to be drawn, fearing they would be used to protect managers from accountability for the success of the IBM company as a whole. It was not until the explosive growth of IBM under Tom Watson, Jr. that it became necessary to build more structure and decentralization into the company.[18]

As noted in the discussion of product spinoff in Chapter 3, there is nothing automatic about the process of moving technology across divisions. Technology sharing in the modern firm takes place most directly in two situations: through long-term research to open up new opportunities, and through vertical integration—either real, with technology suppliers internal to the firm, or virtual, through alliances with vendors. We consider each in turn.

Corporate research.   When firms grow to sufficient size—usually about $1 billion in sales—they can afford a central or corporate research laboratory committed to supporting the operating divisions with research services.[19] These services include:

- exploration of new technical ideas that broaden strategic choices;
- testing the limits to evolutionary improvement of existing technologies, and thus guiding technology strategy;
- providing a window on the world of science and engineering;
- creating conditions for collaboration with universities, national

[17]Robert J. Flaherty, "Harris Corporation's Remarkable Metamorphosis," *Forbes,* May 26, 1980, p. 45.

[18]Thomas J. Watson, Jr. and Peter Petre, *Father, Son & Co.: My Life at IBM and Beyond* (New York: Bantam Books, 1990), p. 151.

[19]The figure of $1 billion is based on four assumptions: (1) very few firms spend more than 10 percent of sales on R&D; (2) few spend more than 10 percent of development expense (committed to planned revenue) on research (an expense justified only by expanded future options); (3) the minimum critical mass for a corporate research laboratory is about 50 people; and (4) research workers with their support each cost about $200,000 per year. Thus one may expect to find the conventional corporate research division only in a subset of the *Fortune* 500 companies. Exceptions, of course, are firms that sell research services: Bolt Beranek and Newman, Arthur D. Little, and Battelle Memorial Institute, for example.

laboratories, and the scientists of other firms to enhance access to new technical ideas and the capacity to absorb them; and

- providing research support for solving unanticipated technical problems that arise in design, development, or production.

These services must be shared across the corporation because of the high entry cost of such a capability, the fungibility of the skills of scientists and research engineers, and the difficulty of measuring concrete returns to the firm. Thus the research function fits all the conditions of dual use (i.e., multiple use for all lines of business), subject only to access restrictions imposed by military customers.

Firms with strong corporate research arms are, of course, in an excellent position to profit from government funding of universities or multipurpose federal labs in science and engineering. Federal technology transfer legislation encourages government laboratories to make unclassified technology available to firms (see Chapter 3), and while the rhetoric emphasizes small business, it is the large multinational corporation with a superb central research laboratory—IBM, AT&T, or Du Pont, for example—that is best positioned to take full advantage. The corporate laboratory is, therefore, an effective diffusion mechanism for U.S. industry. It reaches few firms, but many people and a significant fraction of all manufacturing value-added.

Technology sharing through vertical integration.   IBM Corporation operates the largest microelectronics production company in the world, in terms of dollar volume. Called the IBM Technology Group, it manufactures only for internal use and does not compete with the merchant semiconductor industry. Most other U.S. computer manufacturers buy their integrated circuits from U.S. and foreign semiconductor manufacturers, and may enjoy special relationships with Motorola or Intel Corporation for the collaborative design and production of chips that embody proprietary architectures. In either case, whether vertical integration is achieved internally or externally, the technology supplier may serve all units of the firm. Chips are one example; others might be the design and production of special manufacturing tools, or software. At Hughes, much of the design and production of solar arrays for electric power and traveling wave tubes for signal amplifiers is common to civilian and military businesses.

The common internal supplier of technology may be a unit established with military support. Raytheon, for example, was not one of the original VHSIC contractors, but subsequently was encouraged by the Pentagon to establish a qualified VHSIC facility for manufacturing submicron custom chips. It also has a defense-supported laboratory specializing in gallium arsenide components for extremely high-frequency devices and optical communications. In both cases, the corporate capability provides support throughout the firm.

ORGANIZING FOR DUAL USE

Many firms have weak incentives to work hard at internal technology sharing. Some have little interest in sales to the government, or even actively avoid it. Others, like IBM, have such large civil sales that even a substantial defense division represents a modest fraction of the firm's technical resources. However, Raytheon and Rockwell are examples of firms that are comfortable running two loosely coupled lines of business, one in civil markets and the other in defense. Their managements are selective in their attempts to leverage technology across this market boundary, but they have the experience to do so when it makes sense.

With continuing decline in defense relative to the nation's overall technological effort, the DoD will be increasingly dependent on commercial sources of technology and of innovative and productive capacity. Even in the aerospace industry, which has the largest concentration of government demand at the industrywide level (if not the individual sectors analyzed in this chapter), worldwide airline purchases substantially exceed those of the military. In the case of transport aircraft and the engines that propel them, and in satellite communications as well, leading firms (notably Boeing in aircraft, GE and Pratt & Whitney in engines, Hughes in satellites) have been able to specialize their design, marketing, and management for each market while leveraging a substantially common technology base. Other companies—Rockwell, Harris, GM, and Ford—have deliberately made acquisitions of defense or space firms in order to gain competence in new technologies. Middle-sized, technology-defined companies such as Lord have less choice. With business strategy based on a technological niche, growth depends on broadening the markets they serve—by including defense, for instance.

We have seen that the defense-intensive industrial sectors represent a significant, if minority, share of U.S. manufacturing; they are at the high-tech end of the spectrum, driven by government R&D investments that are 3 to 4 times as great (as a fraction of sales) as in civil markets. Second-tier industrial firms play a major role in the diffusion of technology in the U.S. economy, both from high-risk federal programs into civil applications, and from leading civil industries, such as computers, into military systems. U.S. policy, however, has given much more emphasis to mission projects than to diffusion. Second-tier firms have remained largely invisible in Washington debates about competitiveness, eclipsed by both small business and large.

If government expects the technology it generates in pursuit of its missions to create competitive advantages in civil markets, it must pursue active policies to promote technology diffusion. These policies should reflect a realistic understanding of how firms are organized and managed. This understanding must begin with basic facts:

- Defense's share of national economic activity and of manufacturing—5.5 percent of GNP and 8.7 percent of durable goods—is substantially smaller than its 31 percent share of the nation's R&D.
- At the corporate level and the industry-sector level, defense work is mixed with much larger amounts of commercial business. More than four-fifths of revenues in the 50 largest defense manufacturing sectors come from civil sales, and only 9 percent of the business of the largest defense firms consists of defense prime contracts. Taken in aggregate, the defense industry is embedded within a much larger civilian economy. Nevertheless, most defense business is isolated in separate corporate divisions, and many firms have weak incentives to work at technology sharing. Subcontractors and second-tier firms may represent exceptions, but they have been little studied. Corporate research organizations are important instruments for sharing technology within a firm and accessing it externally, but few firms are large enough to afford them.
- As a result, there is nothing automatic about leveraging defense technology investments to benefit commercial production, or vice versa. Government must devise new policies if it wishes to compensate for cuts in defense procurement, or if it wishes the satisfaction of its own technology needs to generate competitive advantage in civil markets.

APPENDIX 6-A:
THE TOP DEFENSE CONTRACTORS

The firms listed in Table 6-A-1 were drawn from the 100 largest recipients of DoD prime contract awards in 1988. From this list the following were deleted:

- Firms that were not among the 100 largest prime contractors in at least two of the previous four years.
- Nonprofit institutions, privately held firms, joint ventures, and foreign-owned firms. These were omitted because of the difficulty of obtaining meaningful sales and civil R&D data.

The 67 firms remaining performed 60 percent by value of all DoD's prime contract awards over $25,000 in 1988 (66 percent for the top 100 firms). For each firm listed in Table 6-A-1, the following data is presented:

Primary line of business.   Dun's Marketing Services classifies companies into a primary line and up to five secondary lines of business. We have used the primary line of business in the 1989 *Million Dollar Directory*, which should reflect 1988 conditions. A few cases required checking press accounts and annual reports. Subtotals for firms in the same primary line of business are shown at the bottom of the table. These subtotals assign the entire firm to the category representing its primary line of business.

Prime contracts.   Prime contract awards (PCAs) and RDT&E awards were found to be highly variable between years. To adjust for these fluctuations, a five-year average was constructed for each firm's DoD awards. In the case of PCAs, this was done by compiling awards for the preceding four years, correcting them for inflation using the Federal Defense Durables Deflator,[1] and averaging. When a firm was not in the top 100 contractors for all five years, the average was taken only over the years in which the firm appeared in the sample. This has the effect of biasing the PCA estimates slightly upward. In the case of mergers, divestitures, and acquisitions, calculations were made as if the firm had been organized throughout the period as it was at the end of 1988—for example, both Burroughs' and Sperry's prime contracts in 1984 were used in constructing the five-year average for Unisys.

RDT&E.   The RDT&E numbers were constructed by a similar five-year averaging process, using an R&D deflator rather than the Federal Defense Durables Deflator.[2] Since RDT&E awards as small as $1 million are reported each year in DoD's listing of the top 500 RDT&E

---

[1] Developed by the Commerce Department, Bureau of Economic Affairs.

[2] The R&D deflator used was developed by Bronwyn Hall of the University of California, Berkeley.

contractors, each year in which no data were available was counted as zero.

Civil sales. While defense sales are highly variable, civil sales are generally much more stable year to year. Since reliable sales figures for individual corporate divisions are hard to assemble, we felt that the inaccuracy that would be introduced by trying to correct for changing corporate structure would outweigh the gains of constructing five-year averages. Thus 1988 civil sales were estimated by subtracting 1988 prime contract awards from 1988 total sales. Only in two cases was this obviously problematic: Avondale Industries sales in 1988 were essentially equal to their prime contracts, yielding near-zero civil sales; and Singer Corporation—which suffered an 84 percent drop in sales in 1988 as it divested most of its businesses to pay off debt—shows negative civil sales by this approach.

Own R&D. The value of the firm's internally funded R&D in 1988 comes from 10-K filings with the Securities and Exchange Commission. Government-sponsored research should not be included in this figure, although independent research and development (IR&D) expenses— which are considered corporate R&D even though the government reimburses part of their cost—will appear.

A few firms did not report civil R&D data. Most, but not all, were companies (such as Pan Am) that would not be expected to have performed much R&D. These firms are included in the subtotals where data are available, such as for prime contracts. However, they have been excluded from calculations of ratios involving civil R&D. Similarly, Singer Corporation, because it shows negative civil sales, has been excluded from calculations involving civil sales.

TABLE 6-A-1:
Major Defense Contractors Ranked by Prime Contract Awards (PCA)
(smoothed over 5 years)

	Industry	PCA[a] ($ mil)	RDT&E[a] ($ mil)	Total Sales ($ mil)	PCA in 1988 ($ mil)	Rank in PCA for 1988	Civil Sales ($ mil)	Own R&D ($ mil)	PCA as % of Total Sales (%)	RDT&E as % of Total R&D (%)	RDT&E as % of PCA (%)	Own R&D as % of Civil Sales (%)
McDonnell Douglas	Aero	$8,030	$1,413	$15,069	$8,003	1	$7,066	$610	53%	70%	18%	9%
General Dynamics	Aero	7,226	484	9,551	6,522	2	3,029	413	70	54	7	14
General Electric	Gen	6,449	1,073	49,414	5,701	3	43,713	1,155	13	48	17	3
Lockheed	Aero	5,224	1,578	10,590	3,538	8	7,052	536	43	75	30	8
Rockwell International	Aero	4,683	791	11,946	2,184	14	9,762	431	32	65	17	4
General Motors	Auto	4,579	559	121,085	3,550	7	117,535	4,754	4	11	12	4
Boeing	Aero	4,178	1,219	16,962	3,018	10	13,944	751	23	62	29	5
Raytheon	Comm	3,717	542	8,192	4,055	5	4,137	271	47	67	15	7
United Technologies	Aero	3,665	379	18,283	3,508	9	14,775	932	20	29	10	6
Martin Marietta	Aero	3,163	1,417	5,727	3,715	6	2,012	195	61	88	45	10
Grumman	Aero	2,965	706	3,591	2,848	11	743	18	80	98	24	2
Litton Industries	Comp	2,112	144	4,864	2,561	12	2,303	80	48	64	7	3

Company	Category											
Westinghouse Electric	Gen	1,956	469	12,500	2,185	13	10,315	190	16	71	24	2
Tenneco	Oil	1,952	2	13,234	5,058	4	8,176	N/A	19	N/A	0	N/A
UNISYS	Comp	1,932	362	9,902	1,380	16	8,522	713	18	34	19	8
Westmark Systems (civil data for Textron only)	Comm	1,904	249	7,279	1,468	15	5,811	168	25	60	13	3
Honeywell	Instr	1,753	306	7,148	1,365	17	5,783	323	23	49	17	6
IBM	Comp	1,534	441	59,681	1,065	20	58,616	4,419	3	9	29	8
LTV	Misc	1,439	237	7,325	942	21	6,383	23	18	91	16	0
Texas Instruments	Comp	1,272	147	6,295	1,232	19	5,063	494	20	23	12	10
TRW	Comp	1,135	621	6,982	1,250	18	5,732	225	17	73	55	4
ITT	Comm	1,080	169	19,355	769	25	18,586	516	5	25	16	3
Allied-Signal	Aero	994	25	11,909	711	26	11,198	415	8	6	3	4
FMC	Chem	923	78	3,287	862	22	2,425	144	28	35	8	6
Northrop	Aero	916	162	5,797	533	32	5,264	206	15	44	18	4
GTE	Comm	881	245	16,460	423	38	16,037	297	5	45	28	2
Ford Motor	Auto	870	287	92,446	791	23	91,655	2,930	1	9	33	3
Eaton	Auto	797	67	3,468	218	68	3,250	122	20	36	8	4
Singer	Comm	781	155	302	785	24	>0	5	N/A	97	20	N/A

TABLE 6-A-1
(continued):

	Industry	PCA[a] ($ mil)	RDT&E[a] ($ mil)	Total Sales ($ mil)	PCA in 1988 ($ mil)	Rank in PCA for 1988	Civil Sales ($ mil)	Own R&D ($ mil)	PCA as % of Total Sales (%)	RDT&E as % of Total R&D (%)	RDT&E as % of PCA (%)	Own R&D as % of Civil Sales (%)
AT&T	Comm	741	115	51,974	565	30	51,409	2,572	1	4	16	5
Loral	Instr	735	60	1,187	494	35	693	67	51	47	8	10
Gencorp	Auto	707	213	1,891	639	28	1,252	33	36	87	30	3
Hercules	Chem	536	94	2,802	499	33	2,303	74	19	56	18	3
Teledyne	Aero	504	105	4,523	469	36	4,054	79	11	57	21	2
Chevron	Oil	504	0	25,196	328	54	24,868	208	2	0	0	1
Motorola	Comm	480	105	8,250	381	44	7,869	665	6	14	22	8
Harris	Comp	456	116	2,063	371	45	1,692	117	21	50	25	7
Exxon	Oil	449	0	79,557	282	59	79,275	551	1	0	0	1
Pan Am	Misc	441	4	3,569	357	49	3,212	N/A	12	N/A	1	N/A
Morton Thiokol	Aero	438	80	2,316	392	42	1,924	45	19	64	18	2
Atlantic Richfield	Oil	397	0	17,626	303	56	17,323	114	2	0	0	1
Harsco	Auto	377	26	1,279	496	34	783	12	33	69	7	1

Company	Category												
Emerson Electric	Gen	325	50	6,651	253	63	6,398	204	5	20	15	3
Oshkosh Truck	Auto	322	0	353	260	62	93	7	78	0	0	8
Computer Sciences Corp	Comp	317	67	1,304	368	46	936	N/A	25	N/A	21	N/A
Morrison Knudsen	Const	307	0	1,909	170	82	1,739	N/A	15	N/A	0	N/A
Chrysler	Auto	303	23	35,473	211	70	35,262	881	1	3	8	2
Penn Central	Misc	300	20	1,547	329	53	1,218	10	20	67	7	1
Avondale Industries	Misc	294	0	570	580	29	0	N/A	100	N/A	0	N/A
E-Systems	Comm	282	28	1,439	262	60	1,177	N/A	19	N/A	10	N/A
Control Data	Comp	275	23	3,628	315	55	3,313	336	8	6	8	10
Dyncorp	Const	268	4	556	421	39	135	N/A	67	N/A	2	N/A
Coastal Corp	Oil	266	0	8,186	213	69	7,973	N/A	3	N/A	0	N/A
Science Applications Intl	Misc	265	102	863	344	51	519	7	34	94	38	1
Mobil	Oil	245	0	48,198	302	57	47,896	230	1	0	0	0
Olin	Chem	243	20	2,308	331	52	1,977	58	11	26	8	3
Sequa	Chem	223	24	1,713	220	67	1,493	17	13	58	11	1
Amoco	Oil	202	0	21,150	188	75	20,962	272	1	0	0	1
Zenith Electronics	Misc	196	0	2,686	248	65	2,438	101	7	0	0	4
Texaco	Oil	190	0	33,544	131	98	33,413	170	1	0	0	1

TABLE 6-A-1
(continued):

	Industry	PCA^a ($ mil)	RDT&E^a ($ mil)	Total Sales ($ mil)	PCA in 1988 ($ mil)	Rank in PCA for 1988	Civil Sales ($ mil)	Own R&D ($ mil)	PCA as % of Total Sales (%)	RDT&E as % of Total R&D (%)	RDT&E as % of PCA (%)	Own R&D as % of Civil Sales (%)
Hewlett-Packard	Comp	189	0	9,831	149	90	9,682	1,019	2	0	0	11
Figgie International	Gen	183	111	1,200	129	100	1,071	15	15	88	61	1
Eastman Kodak	Misc	179	2	17,034	186	76	16,848	1,147	1	0	1	7
United Industrial Corp	Gen	174	30	315	165	85	150	7	54	81	17	5
Kaman	Aero	161	53	767	184	79	583	8	22	87	33	1
Digital Equipment Corp	Comp	151	6	11,475	175	81	11,300	1,307	1	0	4	12
Sundstrand	Aero	149	2	1,477	185	78	1,292	115	10	2	1	9

Subtotals	Number of firms	PCA^a ($ mil)	RDT&E^a ($ mil)	Total Sales ($ mil)	PCA in 1988 ($ mil)	Rank in 1988	Civil Sales ($ mil)	Own R&D ($ mil)	DoD Sales/ Total Sales	RDTE/ Total R&D	RDTE/ PCA	Own R&D/ Civil Sales
Aero: Aerospace SIC 372 and 376	14	42,296	8,414	118,509	35,810		82,699	4,753	34%	64%	20%	6%
Auto: Automotive SIC 3011 and 371	7	7,955	1,175	255,995	6,165		249,830	8,738	3	12	15	3
Chem: Chemicals SIC 28	4	1,925	216	10,110	1,912		8,198	293	19	42	11	4

Comm: Communications SIC 3666 and 4811	8 (6)[b]	8,708	113,251	1,608	9,866	4,494	105,026	8[c]	24[b]	16	4[b]
Comp: Computers, Electronics, and Software SIC 357, 367, and 737	10 (9)[d]	8,866	116,025	1,927	9,373	8,709	107,159	8	18[d]	21	8[d]
Const: Construction Contractors SIC 16, 17, and 8311	2 (0)[d]	591	2,465	4	575	0	1,874	23	N/A[d]	1	N/A[d]
Gen: General Industrial Equipment SIC 351-356, 358-359, and 361-362	5	8,433	70,080	1,733	9,087	1,572	61,647	13	52	19	3
Instr: Instruments SIC 381 and 382	2	1,859	8,335	366	2,488	391	6,476	28	48	15	6
Oil: Oil and Natural Gas SIC 1311, 2911, 492, and 5172	8 (6)[d]	6,805	246,692	2	4,205	1,545	239,887	2	0[d]	0	1[d]
Misc: Miscellaneous Includes 7 SIC categories, each represented by only one firm	7 (5)[d]	2,986	33,593	365	3,114	1,287	30,617	9	22[d]	12	5[d]
TOTAL From All Sectors	67 (59)[d]	82,135	975,056	15,810	90,884	31,782	893,413	9[b]	33[d]	17	4[d]

Notes and Symbols:
N/A—civil R&D data not available
SIC—Standard Industrial Classification (see Appendix 6-B)
[a]Smoothed over five years; see text
[b]Excluding Singer Corporation and E-Systems
[c]Excluding Singer Corporation
[d]Excluding firms reporting no civil R&D data

197

APPENDIX 6-B:
THE 50 INDUSTRIAL SECTORS WITH THE MOST DEFENSE
SHIPMENTS

This appendix presents data from the Commerce and Defense Departments on the 50 four-digit SIC (Standard Industrial Classification) sectors with the highest defense shipments. The sectors listed are those that had the highest "total direct and indirect DoD expenditures" in 1987, according to the Defense Economic Impact Modeling System (DEIMS), including both direct shipments of goods and services to DoD and indirect shipments to DoD contractors for use in goods or services that they in turn sell to DoD.

Within the manufacturing sector of the economy, the Standard Industrial Classification System distinguishes 20 major categories, identifying each with a two-digit number: Industrial and Commercial Machinery and Computer Equipment, for example, is SIC 35; Transportation Equipment is SIC 37. Each two-digit category is subdivided into smaller categories identified with three-digit codes; and each three-digit category into four-digit subcategories. The entire manufacturing sector is covered by some 450 four-digit SIC codes. Note that although the SIC definitions were altered in 1987, these tables use the 1972–1987 SIC structure.

For each of the 50 four-digit industrial sectors identified by DEIMS as having the most defense shipments in 1987, the tables in this appendix present total shipments and total value-added from the Commerce Department's *1987 Annual Survey of Manufacturers,* along with DEIMS estimates for total defense shipments.[1] (Value-added equals the value of shipped products, or shipments, minus the prices of the inputs used to make those products.) The ratio of value-added to shipments is shown for each sector.

Note that the total defense shipments of $165 billion given for the 50 industrial sectors listed here exceeds by more than a factor of two the $80 billion spent on defense procurement in 1987. How can this be? In adding up defense shipments from more than one industrial sector, many products are counted more than once. Intermediate goods produced in one sector that are later incorporated in products from another sector are counted when first produced and again when the

---

[1] DEIMS data are from Department of Defense, Washington Headquarters Services, Directorate for Information Operations and Reports, *Projected Defense Purchases: Detail by Industry and State, Calendar Years 1986 through 1991* (Washington, DC: Department of Defense).

systems into which they have been included are sold. Thus when Northrop buys avionics from IBM to build into a B-2 bomber, the value of the electronics will appear not only in IBM's shipments but also in Northrop's. Therefore, when electronics and aviation shipments are added, the value of the avionics package will be counted twice.[2]

To avoid this double-counting we must use *value-added* data, in which the money paid IBM for the avionics would be subtracted from the price of the bomber. (Using the value-added data, we see that these 50 sectors account for almost 40 percent of U.S. manufacturing value-added for 1987.) Unfortunately, we cannot determine value-added for defense and commercial production separately unless we assume that the ratio of value-added to shipments is the same for each.

The tables also present the ratio of DoD shipments to total shipments. In two cases, this ratio is above 100 percent. There are several possible explanations, which contribute to uncertainties for all of the sectors listed:

- DEIMS codes do not correspond exactly to SIC codes, and errors could be introduced in our procedure of allocating a DEIMS category's defense shipments among its component SIC codes in proportion to the total shipments of those SIC codes.
- Imported goods enter into the calculation of defense purchases but not the domestic shipments. However, it seems unlikely that this would be a very significant correction in sectors overwhelmingly dominated by defense.
- The DEIMS model may be overly optimistic in its estimates of military demand from various industrial sectors, inasmuch as it is used by DoD to demonstrate its importance in the economy.

Finally, for each sector we identify whether or not it is considered to be high-tech. The classification criteria used are based primarily on the technical skill mix in employment.[3] In all, 100 of the 451 SIC codes are identified as high-tech.

Note that the DEIMS data do not permit estimating what fraction of all defense production is performed by the 50 industrial sectors listed.

---

[2] Another reason why defense shipments exceed the procurement budget is that DoD buys many goods and services with funds other than those contained in the procurement budget category. For example, expendable supplies are bought largely with operations and maintenance funds, and buildings from the military construction account.

[3] Ann Markusen, Peter Hall, and Amy Glasmeier, *High-Tech America* (Boston: Allen and Unwin, 1986).

To calculate this we would need to divide the value-added for defense production in these 50 sectors by all defense value-added, which is not available from these data. But DEIMS does show that 61 percent of all direct defense purchases are accounted for by 10 industries alone, indicating a substantial concentration of DoD purchases among the top 50 industrial sectors.[4]

In the following tables we present the top 50 industrial sectors twice: once in order of DoD shipments, and again in order of DoD share of sector output.

[4] Economic Analysis Division, Office of the Director, Program Analysis and Evaluation; Office of the Secretary of Defense, "Defense Purchases: An Introduction to DEIMS," undated (probably 1984), p. 22.

TABLE 6-B-1:
*The 50 Largest Defense Industrial Sectors,
Sorted by DoD Shipments*

SIC	Industry	DoD Shipments '87 ($ mil)	Total Shipments '87 ($ mil)	DoD Share 1987	Value-Added '87 ($ mil)	Value-Added/ Shipments '87	High-Tech?
3662	Radio/TV communication equipment	$33,649	$55,009	65%	$36,432	66%	YES
3721	Aircraft	16,999	40,038	42	16,746	42	YES
3728	Aircraft equipment, n.e.c.	9,754	17,923	54	11,779	66	YES
3761	Guided missiles & space vehicles	9,018	21,566	42	15,000	70	YES
3731	Ship building & repairing	8,044	8,504	95	5,213	61	NO
3825	Instruments to measure electricity	6,617	7,718	86	5,026	65	YES
3724	Aircraft engines & engine parts	6,022	20,244	30	11,697	58	YES
3679	Electronic components, n.e.c.	5,234	22,503	30	12,280	54	YES
2911	Petroleum refining	5,223	118,507	4	14,128	12	YES
3573	Electronic computing equipment	5,155	59,195	10	31,287	46	YES
3312	Blast furnaces & steel mills	4,428	41,141	11	16,963	41	NO
3674	Semiconductors & related devices	4,161	19,578	21	13,085	67	YES
3483	Ammunition except small arms, n.e.c.	3,662	3,978	92	2,537	64	YES
3711	Motor vehicles and car bodies	2,817	133,343	2	36,066	27	NO

TABLE 6-B-1
(continued):

SIC	Industry	DoD Shipments '87 ($ mil)	Total Shipments '87 ($ mil)	DoD Share 1987	Value-Added '87 ($ mil)	Value-Added/ Shipments '87	High-Tech?
3079	Misc. plastic products	2,734	61,594	5	30,990	50	NO
2869	Ind. organic chemicals, n.e.c.	2,492	42,153	6	17,934	43	YES
3795	Tanks and tank components	2,320	2,522	92	969	38	YES
3489	Ordnance and accessories, n.e.c.	2,227	1,678	133	1,425	85	YES
3599	Machinery, except electrical, n.e.c.	2,052	15,581	14	10,765	68	NO
3714	Motor vehicle parts and accessories	2,014	60,873	3	25,957	43	NO
3811	Engineering and scientific instruments	1,652	3,973	42	2,590	63	YES
2821	Plastics material and resins	1,577	26,248	6	10,912	42	YES
2819	Ind. inorganic chemicals, n.e.c.	1,561	12,980	12	7,356	57	YES
3823	Process control instruments	1,493	4,788	31	3,200	67	YES
3832	Optical instruments and lenses	1,396	5,319	28	3,267	62	YES
3769	Space vehicle equipment, n.e.c.	1,393	1,142	122	838	73	YES
3621	Motors and generators	1,367	6,927	20	3,875	56	YES
3537	Industrial trucks and tractors#	1,162	2,392	50	934	37	YES

3764	Space propulsion units and parts	1,102	3,537	31	2,314	65	YES
3544	Special dies, tools, jigs, fixtures	1,095	7,598	14	5,348	70	YES
3334	Primary aluminum	1,091	5,017	22	1,904	38	NO
3511	Turbines and turbine generator sets	1,031	3,447	30	1,973	57	YES
3661	Telephone and telegraph apparatus	1,031	16,101	7	8,823	50	YES
3357	Nonferrous wire drawing and insulating	1,016	10,639	10	4,156	39	NO
3541	Machine tools, metal cutting type	1,016	3,248	31	1,702	52	YES
3678	Electronic connectors	904	3,910	23	2,450	63	YES
3841	Surgical and medical instruments	896	7,626	12	5,125	67	YES
3622	Industrial controls	879	5,150	17	3,127	60	YES
2621	Paper mills, except building paper	859	28,885	3	13,687	47	NO
3829	Measuring and controlling devices, n.e.c.	837	2,721	32	1,775	63	YES
3339	Primary nonferrous metals, n.e.c.	787	(D)	(D)	(D)	(D)	NO
3842	Surgical appliances and supplies	786	8,754	9	5,604	64	YES
3469	Metal stampings, n.e.c.	782	8,207	10	4,240	52	NO
3471	Plating and polishing	768	3,876	20	2,642	68	NO
3519	Internal combustion engines, n.e.c.	755	10,980	7	5,045	46	YES
3452	Bolts, nuts, rivets, and washers	754	5,074	15	2,990	59	NO

TABLE 6-B-1
(continued):

SIC	Industry	DoD Shipments '87 ($ mil)	Total Shipments '87 ($ mil)	DoD Share 1987	Value-Added '87 ($ mil)	Value-Added/ Shipments '87	High-Tech?
3353	Aluminum sheet, plate, and foil	715	9,497	8	1,790	19	NO
3531	Construction machinery#	681	12,735	5	5,757	45	YES
3861	Photographic equipment and supplies	673	19,505	3	13,059	67	YES
3822	Automatic controls	653	2,197	30	1,405	64	YES
	Total for 50 Industries	$165,335	$996,121	17%	$444,167	44%	36/50
	All 451 Manufacturing Sectors	N/A	$2,480,236	N/A	$1,166,555	47%	100/451

Symbols:
preliminary data; since revised by Census
n.e.c.   not elsewhere classified
(D)   not disclosed for confidentiality reasons
N/A   not available

204

TABLE 6-B-2:
The 50 Largest Defense Industrial Sectors,
Sorted by DoD Share of Sector Output

SIC	Industry	DoD Shipments '87 ($ mil)	Total Shipments '87 ($ mil)	DoD Share 1987	Value-Added '87 ($ mil)	Value-Added/ Shipments '87	High-Tech?
3489	Ordnance and accessories, n.e.c.	$2,227	$1,678	133%	$1,425	85%	YES
3769	Space vehicle equipment, n.e.c.	1,393	1,142	122	838	73	YES
3731	Ship building and repairing	8,044	8,504	95	5,213	61	NO
3483	Ammunition except small arms, n.e.c.	3,662	3,978	92	2,537	64	YES
3795	Tanks and tank components	2,320	2,522	92	969	38	YES
3825	Instruments to measure electricity	6,617	7,718	86	5,026	65	YES
3662	Radio/TV communication equipment	33,649	$55,009	61	36,432	66	YES
3728	Aircraft equipment, n.e.c.	9,754	17,923	54	11,779	66	YES
3537	Industrial trucks and tractors#	1,162	2,392	49	934	39	YES
3721	Aircraft	16,999	40,038	42	16,746	42	YES
3761	Guided missiles and space vehicles	9,018	21,566	42	15,000	70	YES
3811	Engineering and scientific instruments	1,652	3,973	42	2,590	65	YES
3541	Machine tools, metal cutting type	1,016	3,248	31	1,702	52	YES
3823	Process control instruments	1,493	4,788	31	3,200	67	YES

TABLE 6-B-2
(continued):

SIC	Industry	DoD Shipments '87 ($ mil)	Total Shipments '87 ($ mil)	DoD Share 1987	Value-Added '87 ($ mil)	Value-Added/ Shipments '87	High-Tech?
3764	Space propulsion units and parts	1,102	3,537	31	2,314	65	YES
3829	Measuring and controlling devices, n.e.c.	837	2,721	31	1,775	65	YES
3511	Turbines and turbine generator sets	1,031	3,447	30	1,973	57	YES
3724	Aircraft engines and engine parts	6,022	20,244	30	11,697	58	YES
3822	Automatic controls	653	2,197	30	1,405	64	YES
3832	Optical instruments and lenses	1,396	5,319	26	3,267	61	YES
3679	Electronic components, n.e.c.	5,234	22,503	23	12,280	55	YES
3678	Electronic connectors	904	3,910	23	2,450	63	YES
3334	Primary aluminum	1,091	5,017	22	1,904	38	NO
3674	Semiconductors and related devices	4,161	19,578	21	13,085	67	YES
3471	Plating and polishing	768	3,876	20	2,642	68	NO
3621	Motors and generators	1,367	6,927	20	3,875	56	YES
3622	Industrial controls	879	5,150	17	3,127	61	YES
3452	Bolts, nuts, rivets, and washers	754	5,074	15	2,990	59	NO

SIC	Industry					
3544	Special dies, tools, jigs, fixtures	1,095	7,598	14	5,348	70	YES
3599	Machinery, except electrical, n.e.c.	2,052	15,581	13	10,765	69	NO
2819	Ind. inorganic chemicals, n.e.c.	1,561	12,980	12	7,356	57	YES
3841	Surgical and medical instruments	896	7,626	12	5,125	67	YES
3312	Blast furnaces and steel mills	4,428	41,141	11	16,963	41	NO
3357	Nonferrous wire drawing and insulating	1,016	10,639	10	4,156	39	NO
3469	Metal stampings, n.e.c.	782	8,207	10	4,240	52	NO
3842	Surgical appliances and supplies	786	8,754	9	5,604	64	YES
3573	Electronic computing equipment	5,155	59,195	9	31,287	53	YES
3353	Aluminum sheet, plate, and foil	715	9,497	8	1,790	19	NO
3519	Internal combustion engines, n.e.c.	755	10,980	7	5,045	46	YES
3661	Telephone, telegraph apparatus	1,031	16,101	6	8,823	55	YES
2821	Plastics material and resins	1,577	26,248	6	10,912	42	YES
2869	Ind. organic chemicals, n.e.c.	2,492	42,153	6	17,934	43	YES
3531	Construction machinery#	681	12,735	5	5,757	45	YES
3079	Misc. plastic products	2,734	61,594	4	30,990	50	NO
2911	Petroleum refining	5,223	118,507	4	14,128	12	YES

TABLE 6-B-2
(continued):

SIC	Industry	DoD Shipments '87 ($ mil)	Total Shipments '87 ($ mil)	DoD Share 1987	Value-Added '87 ($ mil)	Value-Added/ Shipments '87	High-Tech?
3861	Photographic equipment and supplies	673	19,505	3	13,059	67	YES
3714	Motor vehicle parts and accessories	2,014	60,873	3	25,957	43	NO
2621	Paper mills, except building paper	859	28,885	3	13,687	47	NO
3711	Motor vehicles and car bodies	2,817	133,343	2	36,066	27	NO
3339	Primary nonferrous metals, n.e.c.	787	(D)	(D)	(D)	(D)	NO
	Total for 50 Industries	$165,335	$996,121	17%	$444,167	45%	36/50
	All 451 Manufacturing Sectors	N/A	$2,480,236	N/A	$1,166,555	47%	100/ 451

Symbols:
preliminary data, since revised by Census
n.e.c.   not elsewhere classified
(D)   not disclosed for confidentiality reasons
N/A   not available

208

7

Cross-National Comparisons
J. Nicholas Ziegler*

Interest in the relationship between military and commercial technologies is by no means limited to the United States. All countries seek to balance military and economic priorities, but they do so through a diverse assortment of institutional mechanisms. The dual-use issues analyzed in this volume therefore illuminate one of the central questions in comparative politics: the relationship between military power and economic prosperity.

This chapter surveys institutions for allocating R&D resources to civil and military purposes in four countries—France, Germany, Sweden, and Japan. These countries illustrate the diversity of arrangements found in the industrial democracies for linking commercial and military objectives in technology development.

Prospects for dual use are also of vital concern outside the advanced industrial world. Efforts at reform in what was the Soviet Union hinge partly on the ability to "convert" a huge defense sector to civil production. Among the newly industrializing countries, Brazil and Indonesia have attempted to build indigenous high-tech industries from heavily subsidized defense sectors. Such ambitions reflect a great deal of faith

*Assistant Professor, Sloan School of Management, MIT. The author gratefully acknowledges the comments of the authors of this volume, as well as Michael Chinworth, Ethan Kapstein, Richard Samuels, Todd Watkins, and Steve Weber. Research for this chapter was supported by the Center for Science and International Affairs, Harvard University, and the Sloan School of Management, MIT. Statements of fact and interpretation reflect the views of the author.

in the effectiveness of dual-use relationships. It is important to ask whether the experience of the industrial democracies lends support to such a development strategy.

As Table 7-1 indicates, the four countries compared here differ widely. France spends almost as much as the United States on defense R&D as a fraction of total R&D (though much less in absolute terms). As a result, France has sought to use defense as a foundation for commercial competitiveness. Germany and Japan exhibit world-class performance in technology-intensive sectors with much lower levels of spending on defense R&D. Government agencies in both countries use a variety of other mechanisms for supporting technology development. Sweden is an intermediate case. It combines a considerable commitment to self-sufficiency in defense with remarkable success in commercial fields.

This chapter focuses on the links between public policy and commercial strategies, treating incentives and barriers to dual use at the level of the firm through abbreviated cases. While it is natural to seek lessons that can be applied to the U.S. case, it is necessary to recall that institutional arrangements for R&D are informed by political traditions and constrained by industrial profiles that are specific to each country's historical experience. In particular, the much greater levels of defense spending in the United States make its situation fundamentally different from the situation in any of these four countries.

Like other chapters in this volume, this one suggests that the concept of spinoff is far too narrow to capture the many possible ways of managing the links between military and commercial technology development. As indicated in Table 7-1, a range of factors affects a country's ability to manage the development of dual-use technologies. First, different countries have different degrees of commitment to military self-sufficiency, and, consequently, to the military side of the dual-use relationship. Second, the state may play a more or less activist role in guiding economic development. Third, the portion of public resources committed to defense R&D may vary considerably. Finally, defense is stuctured differently among countries, with important implications for dependence on export markets.

FRANCE

Of the four countries examined in this chapter, France has most assiduously promoted economic development as an ingredient in military and political power. As Jean-Baptiste Colbert, the royal comptroller under Louis XIV, said, "Trade is the source of finance and finance is the

TABLE 7-1:
Dual Use in Five Countries (1988)

	Commitment to Self-Sufficiency in National Security	State's Role in Economic Development	R&D as Percentage of GDP	R&D Financed by Government (%)	Defense R&D as Percentage of Government R&D	Defense Budget ($ bil)	Defense R&D ($ bil)	Defense Industrial Structure	Export Dependence of Defense Industry
France	High	Active	2.3%	50.6%	37.4%	$ 31.9	$ 3.9	Limited competition among a few state-oriented firms	High
Sweden	High	Consensual	2.9	36.9[a]	24.1	4.9	0.4	Limited competition among commercially oriented firms	Moderate
FRG	Low	Regulatory	2.8	33.9	12.4	29.2	1.1	Moderate competition among commercially oriented firms	Low
Japan	Low	Promotional	2.6	21.5[a]	3.5[b]	28.4	0.4[b]	Managed competition among commercially oriented firms	Very Low
United States	High	Limited	2.7	48.8	67.8	292.0	40.1[c]	Wide competition among range of firms	Low

Sources: R&D expenditures and appropriations from OECD, Main Science and Technology Indicators (1990/Number 1) with adjustments as noted in Table 4-1, Chapter 4 of this volume. Defense budget figures (including currency conversions) from International Institute for Strategic Studies, The Military Balance (various years).
[a] 1987 figures.
[b] Based on trade press sources and unpublished OECD estimates for 1987.
[c] Based on OECD, unadjusted for IR&D spending.

vital nerve of war."[1] The establishment of the Académie des Sciences in 1666 testified to the view that scientific knowledge itself should be cultivated in the service of state power.

Much the same principles of state-led growth and state-supported science remained prominent in the post-World War II period. The links between science and autonomy were clearest in France's program for developing an independent nuclear capability. In the past two decades, state support for defense-related industries has remained high as French governments tried various strategies for aiding French firms in international competition.

Commitment to industrial and military autonomy has fostered a dual industrial structure. Sectors such as steel, oil, electronics, and aerospace are dominated by large, technologically advanced firms that receive a great deal of aid from the state, particularly from the Ministry of Defense. These state-oriented firms tend to avoid competition by focusing within their chosen segments. The best known—Dassault in jet fighters, SNECMA in jet engines, Thomson-CSF in defense electronics, and Aerospatiale in military helicopters—have achieved nearly unrivaled market power within France. In contrast to these industries, sectors like textiles or mechanical engineering are populated by small, comparatively backward firms that have difficulty participating in state-sponsored efforts for technological advance.

Administrative Resources

The state's long-standing role means that publicly funded facilities for research and development have evolved on a "mission-oriented" basis toward specific purposes designated by the central state administration.[2] The Centre National de la Recherche Scientifique (CNRS), founded in 1939, insulates basic research to some degree from political interference.[3] Applied research takes place in a variety of specialized installations such as the Atomic Energy Commissariat (Commissariat à l'Energie Atomique, CEA), the National Center for Space Studies

---

[1] Quoted in Edward Mead Earle, "Adam Smith, Alexander Hamilton, Friedrich List: The Economic Foundations of Military Power," in Peter Paret, ed., *Makers of Modern Strategy: From Machiavelli to the Nuclear Age* (Princeton: Princeton University Press, 1986), p. 17.

[2] The term "mission-oriented" is elaborated in Henry Ergas, "Does Technology Policy Matter?" in Bruce Guile and Harvey Brooks, eds., *Technology and Global Industry* (Washington, DC: National Academy Press, 1987).

[3] Robert Gilpin, *France in the Age of the Scientific State* (Princeton: Princeton University Press, 1968), p. 159. See also Pierre Papon, *Le Pouvoir et la science en France* (Paris: Centurion, 1979), pp. 34–39.

(Centre National d'Etudes Spatiales, CNES), the National Center for Telecommunications Engineering (Centre National d'Etudes des Télécommunications, CNET), and the National Bureau for Aeronautical Engineering and Research (Office National d'Etudes et de Recherches Aérospatiales, ONERA). With the exception of the largely autonomous CEA, these *grands organisms* are generally supervised by one of the ministries.

Two higher-level administrative bodies coordinate these facilities. After the founding of the Fifth Republic in 1958, a scientific council called Délégation Générale à la Recherche Scientifique et Technique (DGRST) was attached to the office of the prime minister.[4] The DGRST exerted considerable control over the civil R&D budget until the early 1980s, when it was dissolved into a new ministerial organization as part of the Socialist government's effort to link research and industrial policy. Defense-related R&D is coordinated in the Ministry of Defense through the Délégation Générale pour l'Armement (DGA), created in 1961 from a series of separate procurement bureaus that had served the military services.[5]

Both the DGRST and the DGA fitted into an administrative recipe for state-led economic growth that combined a long-term time horizon, represented by the Commissariat du Plan, and a coordinated set of macroeconomic policies, imposed by the powerful Finance Ministry.[6] The DGRST advanced long-term research priorities through its links to the prime minister and the specialized committees of the plan.[7] The DGA wielded particular weight in annual budget questions because its cause was represented by the powerful Ministry of Defense. These administrative arrangements supported an industrial policy that aimed at promoting "national champion" firms in key sectors.

As the advanced economies grew increasingly interdependent in the 1970s, the desirability of state-led planning came under periodic question. In the 1970s, Giscard d'Estaing's program for "industrial redeployment" called on large firms to wean themselves from state support and adjust to the constraints of a world marketplace.[8] The government's agenda turned sharply toward monetary policy, and the

---

[4] Papon, *Le Pouvoir et la science*, pp. 49ff.

[5] Pierre Dussauge, *L'industrie française de l'armement* (Paris: Economica, 1985), pp. 29–30.

[6] Peter Hall, *Governing the Economy: The Politics of State Intervention in Britain and France* (New York: Oxford University Press, 1986), pp. 171–172, 187–189.

[7] Gilpin, *France in the Age of the Scientific State*, p. 226.

[8] Suzanne Berger, "Lame Ducks and National Champions: Industrial Policy in the Fifth Republic," in W.G. Andrews and S. Hoffmann, eds., *The Fifth Republic at Twenty* (Albany: State University of New York Press, 1981), pp. 292–310.

influence of the Commissariat du Plan declined accordingly. The Socialist government revived the planning impulse in 1981, but quickly found it impossible to implement an autarkic economic policy in an interdependent world.[9]

While French officials have improvised on arrangements and strategies for promoting civil industry, there has been no similar effort to dilute the state's role in the defense sector. The DGA remains a highly centralized locus of control whose decisions ramify widely through the French economy. Altogether, it employs more than 70,000 persons, of whom 48,000 are primarily involved in the agency's industrial missions. The DGA has two divisions, for acquisition programs and international relations, respectively. The Delegate for Armament Programs (Délégué aux Programmes d'Armements, DPA) directs the research, development, production, financing, testing, and delivery of weapons systems. It has probably exercised more influence over French industry than any other part of the French state.[10] The Delegate for International Relations (Délégué aux Relations Internationales, DRI) is the administrative agency charged primarily with international arms sales.

Within the DGA, the Directorate for Research, Studies, and Technology (Direction des Recherches, Etudes, et Technique, DRET) coordinates military R&D. With a staff of 2,000, the DRET has responsibilities that include scanning the public and private research communities; overseeing ONERA, the national aerospace laboratory; organizing special task forces; monitoring research contracts issued by the technical directorates; and maintaining links to the research activities of universities, civilian labs, and military schools.

France allocates a greater proportion of its R&D effort to defense than any of the industrialized democracies except the United States and Great Britain. Military R&D represents roughly one-third of government expenditures on R&D, and about 20 percent of the country's total R&D. Despite this commitment of resources, the DGA has been forced to concentrate on a few lines of technological development. Expenditures for military R&D equal roughly 25 percent of the amount budgeted for acquisitions.[11] Although this figure represents a high degree of R&D intensity, it is considerably less than in the United States,

[9] For details on the administrative changes brought to the DGRST and its successor agencies during the 1980s, see the *Bottin Administratif* (Paris: Dodin Bottin, 1977), p. 107; (1982), pp. 697–698; (1983), pp. 626, 632; (1989), p. 725.

[10] Edward Kolodziej, *Making and Marketing Arms: The French Experience and Its Implications for the International System* (Princeton: Princeton University Press, 1987), pp. 260–261.

[11] Dussauge, *L'industrie française de l'armement*, p. 41; and "La loi de programmation militaire 1987–1991," *Régards sur l'Actualité*, no. 137 (January 1988), p. 31.

where defense R&D expenditures in the 1980s were equal to nearly half the level of the acquisitions budget.[12]

Industrial Profile

In keeping with its centralized political structure, France's arms industry is highly concentrated. Whether organized as state-administered arsenals or as commercial firms, a few dominant producers in each weapons category enjoy close, bilateral ties with French procurement agencies.

For the army and the navy, weapons systems are produced by a centuries-old network of state-owned arsenals. Tanks and other ground equipment are made in ten manufacturing centers that comprise the GIAT (Groupement Industriel des Armements Terrestres) within the Directorate for Ground Armament. The GIAT employs approximately 17,000 people and coordinates business with a cluster of private firms. Ships are built in state-controlled arsenals under the Directorate for Naval Construction (Direction des Constructions Navales, DCN). The naval system employs over 30,000 persons and also cooperates with private firms.[13]

In recent years, the arsenals have looked to foreign sales to temper their dependence on the French military. In some years, the GIAT has shipped as much as 40 percent of its matériel to foreign arms purchasers. On occasion, the GIAT has developed armored vehicles for the export market alone and even sacrificed the French army's development schedules to make foreign sales.[14]

France's main aerospace and electronics suppliers are organized as commercial enterprises, but their relations to the DGA are so similar to those of the GIAT and the shipyards that they are often called arsenal firms. On a sectorwide basis, they gain half to three-fourths of their revenues from military sales. At the firm level, military business can be heavily skewed toward export sales, particularly to the Middle East and Africa. The best example is Dassault-Bréguet, the French national champion in fighter aircraft. Though it was never fully nationalized, 90 percent of Dassault's revenues came from military sales through the early 1980s, of which about 70 percent were exports. Only recently has the firm expanded its sales of business jets to 30 percent of annual

---

[12] See Chapter 4 in this volume; also Jacques Gansler, *Affording Defense* (Cambridge, MA: MIT Press, 1989), p. 215.

[13] Kolodziej, *Making and Marketing Arms*, pp. 233–235, 247–248; Dussauge, *L'industrie française de l'armement*, p. 34.

[14] Kolodziej, *Making and Marketing Arms*, pp. 233, 273.

revenues. Aerospatiale (Société Nationale Industrielle Aérospatiale, SNIAS), a state-controlled firm, receives about 60 percent of its revenues from military projects, more than half from foreign purchasers. As France's partner in the Airbus consortium, Aerospatiale has benefited from the civil aircraft market and from strength in civil helicopters.[15]

Thomson, a large holding company in electronics, was wholly nationalized by the Socialists in 1982. Avionics and ground armaments comprised 30 percent of sales in the early 1980s, more than half exported. Thomson's military businesses grew quickly through the 1980s, and by 1987 came to well over 40 percent of total revenues. Matra, a smaller firm initially focused on ordnance and military electronics, acquired operations in transportation and industrial control systems while growing twentyfold in the 1970s. The firm's military business has recently ranged between 40 percent and 60 percent of sales, more than two-thirds for export.[16]

Recent Developments

The strengths and weaknesses of the French approach are illustrated by the recent experiences of Thomson and Aerospatiale. Both are major state-owned defense suppliers with extensive international business in civil products, yet they have managed the ties between their military and nonmilitary activities in very different ways.

After a decade of restructuring, Thomson has centered its activities on defense electronics and consumer electronics. In the process, the firm has sold operations in telecommunications and medical electronics, and merged its semiconductor operations in a joint venture with the Italian firm, Società Generale Semiconduttori (SGS). Among its acquisitions has been General Electric's consumer electronics operations, including RCA.[17] These changes have transformed Thomson from a conglomerate of numerous fiefdoms to a company with interests

[15] For information on defense sales and export dependence, see Kolodziej, *Making and Marketing Arms*, pp. 177, 203, 222, 233, 273; and *French Company Handbook 1989* (Paris: International Business Development, 1989).

[16] Data on firm-level dependence on military business are drawn from Kolodziej, *Making and Marketing Arms*, pp. 202–205; Dussauge, *L'industrie française de l'armement*, pp. 19–23, 141–144; and the *French Company Handbook*.

[17] Articles on Thomson's restructuring include "Alliance franco-italienne dans les semiconducteurs," *Le Monde*, April 30, 1987; "Riding a National Champion," *Financial Times*, January 30, 1989; "Buying Spree Boosts Thomson-CSF's Empire," *Electronics Business*, November 27, 1989.

in defense and consumer electronics bridged by the joint semiconductor venture with SGS. Still, the logic of Thomson's strategy depends on acquiring market share rather than managing technologies with applications in both civil and military markets.

Thomson officials stress financial rather than technological links between military and commercial sales; technologically based synergies among operating divisions appear limited. The semiconductor venture, Thomson-SGS Microelectronics, has become a merchant producer that supplies Thomson's other divisions on a conventional make-or-buy basis. Thomson seeks price and quality for products like televisions through huge production runs in globally dispersed factories. In contrast, the company's defense-related branch draws 80 percent of its revenue from military orders and specializes in customized systems, including field communications for the U.S. Army and a submarine detection system for Saudi Arabia.[18] The division has been described as a world of its own, whose traditional managers have only reluctantly accepted the more international orientation of Thomson's chief executive, Alain Gomez.[19]

By building market share in two highly competitive segments, Thomson has acted on the view that global economies of scale remain critical in electronics. Even if the consumer business no longer benefits much from synergy with defense systems, consumer markets provide revenues that help cover R&D expenses for both divisions. At the same time, as a leading advocate for European standards in high-definition television (HDTV), Thomson has positioned itself to gain from anticipated links between HDTV componentry and high-resolution military systems.[20] To date, however, Thomson's growth has reflected effective management of separate businesses, without significant technology transfer between civil and military operations.

Aerospatiale has tried much harder to link its military and commercial activities, combined within each of the main operating divisions. The aircraft division, France's participant in the Airbus consortium, is only about 20 percent dependent on military sales, while the helicopter unit splits its sales almost equally between civil and military markets. The space and ballistics division sells 80 percent of its output to the

---

[18] Janice McCormick and Nan Stone, "From National Champion to Global Competitor: An Interview with Thomson's Alain Gomez," *Harvard Business Review* (May–June, 1990), pp. 126–135; "Les Alchimistes de Corbeville" and "Thomson: le Maître des Armes," both in *Le Nouvel Economiste*, April 3, 1987, pp. 22ff.

[19] For managerial attitudes within Thomson, see *Le Nouvel Economiste*, April 3, 1987; and *Electronics Business*, November 27, 1989.

[20] "Keeping Cool in a High-Stakes Game," *Financial Times*, July 16, 1990.

military. Only the tactical missile division sells exclusively to military customers.[21]

There are several indications that much of the firm's success stems from effective leveraging of military R&D. Through the early 1980s, the French military provided two-thirds of Aerospatiale's entire R&D budget, which amounted to 20 percent of the firm's revenues, reaching 29 percent in 1987.[22] A new corporate laboratory focused on areas of generic interest to the group, including materials, data processing, and production technologies. The helicopter division best exemplified the firm's ability to bridge military and civil markets, selling to more than 100 countries. Both its civil and military helicopters have held speed and altitude records, and recently the firm has been the leading supplier of civil helicopters to North America.[23]

The achievements of firms like Thomson and Aerospatiale continue to rest on government policies, including subsidies that have sometimes been massive. Airbus Industrie has established itself as a serious competitor to Boeing and McDonnell Douglas for civil air transports, but critics note that billions of dollars of initial capitalization came from the sponsoring governments (primarily France, Germany, and Britain). Airbus now claims to be profitable on operations, but the initial investments will probably never be repaid. Thomson, though large, continues to lose money on semiconductors and consumer electronics. With the European Community seeking to limit state subsidies for commercial industry, the future for France's dual-use firms seems as unsettled today as in the 1960s, when De Gaulle's ambitious *Plan Calcul* sought to create a national champion in the information industry.

With Europe 1992 promising to open French markets to more imports, and with Brussels working to pry Paris away from subsidies, executives of French firms as well as officials in the French government will need to learn new ways. The end of the cold war, accompanied by greater competition in export markets for armaments, may make this transition still more difficult by limiting this source of revenues.

SWEDEN

At first glance, Sweden and France appear to share important characteristics. Both have predicated national autonomy on a self-sufficient defense industry. In Sweden, however, the principle of self-sufficiency

---

[21] Dussauge, *L'industrie française de l'armement*, p. 19.

[22] Ibid., p. 46.

[23] *French Company Handbook 1989.*

rests on rather different assumptions. National autonomy is perceived in terms of territorial integrity, whereas in France, autonomy is viewed as an extension of historical greatness. Sweden has consequently pursued defense policy and economic development in tandem, as roughly equal goals of state action.

Sweden's defense industrial capabilities reside in a group of technologically sophisticated firms that have followed specialized product strategies with considerable success in international markets. Although controlled by a small group of private investors, firms such as Saab-Scania, Volvo, Ericsson, and Bofors are counted among the country's major social institutions. They are expected to fulfill a broad range of social priorities, such as regional development and full employment, as well as military self-sufficiency. Because Sweden's definition of self-sufficiency emphasizes territorial integrity rather than technological autarky, these firms have relied on external technology sourcing much more than their French counterparts.[24]

Administrative Resources

Many of the Swedish state's instruments for economic development have appeared since the 1950s, as the economy shifted from agriculture and forest products toward manufacturing and services. Owing to a small domestic market, Swedish firms typically exported as much as half of their output and exports soon accounted for roughly one-third of the country's GNP.

Sweden's recent efforts at industrial restructuring have been guided by a deeply rooted principle of balanced adjustment that stems from the historical compromise between Swedish labor and management. In contrast to France, where politics has centered within a monolithic bureaucracy, politics in Sweden has revolved around regular patterns of structured consultation. The modern version of this structured consultation appeared when Swedish industrialists learned to negotiate with the Swedish working class in the 1930s, once the latter's representatives in the Social Democratic party could no longer be excluded from political power. From the 1930s onward, the Swedish state's legitimacy has rested on its ability to distribute the proceeds of economic growth among industrial owners and organized labor groups.[25]

[24] On Swedish industry, see "How Sweden Became Europe's Industrial Powerhouse," *Business Week*, August 18, 1986, pp. 100A–100B; "The Swedish Economy," supplement in *The Economist*, March 3, 1990; and "The Stateless Corporation," *Business Week*, May 14, 1990, p. 103.

[25] On Sweden's historical compromise, see Hugh Heclo and Henrik Madsen, *Policy and*

Since the Social Democratic party governed from 1932 through 1976, Sweden's institutional arrangements for industrial promotion have been stamped by the need for balanced adjustment. In 1967, a Ministry of Industry and a state-owned investment bank were created to help ailing industries. One of the ministry's tasks was to distribute public R&D funds, which had previously been disbursed primarily by the Ministry of Education and the Ministry of Defense.[26]

In addition to these instruments for focused industrial intervention, Sweden also developed broad policies for regional development and labor-market adjustment. With a territorial area equivalent to France's but a population one-seventh as large, Sweden was always susceptible to regional imbalances. Since the 1960s, the state has encouraged firms to invest in depressed areas through loans and grants. After the Social Democratic government gave way to a center-right coalition in the latter part of the 1970s, regional development policies targeted small and medium-size firms. The Social Democratic years also left Sweden with an extensive array of labor-market policies—including relocation grants, job-placement networks (private employment agencies are prohibited), selective training and retraining programs, and temporary income subsidies.[27]

Without this broad panoply of development supports, it is doubtful that Sweden could base its policy of armed neutrality on the principle of self-sufficiency. Although Sweden spends over 4 percent of its GNP on defense, this amounted to only $4.9 billion in 1988 (see Table 7-1). The doctrine of armed neutrality confronts severe constraints in such a small country. The military has frequently paid a premium to have major weapons systems developed by Swedish firms. Like France, Sweden has been obliged to pursue an increasingly selective approach to weapons R&D—emphasizing target-acquisition systems, electronic countermeasures, detection systems, and other technologies of particular importance for the defense of a large Nordic territory.[28]

Additional constraints stem from Sweden's overall security doc-

Politics in Sweden: Principled Pragmatism (Philadelphia: Temple University Press, 1987), pp. 9ff.; and Andrew Martin, "The End of the 'Swedish Model'?" *The Bulletin of Comparative Labour Relations,* vol. 16 (1987), pp. 93–128.

[26] Bernard Udis, *The Challenge to European Industrial Policy: Impacts of Redirected Military Spending* (Boulder, CO: Westview Press, 1987), p. 85.

[27] For Swedish economic development policies, see Heclo and Madsen, *Policy and Politics in Sweden;* Martin, "The End of the 'Swedish Model' "; OECD, *Reviews of National Science and Technology Policies: Sweden* (Paris: OECD, 1987); and Udis, *The Challenge to European Industrial Policy,* pp. 88ff.

[28] See Inga Thorsson, *In Pursuit of Disarmament: Conversion from Military to Civilian Production in Sweden* (Stockholm: Allmänna Förlaget, 1984), vol. I, pp. 95–96.

trine, which emphasizes international cooperation and global economic development as well as purely military defense. Commitments to neutrality and to third-world economic development impose strict limitations on military exports.[29] With few exceptions, the government has approved military sales only to other neutral countries in the industrialized world. Unlike French firms, therefore, Swedish defense suppliers cannot amortize the costs of new weapons systems by cultivating ever wider export markets.

The need to finance military R&D through domestic revenues puts Swedish officials under great pressure to pursue military procurement and industrial development as mutually reinforcing goals. Procurement contracts are used to enhance labor-market stability and regional economic vitality. Proponents as well as opponents of armed neutrality support industrial restructuring, because they see civil industry as a repository of critical skills and capacities that the military cannot support on its own.[30]

The Defense Matériel Administration (Försvarets materielverk, FMV) oversees procurement. The FMV supports the supreme commander and the staffs of the three military services, which work closely with industrial firms during the acquisition cycle. The FMV must hold Swedish suppliers to world standards for cost and performance, unless explicitly allowed by civilian authorities to favor domestic firms. In practice, the FMV has usually bought 85 to 90 percent of its matériel from Swedish producers, who in turn purchase an estimated 15 to 20 percent of their inputs from abroad.[31]

Like other aspects of procurement, defense R&D is embedded in a broader institutional context. Swedish industry finances nearly two-thirds of the country's R&D from its own resources. For public R&D, Sweden, unlike France, has no central ministry and few government laboratories. Instead each ministry is responsible for its own R&D. On the view that publicly financed research should be carried on primarily within the universities, more than 30 percent of government R&D funds flow through the Ministry of Education. During the 1980s, the Defense Ministry controlled 22 to 27 percent of public R&D funds, spent largely in industry. The share allocated by the Ministry of Indus-

---

[29] Ibid., pp. 49, 62–63, 112ff.

[30] As one defense planner put it, "There is a great need for restructuring so we'll be able to keep the competence that is absolutely necessary in a war situation." In *Sweden Now,* February 1989, p. 20.

[31] This paragraph is based on Thorsson, *In Pursuit of Disarmament,* vol. I, pp. 72ff., 92–94, 167ff., and vol. I, part 1b.

try, which in some years exceeded defense R&D, dropped to 16.5 percent in 1986–1987.

Measured as a fraction of all public R&D spending, government disbursements for defense R&D (27 percent in 1987) lie between the French and German figures. It is difficult to compare defense R&D and equipment expenditures, partly because Swedish defense budgets include R&D in nonmilitary categories. In 1982–1983, however, the FMV reported spending $180 million (1,134 million Swedish kroner, SEK) on R&D and $886 million (5,566 million SEK) on equipment for a ratio of roughly 20 percent. Of FMV's R&D disbursements for 1982–1983, industry received $140 million (882 million SEK), or about three-quarters, with a dozen firms accounting for almost all of that amount.[32]

Although most defense R&D is performed by industry, the military does have its own facility, the National Defense Research Institute (Försvarets Forskningsanstadt, FOA). The FOA concentrates on applied studies that complement development done by the equipment suppliers. This can be particularly important during the course of a major acquisition program, when preliminary test results, unforeseen technical issues, and revised cost estimates lead to midstream corrections. Smaller research programs are conducted by the FMV itself, the service staffs, the bureau for civil defense, and other agencies. In 1983, the FOA employed 1,385 persons, while the defense agencies combined had about 2,100 employees engaged primarily in R&D. One of Sweden's few mission-oriented laboratories, the National Aeronautical Research Institute, receives approximately $12 million (80 million SEK) for applied R&D from the armed forces as well as the aerospace companies. These agencies and institutes give Sweden the capability to direct funds and manpower to specific research problems. They are, however, among the relatively few Swedish analogs to the French *grands organisms*.

Industrial Profile

Sweden's defense industry is unique. Though highly concentrated, it is generally regarded as well integrated with the country's overall industrial base. The ten largest defense contractors supply 80 percent

---

[32]Calculated from information in Thorsson, *In Pursuit of Disarmament*, vol. I, pp. 97, 159, 164, 165–166, using the exchange rate of $1 to 6.282 Swedish kroner.

of the defense industry's output.[33] Even the largest contractors are subsidiaries of larger groups that are dominated by civil business lines.

Given Sweden's size, there could hardly be more than one or two producers for any major system category. Both Volvo and Saab-Scania build military trucks, but only Hägglund & Söner makes tracked vehicles, including tanks (with recent models involving Bofors as well). The ordnance suppliers include Bofors for large guns and munitions and the National Defense Factories Group (Förenade Fabriksverken, FFV) for smaller arms. Naval vessels are built by two state-owned firms: Kockums (submarines) and Karlskronavaret (surface ships).

Sweden is one of the smallest countries to attempt independent development of advanced combat aircraft. The core of the country's aerospace industry is Saab-Scania, prime contractor for the JA 37 Viggen interceptor and the newer JAS 39 Gripen fighter. Volvo Flygmotor adapts aircraft engines to Swedish needs from foreign designs. Bofors and Saab-Scania build missiles. Four firms—Ericsson Radio Systems, Philips Elektronikindustrier, Bofors Aerotronics, and SATT—supply electronics technologies that are central in Swedish efforts to limit superpower surveillance.[34]

The difference between the French and Swedish approaches to self-sufficiency extends to the organization and business focus of firms. Almost all French defense firms behave much like arsenals, while seeking economies of scale and scope through exports to other military customers. In Sweden, by contrast, even some of the arsenal-type suppliers have been diversifying into commercial markets. Although founded as an armaments workshop, the country's oldest weapons group, FFV, has been supervised by the Ministry of Industry since 1970. In 1983, 70 percent of its revenues came from defense contracts, 30 percent came from sales of industrial equipment, automotive components, commercial aircraft maintenance services, and sporting firearms. The country's tank and tracked vehicle supplier, Hägglund, earned only 10 to 40 percent of its revenue from military orders through the 1970s—representing only 2 percent of consolidated revenues for the parent group, ASEA. Bofors, the private firm most dependent on military sales, saw defense projects increase from 30 percent of revenues in the early 1970s to 50 percent in the early 1980s.[35]

The commercial orientation of Swedish defense contractors is clearest in aerospace and electronics. Saab-Scania, Sweden's largest military

---

[33] See Gansler, *The Defense Industry*, p. 245; and Thorsson, *In Pursuit of Disarmament*, vol. Ib, p. 18.

[34] Thorsson, *In Pursuit of Disarmament*, vol. I, pp. 99–107.

[35] Ibid., pp. 116–119.

contractor, obtains no more than 8 percent of sales from defense. Even as deliveries of the Gripen began, increasing sales of Saab's 340 commuter airliner enabled the company's Aircraft Division to reduce its military dependence from 90 percent in 1983 to 48 percent in 1989. In the other division with military business—Saab-Scania Combitech—defense represented roughly 55 to 60 percent of sales through the 1980s. Volvo Flygmotor's commercial and military businesses both expanded during the 1980s, with defense representing 50 to 60 percent of sales. For the parent group, AB Volvo, military contracts came to less than 1.5 percent of worldwide sales during most of the 1980s. Ericsson's military electronics work comprises roughly half the revenues of the radio systems group but well under 10 percent of Ericsson's consolidated sales. Defense contracts won by the Philips subsidiary in Sweden account for about half of the control division's work, and 20 percent of Philips Sweden's overall sales.[36]

Recent Developments

Sweden's policies of military self-sufficiency and export restrictions have been severely tested since 1986, with the cost and complexity of modern weapons systems straining resources all around. Like their French counterparts, Swedish defense firms have responded in a variety of ways. Yet the Swedish political establishment has—unlike its French counterpart—encouraged firms to diversify from defense to commercial activities.

Bofors, the privately owned Swedish firm that has most clearly followed an arsenal strategy, made a strategic choice to remain in the defense business—if necessary by seeking greater ordnance sales in foreign markets—in the 1970s. The firm also sought to strengthen its chemicals business, merging in 1984 with Kema Nobel under the new name Nobel Industries of Sweden.[37] Although merger with Kema Nobel lowered the share of military revenues in the new group's consolidated revenues, military activities remained dominant in the former Bofors Ordnance Division, which gained new capabilities in explosives and advanced materials from Nobel's other divisions. The Ordnance Division became an increasingly integrated defense systems house, but it has shown no signs of exploiting dual-use links to extend its technological expertise to civil applications. The division is still known by its

[36] Ibid., pp. 118, 122–123, and company reports, 1987–1988.
[37] Thorsson, *In Pursuit of Disarmament*, vol. II, pp. 118–122.

old name, Bofors. Since its profit margins consistently exceed those of other business lines, group management has little incentive to dilute the Ordnance Division's cohesion or identity.

The limitations of the strategy have become apparent in recent years. As Bofors focused its technological capabilities on defense systems, it was forced to look further afield for markets. In 1987, the firm's relations with the Swedish government were soured by public accusations that it had sold arms illegally to Iran and Iraq and made questionable payments to gain a major contract for howitzers from India. While some officials in Sweden's defense establishment suggested that Bofors should give up its ordnance business altogether, the firm recently absorbed Ericsson's command and control interests within its own electronics operations and has successfully tested its new RBS-70 missile. To remain profitable, Bofors Ordnance had to cut employment by almost a third, from 4,800 in 1989 to 3,000 in mid-1990.[38]

Saab-Scania adopted a very different strategy in the 1980s. In theory, the company was well positioned to seek a dominant role in integrated defense systems. In practice, management decided on an explicit dual-use strategy by joining the firm's technology-intensive ventures and informal skunk works in a new multitechnology business group called Saab-Scania Combitech. The objective was to coordinate the group's expertise in electronics and provide a technology house for both internal and external development projects.[39] Combitech units include automation, production control, measuring systems, aerospace instruments, electronic circuit boards, marine electronics, automotive electronics, missiles, and training systems. Instead of focusing on defense products as Bofors did, Saab-Scania concentrated its existing capabilities in order to apply them to a range of civil and military markets. The aim was to "combine technological economies of scope with (joint) economies of scale" and to promote attentive management of technology transfer. It is difficult to assess the consequences of such measures without tracing specific products and components, but outside observers as well as company reports suggest that technical communication among Combitech companies has been strengthened.[40] Combitech grew from under 1,700 employees in 1983 to 2,600 in

---

[38] *New York Times*, January 24, 1990, p. D5.

[39] Ove Granstrand and Sören Sjölander, "Managing Innovation in Multi-Technology Corporations," *Research Policy*, vol. 19, no. 1 (February 1990), pp. 35–60; Saab-Scania *Annual Report*, 1988, p. 35.

[40] See Granstrand and Sjölander, "Managing Innovation," p. 46.

1988, when it was among Saab-Scania's most profitable divisions, returning over 20 percent on capital.

Despite their past successes, the longer-term future seems cloudy for Sweden's large firms. Like many other European multinationals, they have suffered badly in international markets in recent years. The automobile divisions of Volvo and Saab have both sought refuge in alliances with larger foreign companies. Sweden's labor force, well educated and trained, is also well supported—one reason why Swedish firms sometimes face costs above those of their competitors. Though the coming entry into the European Community promises to alter the context for defense policies—and although Sweden's multinationals have proven remarkably adept at negotiating difficult times in the past—the country may well find that self-sufficiency in defense is no longer an achievable policy in a global economy.

GERMANY

Unlike France and Sweden, which attach high value to military self-sufficiency, the Federal Republic of Germany has subordinated military autonomy to membership in the Western alliance. Constitutional restrictions limited the West German military to a defensive force posture, while political considerations made multilateral projects more acceptable than an autarkic orientation. These constraints discouraged the Federal Republic from adopting a high-profile role in military technology. Additional constraints grew out of the history of misuse of science and technology during the Nazi period. The postwar constitution guaranteed freedom of teaching and research, and public opinion resisted giving the armed forces significant influence over institutions for science or technology.

Even before World War II, the German state's influence over technology development was balanced by powerful industrial interests. The concentration of German industry led the economist Rudolf Hilferding to formulate his theory of "organized capitalism" in the 1920s. German officials generally tried to shape economic outcomes by regulating the framework of competition. Occasionally, public authorities attempted to dictate outcomes more directly, most notably in programs for developing synthetic substitutes for natural raw materials.

After World War II, the regulatory approach became dominant. In the early postwar years, economic policy reflected the doctrine of ordo-liberalism (*Ordnungspolitik*), popularized in Ludwig Erhard's campaign for the "social-market economy."[41] The doctrine of ordo-liberalism called for measures and institutions that would "impart to competition the framework, rules, and machinery of impartial supervision, which a competitive system needs as much as any game or match if it is not to degenerate into a vulgar brawl."[42] This framework-setting approach aimed at a healthy mix of large firms with small and medium-sized businesses. It assigned primary responsibility to the state for setting the conditions of competition and it limited sector-specific interventions, relying instead on incentives available to a large universe of firms.

This historical experience gave the Federal Republic's defense industry its particular characteristics. Germany has long been known for industrial giants like Daimler-Benz, Siemens, and Krupp. In the postwar decades, large diversified firms secured their futures from civil markets and drew a minor portion of their revenues from military work. Equally important were the numerous small and medium-sized companies that provided a great deal of specialized know-how to the larger systems houses. These long-term relationships between large firms and small or medium-sized firms helped industry adjust to turbulent markets, accounting for much of the country's economic vitality in the postwar period.[43]

Administrative Resources

Over the years, German technology policy has shifted between a regulatory or framework-setting approach and a more interventionist or mission-oriented approach.[44] In the 1950s and 1960s, the government

[41] The notion of *Ordnungspolitik* is discussed in Christopher Allen, "The Underdevelopment of Keynesianism in the Federal Republic of Germany," in Peter Hall, ed., *The Political Power of Economic Ideas* (Princeton: Princeton University Press, 1989), pp. 263–289. It was the intellectual foundation for the policies of Ludwig Erhard, West Germany's first minister of economics, whose notion of the "social market economy" was intended to reconcile market efficiency and social priorities. On the regulatory bent in the German policymaking style, see K.H.F. Dyson, "West Germany: The Search for a Rationalist Consensus," in J. Richardson, ed., *Policy Styles in Western Europe* (London: Allen and Unwin, 1982), pp. 17–46.

[42] Wilhelm Röpke, quoted in Allen, "The Underdevelopment of Keynesianism," p. 281.

[43] This view is most forcefully advanced in Gary Herrigel, "Industrial Order and the Politics of Industrial Change: Mechanical Engineering," in Peter Katzenstein, ed., *Industry and Politics in West Germany* (Ithaca: Cornell University Press, 1989), pp. 185–220.

[44] Otto Keck, "West German Science Policy Since the Early 1960s: Trends and Objectives," *Research Policy*, vol. 5, no. 2 (1976); pp. 116–157, Volker Hauff and Fritz Scharpf, *Technologiepolitik als Strukturpolitik* (Frankfurt: Europäische Verlagsanstalt, 1975); *Bundesbericht Forschung 1984*; and interviews conducted at the Bundesministerium für Forschung und Technologie, summer 1990.

promoted civilian nuclear power, aerospace, and other technologies that could help the country reestablish its role in world politics. The establishment of a Ministry for Research and Technology (*Bundesministerium für Forschung und Technologie*, BMFT) in 1972 signified the Social Democrats' desire for more forward-looking policies to spearhead adjustment to anticipated "structural" changes in the international economy. The Social Democrats directed resources to data processing and other industrial technologies, as well as nuclear power and aerospace. From the late 1970s, this anticipatory tilt has been gradually tempered. According to the "subsidiarity principle" articulated by the Christian Democratic government in May 1983, the state was to moderate its R&D investments and encourage entrepreneurial initiative by returning to favorable "framework conditions" for research, development, and innovation. In the late 1980s, elements of the earlier emphasis on large projects returned as over two-thirds of the annual increments in BMFT's budget were allocated to aerospace projects.

The BMFT disburses 55 to 60 percent of federal R&D resources, but it has always worked within a dense network of other agencies. Despite a low profile, the Defense Ministry's R&D disbursements grew from 15 percent to more than 20 percent of federal R&D outlays through the 1980s. Remaining federal funds for R&D are allocated through the Economics Ministry (7 to 9 percent), which oversees programs subsidizing R&D personnel in small firms, and the Ministry of Science and Education (8 to 9 percent), which supports R&D in educational institutions.[45]

The federal ministries together disburse only about one-fourth of the country's total R&D funds. The regional (*Länder*) authorities—which exercise primary authority over education in Germany's federal system—account for another 13 to 15 percent of the total. Industry itself finances between 55 and 65 percent of all R&D.[46]

The performance of R&D in the Federal Republic is entrusted to three groups of laboratories, which are heavily supported by the BMFT with ancillary funding from the *Länder* in which they are located. These include the Max Planck Institutes for basic research and the Fraunhofer Institutes for applied and industrial research. A lesser-known set of facilities, the so-called large research organizations (*Großforschungseinrichtungen*) were mostly founded in the 1950s and 1960s to promote

---

[45] *Bundesbericht Forschung 1988*, Table II/4, p. 68; and *Faktenbericht 1990*, Table I/3, p. 43.

[46] These figures, from *Bundesbericht Forschung 1988*, Table II/2, p. 60, show somewhat greater public expenditures than figures reported by the OECD (and given in Table 7-1). See also *Facts and Figures 1990; Update of the Report of the Federal Government on Research 1988*, Tables I/2 and I/4, pp. 29, 43.

research in civil nuclear power and aerospace. There are now thirteen such facilities whose concerns have been extended to fields including polar research and elementary particle physics. Together these three groups of labs conducted 10 percent of the country's R&D in 1987, while institutions of higher education performed another 12.5 percent. The greatest part of the Federal Republic's R&D—71 percent in 1987—is conducted by industry.[47]

This broad range of institutions gives the Ministry of Defense ways to finance R&D without managing its own laboratories. Owing to the widespread view that military goals should not influence scientific research, the Ministry of Defense operates no in-house facilities for R&D. In the 1950s, the ministry had difficulty finding external institutions to perform its research until several Fraunhofer Institutes, which were under financial pressure, proved amenable partners.[48] Six Fraunhofer Institutes now receive basic budget finance support (*Grundfinanzierung*) from the Ministry of Defense for work in solid-state applications, ballistics, materials, and related fields. The ministry also supports a series of institutes in the applied natural sciences (*Forschungs-Gesellschaft für angewandte Naturwissenschaften*) that work in communications and related areas. The German Aerospace Research Establishment (*Deutsche Forschungsanstalt für Luft- und Raumfahrt, e.V.,* DLR) is the only large research organization that receives budget contributions from the Ministry of Defense as well as from the BMFT. All of these external facilities receive roughly three-fourths of their defense funding through grants, with the remainder consisting of contracts for specific projects.[49]

While these relationships enable the Ministry of Defense to sponsor research without encroaching on the BMFT's general responsibility for science and technology, its subordinate role means the ministry must depend heavily on technological scanning capabilities, which are reinforced by the Fraunhofer Institute for Scientific and Technical Trend Analysis. The ministry coordinates its research priorities periodically

---

[47] A miscellaneous group of state-owned institutions—mostly libraries and archives together with international organizations funded from German sources—account for the remainder. For figures and a list of the *Großforschungseinrichtungen,* see the *Bundesbericht Forschung 1988,* pp. 62 (Table II/3), 296ff., and 352 (Table VII/3).

[48] For nonuniversity research institutions, including the Fraunhofer Society, see Hans-Willy Hohn and Uwe Schimank, *Konflikte und Gleichgewichte im Forschungssystem; Akteurkonstellationen und Entwicklungspfade in der staatlich finanzierten außeruniversitären Forschung* (Frankfurt/M: Campus, 1990).

[49] Interviews conducted at the Federal Ministry for Defense, Bonn, Summer 1990; *Bundesbericht Forschung 1988,* pp. 198–199, 298; *Rheinischer Merkur,* May 6, 1989, reprinted in the *German Tribune,* May 28, 1989.

with those of the BMFT by issuing a comprehensive research and technology document, *Forschungs- und Technologiekonzept*, supplemented
with annual guidelines. By publicizing its long-term view of military
priorities, the ministry encourages defense firms to commit their own
funds to complementary R&D activities.

These arrangements provide a series of organizational buffers that
enable the Ministry of Defense to meet its research needs while remaining at arm's length from performing institutions. Although they
have emerged through historical circumstance rather than deliberate
design, these organizational buffers ensure that defense R&D remains
consistent with the Federal Republic's regulatory approach to policy.
Rather than conducting its own in-house research and transferring the
results directly to industry, the Ministry of Defense provides general
support to those institutions that are positioned to scan research options and pursue promising possibilities.

This indirect relationship has not prevented German politicians
from following the U.S. example by invoking commercial spinoffs to
justify military spending for R&D.[50] Nonetheless, historical circumstances in the Federal Republic have produced a set of linkages between
commercial and defense technologies significantly different from those
in the United States. Through the 1980s, the Federal Republic's institutional arrangements made civil research results available to military
projects ("spin-on") at least as readily as they oriented basic research
toward military goals with subsequent transfer to civil applications.[51]

Research overseen by the Defense Ministry's Armaments Division
comprises the first step of the acquisition cycle. The Armaments Division (about 400 persons) must come to agreement with the service
staffs on the "guiding concepts" for weapons programs, these concepts
themselves subject to interservice "correlation" by the general staff. It
is only at the end of the concept phase and beginning of the definition
phase that the services generate tactical and technical requirements for
a particular weapons system. If this phase leads to a satisfactory set of
time/cost/performance requirements, then a prime contractor is chosen
and day-to-day responsibility for the project is shifted to a specialized
procurement agency, the Federal Office for Military Technology and

---

[50] The spinoff rationale was invoked, for example, in 1985, when West German firms
were invited to participate in U.S.-sponsored research for the Strategic Defense Initiative. For
critical analysis of the spinoff rationale in the German context, see, for example, Ulrich
Albrecht, "Spin-off: A Fundamentalist Approach," in Philip Gummett and Judith Reppy,
eds., *The Relations between Defence and Civil Technologies* (Dordrecht: Kluwer, 1988), pp. 38–57.

[51] For acknowledgment of civil-to-military transfers as part of the defense research task,
see *Bundesbericht Forschung 1988*, pp. 198–199.

Procurement (*Bundesamt für Wehrtechnik und Beschaffung,* BWB). This office employs roughly 18,000 people who oversee development, prototype testing, production, and other contractual activities. In theory, separate contractors can be chosen for development and production; but in practice, the developer also produces the system.[52]

As a proportion of public R&D expenditures, defense R&D rose from 10 percent in 1984 to 12.7 percent in 1987—a far smaller proportion than in France or Sweden. By comparison to equipment purchases, however, the Federal Republic's defense R&D figures are surprisingly high. In 1987, the ministry spent $1.6 billion (2.8 billion DM) on defense R&D (including testing) and $6.7 billion (12 billion DM) on equipment—a ratio of 23.5 percent. By 1989, the ratio had risen to 26 percent, and ministry requests for 1990 envisioned a ratio of 30.1 percent.[53] Through this combination of arm's-length institutional support for R&D and direct sponsorship of systems development, the Ministry of Defense obtains technology-intensive equipment without dictating the research agendas of the scientific community or industrial firms. Even though the ministry conducts no in-house research or technology development, weapons procurement has become as R&D-intensive in the Federal Republic as in Sweden or France.

Industrial Profile

Political restrictions on direct exports, especially to non-NATO purchasers, encouraged West German defense firms to join cross-national consortia. As a result, they became adept at managing complex international programs for developing and producing military systems. Aided by the dynamism of the small and medium-size sector, the overall size of Germany's manufacturing economy helped the Federal Republic to preserve more competition in procurement than other medium-sized powers.[54]

Within this general picture, a recent trend toward concentration in

[52] This paragraph is based on interviews conducted at the Armament Division, Federal Ministry of Defense, Bonn, July 1990; and U.S. General Accounting Office, *Fact Sheet for the Chairman, Committee on Governmental Affairs: Weapons Acquisition, Processes of Selected Foreign Governments* (Washington, DC: GAO, February 1986), pp. 53–64.

[53] Figures for 1987 are from the Embassy of the Federal Republic of Germany, Washington, DC; figures for 1989 and 1990 are from *Jane's Defence Weekly,* July 15, 1989, p. 70.

[54] For background on the German defense industry, see, for example, Klaus Schomacker, Peter Wilke, and Herbert Wulf, eds., *Zivile Alternativen für die Rüstungsindustrie* (Baden-Baden: Nomos, 1986); and Burkhardt J. Huck, "Verteidigungsausgaben, Rüstungsplanung und Konversion in der Bundesrepublik Deutschland: Daten und Literaturbericht," Working Paper AP2700 (Ebenhausen, FRG: Stiftung Wissenschaft und Politik, May 1991).

aerospace may foreshadow a basic shift away from the mix of small and large firms that has characterized the Federal Republic's industrial structure. This movement began under the auspices of the automotive giant, Daimler-Benz, in 1985, when the firm took control of the aircraft builder Dornier as well as AEG, the troubled electronics firm. In 1988, Daimler added the country's other major aerospace firm, Messerschmidt-Bölkow-Blohm (MBB), to the group, putting all of its aerospace resources under a single subsidiary called Deutsche Aerospace (DASA).

Several independent suppliers remain for most other types of weapons systems. Tanks and armored vehicles are produced by Krauss-Maffei, Krupp MaK Maschinenbau, and Wegmann & Co. Howaldswerke-Deutsche Werft and Blohm und Voss build ships, while other firms provide design services and propulsion. Daimler's subsidiary, MTU, supplies aircraft engines, as does KHD Luftfahrttechnik. AEG continues under Daimler ownership to supply defense electronics, as do Siemens and several small firms. Rheinmetall makes artillery and Diehl produces smaller arms, with numerous smaller firms also participating in the ordnance business.

Almost all German defense firms have tried to balance their military product lines with civil production. Diversification has been difficult for ordnance suppliers, but Rheinmetall also has subsidiaries in packaging machinery and automotive components. The Diehl group does business in military armaments, sporting guns, semifinished metal products, and precision mechanical equipment.[55]

German aerospace producers have anticipated for some time that commercial markets would provide as much or more growth as military markets. MBB has balanced orders for the Tornado combat aircraft with commercial business through its 38 percent stake in Airbus. MBB's other products include medium-range civil aircraft, space systems, helicopters, and guided munitions. In 1988, the firm reduced its military revenues to 46 percent, from as much as 60 percent in 1985. Within Daimler's new integrated aerospace division, military sales account for just under half of all revenues.[56]

In marked contrast to their French counterparts, German electronics firms do not depend heavily on defense systems. AEG no longer publishes separate figures, but in 1983 the firm received only 18 percent of revenues from military production. Siemens, the electronics and electrical equipment giant, though recently buying additional radar and

---

[55] Rheinmetall and Diehl Annual Reports, 1987, 1988.
[56] MBB Annual Reports; Udis, *The Challenge*, p. 68; *The Wall Street Journal*, April 3, 1990.

military communications capacity from the British firm Plessey, still has defense sales under 5 percent of total revenues.[57]

Recent Developments

As in France and Sweden, defense firms in the Federal Republic have been squeezed between demanding technologies and stiffening competition. Difficulties surfaced in 1986, when the government-owned shipyard, Howaldtswerke-Deutsche Werft, was shown to have supplied technical information on West Germany's diesel submarine, the U-209, to the South African government. The affair highlighted the problems of firms confronted with increasing technological complexity, rising development costs, and fewer orders from the German armed forces. The range of responses is illustrated by comparing Daimler-Benz and Krauss-Maffei.

Daimler has undertaken an ambitious effort to transform itself from an automotive firm into an integrated technology company with a dominant position in defense systems. To preserve its position in the upper end of the auto market and to diversify into aerospace, Daimler's management decided several years ago that defense technologies would have to become central in the company's future. Much like several of the integrated U.S. defense contractors, Daimler sought to capture economies of scale and scope in technology management. To accommodate the hundreds of specialized aerospace engineers gained through its acquisitions, Daimler quickly announced plans for a central R&D department, intended to transfer know-how to the newly broadened range of operating divisions. Although Daimler executives did not forecast growing defense sales per se, they did argue that defense technologies were crucial in aerospace as a whole, where they expected growth in civil as well as military markets.[58]

Daimler's strategy means the firm must manage a much wider set of relationships with the federal government. While the recent acquisitions brought substantial R&D capabilities, they also brought orders subject to political changes. The relaxation of East-West tensions has threatened the planned pan-European fighter program, prompting statements from Daimler's chairman that the government should help shoulder the burden caused by any cutbacks. Daimler's consolidation

---

[57] Calculated from figures in the *New York Times*, March 29, 1987, p. III-4; *International Management*, July 1984, p. 7; and company reports.

[58] "Interview with Edzard Reuter," *Financial Times*, April 23, 1990; Daimler-Benz Annual Report for 1989, pp. 15–16, 44, 50, 53.

of Germany's aerospace firms was sweetened by the federal government's support, particularly its protection against currency risks stemming from Airbus's dollar-denominated business. Government officials
believed that greater size would give the firm enhanced clout in the
international arena, and indeed, shortly after acquiring MBB, Daimler
opened talks on aerospace cooperation with Mitsubishi and Pratt &
Whitney.[59]

Krauss-Maffei followed a different strategy. Prime contractor for
antiaircraft guns and the Leopard tank, the company was heavily dependent on military orders through the early 1980s. As a major tank
order from the Netherlands expired in 1986, Krauss-Maffei found itself
at a turning point. Management did not believe that compensating
revenues could be found in nondefense markets and sought increased
latitude for export sales, especially in markets where other NATO defense companies were free to operate. More recently, Krauss-Maffei
has approached other European tank producers to explore reciprocal
subcontracting relationships that would smooth procurement cycles.
Instead of making new acquisitions and seeking greater scale, like
Daimler, Krauss-Maffei devolved responsibility to its operating units
and explicitly asked them to become more flexible. The company won
new orders for locomotives from Spain as well as Germany and bol-
stered its sales of steel castings for power generators and turbines. A
broad base of skilled production workers enhanced the adjustment
capacities of the divisions for programmable automation technologies
such as machine-tool controllers, process engineering systems, and a
line of injection-molding machines and plastics products. Military revenues declined from 77 percent of sales in 1985 to 53 percent in 1988.[60]
Although it may have happened as much by constraint as by choice,
Krauss-Maffei positioned itself surprisingly well for the possibility that
basic political changes might cause dramatic reductions in orders for
its core military products.

Through their different strategies, both Daimler and Krauss-Maffei
suggest that German firms are better placed than their French and
Swedish counterparts to cope with declining defense expenditures.
German firms have benefited from macroeconomic and regulatory sta-

---

[59] "Daimler-Benz Seeks Guarantee on Jobs If Bonn Abandons EFA," *Financial Times*,
April 23, 1990; "Daimler-Benz's Defense Investments Get Tripped Up," *The Wall Street Jour-
nal*, April 3, 1990; and "Daimler-Benz and Mitsubishi Negotiating Cooperative Plan," *New
York Times*, March 7, 1990.

[60] Krauss-Maffei Annual Reports, 1985, pp. 9, 14, 29, and 1988, pp. 9–10.

bility, a highly qualified labor force, and a balanced mix of small and large firms. Oriented strongly toward commercial markets, German firms have largely avoided the deep dependencies on state aid that typify French defense contractors. Embedded within long-standing industrial and banking networks, German firms tend to have stronger finances than many Swedish companies. Effective technology management accounts for part of the successful German record, but it is reinforced by a policy environment and an industrial structure that are essential elements of a robust manufacturing sector. Integration of the former East German territories may bring unexpected changes, but Germany's manufacturing sectors are likely to remain more competitive than their primary rivals within Europe.

JAPAN

Like the Federal Republic of Germany, Japan emerged from World War II in a position of imposed military dependence on the United States. It, too, had constitutional provisions that limited the military to a defensive-force posture. The war produced widespread suspicion of the military's role in Japanese society, particularly in the realm of scientific research. Political constraints at home and abroad therefore encouraged the Japanese state to adopt a low profile in military research and development.

The parallel with Germany does not, however, extend to economic policy. Whereas the Federal Republic adopted a regulatory posture, the Japanese state has played a more activist role. According to many analysts, state-promoted commercial development has become a surrogate for military power as a means of establishing Japan's role in world affairs. While the precise balance of business-state relations is energetically debated, most observers would agree that public officials have exerted considerable influence over the practices and strategies of major Japanese firms.

Despite the American security umbrella, it was far from inevitable that defense production would occupy a subordinate place in Japan's industrial structure. Only after considerable debate in the mid-1950s did Japan's Ministry of Finance and city banks repulse efforts to give defense firms a privileged place in the country's programs for economic reconstruction. The strategy of "indigenization" that had earlier been applied to military technology was directed toward commercial technology. Long-practiced policy capabilities for identifying and assimilating technologies from other countries were therefore transferred from military to commercial objectives.[61]

In the postwar period, defense production has been concentrated within subsidiaries of the major machinery and electronics companies. Typically, one of the main defense suppliers—such as Mitsubishi or Kawasaki—would coordinate a large-scale weapons project using technology licensed from a U.S. source. In aerospace, where most production has been for the military, the Japanese prime contractor would invariably share fabrication with the two or three unsuccessful bidders.[62] This pattern of coordinated development and reciprocal subcontracting began as an adaptation to the constraints of technological inferiority. In recent years, however, Japanese firms have shown improving capabilities in defense-related technologies as they earlier had in commercial products.[63]

Administrative Resources

The role of government institutions in Japan's economic success remains much debated among specialists. The question is sometimes framed in terms of industrial policy versus market forces. The proponents of industrial-policy explanations focus on the state's ability—especially through the Ministry of International Trade and Industry (MITI)—to promote growth by allocating resources to the most promising sectors while protecting Japan's domestic market from import competition. The proponents of market-based explanations focus instead on macroeconomic factors such as savings, investment, education, and low labor costs that help firms to grow.[64]

---

[61] These debates are analyzed in Richard J. Samuels, "Reinventing Security: Japan Since Meiji," in Ethan Kapstein and Raymond Vernon, eds., *Searching for Security in a Global Economy, Daedalus* (Fall 1991), pp. 47–68.

[62] Richard J. Samuels and Benjamin C. Whipple, "Defense Production and Industrial Development: The Case of Japanese Aircraft," in Chalmers Johnson, Laura d'Andrea Tyson, and John Zysman, eds., *Politics and Productivity: The Real Story of Why Japan Works* (Cambridge, MA: Ballinger, 1989), p. 289.

[63] For examples, see Steven K. Vogel, *Japanese High Technology, Politics, and Power* (Berkeley: BRIE Research Paper No. 2, March 1989), pp. 49, 57; and David Sanger, "Rousing a Sleeping Industrial Giant," *New York Times*, May 20, 1990.

[64] The standard argument for state-guided development is Chalmers Johnson, *MITI and the Japanese Miracle* (Stanford: Stanford University Press, 1982). A range of market-oriented explanations is presented in Hugh Patrick and Henry Rosovsky, eds., *Asia's New Giant* (Washington, DC: Brookings Institution, 1976) and extended to high-tech sectors in Hugh Patrick, ed., *Japan's High Technology Industries: Lessons and Limitations of Industrial Policy* (Seattle: University of Washington Press, 1987).

However, this distinction is considered too simple by many scholars. Several intermediate positions provide more nuanced explanations for the Japanese state's ability to promote growth by preserving rather than undercutting market incentives.[65] For analyzing technology policy, these intermediate views are especially useful. They suggest that the concept of the market is viewed differently in Japan than in the United States. Rather than a mechanism for allocating resources efficiently, the market is understood as a social institution for promoting collective growth.

This approach has helped the Japanese state to encourage industrial research and development with great effect. Although public agencies have played an active role, they have done so through facilitation and incentive rather than brute control over resources. Whether by promoting generic technologies or choosing particular industries for support, Japanese policies seek to shape market forces rather than to overturn them.

Despite Japan's reputation for "proactive" industrial policy, the central government provides only 21 percent of the country's R&D outlays—a smaller proportion than in the United States, France, Sweden, or Germany. Moreover, coordination of R&D policy occurs more at the project and the regional level than at the national level. Public funding for R&D has three main outlets: the Ministry of Education (48 percent); the Science and Technology Agency (25 percent); and MITI (13 percent). Smaller fractions of the R&D budget are disbursed by the Japanese Defense Agency (5 percent), the Ministry of Agriculture (4 percent), and the Ministry of Posts and Telecommunications (2 percent).[66]

While overall responsibility for policy coordination is nominally assigned to the Science and Technology Agency (STA), which is

---

[65]The dichotomy between state and market has been critiqued on various grounds. Ronald Dore stresses Japan's distinctive subcontracting practices in "Goodwill and the Spirit of Market Capitalism," a chapter in Dore's *Taking Japan Seriously: A Confucian Perspective on Leading Economic Issues* (Stanford, CA: Stanford University Press, 1987), while Richard J. Samuels emphasizes the "politics of reciprocal consent" in *The Business of the Japanese State: Energy Markets in Comparative and Historical Perspective* (Ithaca: Cornell University Press, 1987). For a more market-oriented argument that acknowledges the key role of MITI's policies before the 1970s, see Kozo Yamamura, *Policy and Trade Issues of the Japanese Economy* (Seattle: University of Washington Press, 1982). For state-centered explanations that acknowledge MITI's ability to incorporate market incentives within its overall policy strategies, see Marie Anchordoguy, "Mastering the Market: Japanese Government Targeting of the Computer Industry," *International Organization*, vol. 42, no. 3 (Summer 1988), pp. 509–543; and Chalmers Johnson, "How to Think about Economic Competition from Japan," *Journal of Japanese Studies*, vol. 13, no. 2 (1987), pp. 415–427.

[66]These figures from the 1988 budget are reported in Vogel, *Japanese High Technology*.

attached to the office of the prime minister, STA has never gained enough control to coordinate comprehensively. Japanese bureaucrats vie continually for jurisdiction over funding. Thus all moneys for university research remain "sacred" to the Ministry of Education (MOE) and beyond STA's accepted brief. STA is primarily responsible for nuclear power and space and has some responsibility for ocean development and biotechnology. Defense activities remain beyond STA's reach.[67]

Although MITI's share of public R&D disbursements is only 13 percent, the ministry has many ways of influencing industry's research agenda. First, MITI's role in formulating trade and industrial policies means that it draws from the best qualified public servants and commands the attention of Japanese industrialists. Second, through its Agency for Industrial Science and Technology (AIST), MITI operates 16 of its own research institutes, of which the Electrotechnical Laboratory (ETL) is perhaps the best known. Third, since the early 1960s, MITI has promoted cooperative research among industrial firms by providing funding (typically about half), by creating a framework for distributing risk and sharing results, and often by providing administrative services to the joint project.[68]

MITI represents Japan in international trade negotiations, and has often reinforced its technology agenda by protecting the domestic market for new technologies. The combination of generic technology development backed by protection of infant industries has been conspicuous in semiconductors and computers. Even in sectors where MITI officials could not dictate industry's structure or guide its technological agenda (such as autos), restrictions on foreign access to the domestic market were a decisive factor in support of industrial development.[69]

---

[67] Office of Technology Assessment, *Holding the Edge: Maintaining the Defense Technology Base*, vol. 2: Appendices (Washington, DC: U.S. GPO, January 1990), Appendix H, p. 177; Interview with Fumio Kodama, Cambridge, MA, December 9, 1989.

[68] For MITI's policies, see, for example, Vogel, *Japanese High Technology*, p. 18; Henry Ergas, "Does Technology Policy Matter?" pp. 221ff.; George R. Heaton, "The Truth About Japan's Cooperative R&D," *Issues in Science and Technology*, vol. 4, no. 3 (Spring 1988), pp. 32ff.; and Glen Fong, "Follower at the Frontier: International Competition and Collaborative Research in Japan," paper delivered at the American Political Science Convention, San Francisco, August 30–September 2, 1990.

[69] For semiconductors, see Michael Borrus, *Competing for Control: America's Stake in Microelectronics* (Cambridge, MA: Ballinger, 1988); for computers, Anchordoguy, "Mastering the Market"; for automobiles, Michael Cusumano, *The Japanese Automobile Industry: Technology and Management at Nissan and Toyota* (Cambridge, MA: Harvard University Press, 1985), pp. 23–24.

Some writers argue that centralized technology-promotion efforts have been less important than Japan's extensive network of regional laboratories and local technology assistance programs. The past few years have seen a proliferation of science and technology efforts, such as Tsukuba Technology City and the new Kansai Science City. Beyond these celebrated cases, each of Japan's 47 prefectures has at least one laboratory, jointly funded by central and prefectural authorities, but with a research agenda largely determined by local industry. Local chambers of commerce (*shokokai*) support such efforts by allocating capital to small business for retraining and new ventures. Such financing comes partly from the central government, but the *shokokai* remain pointedly independent of central as well as prefectural authorities.[70]

Within these elaborate arrangements for industrial promotion, the Japan Defense Agency (JDA) has until recently played a minor role. Owing to recriminations over the military activities of the scientific community during World War II, a new agency, the STA, was created to coordinate R&D policy, while the military was relegated to marginal status. Even in the most recent years, the JDA has accounted for less than 5 percent of government R&D outlays. These spending levels also imply a very low level of research-intensity in Japan's defense effort. During the 1980s, the ratio of R&D expenditures to equipment expenditures has ranged between 4.5 percent and 7.5 percent, far lower than the corresponding ratios for France, Sweden, or West Germany.[71]

Like MITI, however, the JDA has learned to use its limited resources with maximum effect. JDA conducts research and testing through its Technical Research and Development Institute (TRDI), which operates five research centers.[72] TRDI has secured a growing share of the defense budget through the 1980s—from roughly 1.1 percent in 1983 to 2.1 percent in 1988.[73] More important than absolute resource levels is JDA's ability to mobilize the capabilities of Japanese industry. By subsidizing research in high-risk areas, JDA encourages its suppliers to com-

[70] For regional technology policies in Japan, Ergas, "Does Technology Policy Matter?" pp. 217–218; and David Friedman's detailed account of one *shokokai* in *The Misunderstood Miracle: Industrial Development and Political Change in Japan* (Ithaca: Cornell University Press, 1988), pp. 177–200.

[71] Calculated from figures for 1983 to 1988, cited in Vogel, *Japanese High Technology*, p. 50. For similar figures, see also *Aviation Week and Space Technology*, May 1, 1990, p. 59.

[72] Reinhard Drifte, *Arms Production in Japan: The Military Applications of Civilian Technology* (Boulder: Westview Press, 1986), p.35.

[73] Vogel, *Japanese High Technology*, p. 50. Similar figures are given in Drifte, *Arms Production in Japan*, p. 34.

mit their own funds. The JDA bolsters its signals in two ways. First, coordinating closely with MITI, JDA has often stepped into the breach just as MITI's funding declined. In the late 1970s, for example, when a publicly subsidized consortium to produce the commercial YS-11 aircraft was dismantled, MITI encouraged JDA to increase purchases of military aircraft as a way of providing ongoing support for the aerospace industry. MITI's influence within JDA is reinforced because the director of JDA's procurement bureau is traditionally recruited from MITI's aircraft and ordnance division.[74] Second, TRDI's research programs are closely tied to JDA's purchasing plans. Thus procurement contracts can powerfully reinforce TRDI's ability to guide industry's research agenda.

Seeking to draw on the commercial technology base, Japan's policies have forced an explicit questioning of spinoff mechanisms. Japanese policymakers increasingly call for technologies to be transferred from commercial to military (spin-on) rather than the reverse.[75] More recently, they have begun to speak of technology "fusion" to describe innovation through mutual cross-investment by one industry in technologies developed by another. Though the reality may not yet match the rhetoric, the terminology itself suggests that Japanese policymakers think of technology transfer as a multidirectional process.[76]

Industrial Profile

Japan's major defense suppliers are a stable of extremely large, horizontally diversified firms that compete intensely in both domestic and export markets. Their managerial focus on process innovation is well documented for mass-production, low-margin businesses such as consumer electronics, but also appears crucial for firms like Komatsu, whose heavy earth-moving equipment bears more resemblance to defense products.[77]

The JDA's influence over its suppliers is enhanced by competitive

---

[74] Samuels and Whipple, "Defense Production and Industrial Development: The Case of Japanese Aircraft," pp. 277, 282, n. 14; Michael W. Chinworth, "Strategic Technology Management in Japan: Commercial-Military Comparisons" (Cambridge, MA: MIT Japan Program, Working Paper 89-07, 1989).

[75] For elaboration of this point, see Drifte, *Arms Production in Japan*; Samuels and Whipple, "Defense Production and Industrial Development; the Case of Japanese Aircraft"; Vogel, *Japanese High Technology*; and Samuels, "Reinventing Security: Japan Since Meiji."

[76] Interview with Fumio Kodama, Cambridge, MA, December 9, 1989. See also the special supplement on Japanese technology in *The Economist*, December 2, 1989, p. 5.

[77] Christopher A. Bartlett and U. Srinivasa Rangan, "Komatsu Limited," Harvard Business School Case 9-385-277, 1985.

pressures that oblige firms to establish an early presence in new business lines. These firms have long outgrown their earlier dependent relationships with the Japanese state. Yet, because defense projects effectively finance learning in high-risk areas that could provide important competitive advantages, even the biggest firms have been reluctant to opt out of defense consortia and coproduction agreements.[78]

The largest 10 contractors, many of them active in a wide range of weapons systems, supply more than two-thirds of the country's defense equipment.[79] Given Japan's geopolitical position, air and naval forces have the highest priority. For airframes, the leading contractor is Mitsubishi Heavy Industries (MHI), which builds the F-15 under license and does subcontracting for Boeing and others in its Nagoya air works. The other airframe builders, Kawasaki Heavy Industries (KHI) and Fuji Heavy Industries (FHI), share in licensed production as well as civilian subcontracting. MHI, KHI, and the fourth major aerospace contractor, Ishikawajima-Harimi Heavy Industries (IHI), build jet engines under license. Mitsubishi Electric Company and Fujitsu supply avionics, while Nippon Electric Corporation (NEC) specializes in air defense systems. Toshiba, MHI, and KHI produce American-designed missiles under license while also developing their own technologies. KHI, FHI, and MHI build helicopters under license.[80]

In shipbuilding, KHI and MHI are the leading submarine producers, while IHI, Hitachi Shipbuilding, and Mitsubishi Shipbuilding construct surface ships. MHI is the leading tank producer, while Komatsu, Hitachi, and Japan Steel Works manufacture armored vehicles and artillery. More than two dozen firms supply munitions and light arms.[81]

While these firms compete strenuously, their defense capacity is limited by Japan's procurement budget and a ban on arms exports. Although dual-use technologies can be transferred to the United States under a 1983 agreement, Japanese defense suppliers are otherwise prohibited from exporting components for military end products.[82] Al-

[78] OTA, *Holding the Edge,* Appendix H, p. 178; and Chinworth, "Strategic Technology Management in Japan."

[79] "Spending Up in the Land of the Rising Gun," *Far Eastern Economic Review,* October 13, 1988, pp. 64–68; *Aviation Week and Space Technology,* May 1, 1989, p. 59.

[80] Drifte, *Arms Production in Japan,* p. 60. The U.S. licensors include Boeing, Bell, and Sikorsky. Although Japan has not developed space technologies for military purposes, Mitsubishi Electric, Toshiba, and NEC are all active in constructing commercial satellites, while Nissan has established a subsidiary for rocket propulsion.

[81] Drifte, *Arms Production in Japan;* Vogel, *Japanese High Technology,* pp. 53–62; *Aviation Week and Space Technology,* January 8, 1990, p. 29.

[82] For the significance of this agreement, see Samuels, "Reinventing Security: Japan Since Meiji."

though this export ban has not been seriously challenged, there is little doubt that many defense suppliers would like to see it lifted.[83]

Even by itself, however, domestic demand has created considerable momentum among the Japanese arms producers. MHI, Japan's largest defense supplier, has seen its military sales grow from 7 percent to 15 percent of revenues in 10 years, and company spokesmen would like that figure to increase further. KHI, the country's second-largest defense supplier, saw military sales rise to almost 30 percent of total revenues in the 1980s—partly because the company's shipbuilding division suffered severe decline. Other firms, such as IHI and Nissan, have altered their articles of incorporation to add defense to their major business lines.[84]

Recent Developments

The most significant current project for dual-use management in Japan is unquestionably the FSX "Fighter Support Experimental" aircraft. FSX planning exemplifies the main features of Japanese business-government relations. The project is a prime candidate for technology fusion. Through the main licensing agreement with General Dynamics, MHI will receive manufacturing information and certain design/test data for the F-16 airframe, which after extensive modification is to become the FSX. Under MHI's leadership, Japanese firms plan ambitious new design features including composite wing construction, advanced avionics, onboard phased-array radar, and new armaments.[85]

The program typifies the mixture of collaboration and competition that characterizes much of Japan's high-technology industry. The prime contractor for the FSX is MHI, with extensive participation by KHI and FHI in the airframe, and IHI in the engine. Avionics are assigned to Mitsubishi Electric, NEC, and Fujitsu.

Together, these suppliers have already developed many of the generic technologies to be "spun on" to the FSX. As a major platform for new technologies, the project has become the next logical source

[83] "Spending Up in the Land of the Rising Gun," p. 67.

[84] OTA, *Holding the Edge*, vol. 2, Appendix H, p. 183; "Spending Up in the Land of the Rising Gun," p. 67.

[85] For a nontechnical discussion, see John D. Moteff, "FSX Technology: Its Relative Utility to the United States and Japanese Aerospace Industries," U.S. Congress, Congressional Research Service, Report 89-237 (April 12, 1989). On the political issues in both countries, see Daniel Lev Alexander, "Worst Expectations: The U.S.-Japan FSX Fighter Aircraft Dispute and the Politics of International Weapons Cooperation," Undergraduate Thesis, Harvard University, 1990.

of revenue from technologies on which Japanese firms were already at work. In addition, Japanese firms had sought the opportunity to try large-scale systems integration tasks, where Japan has remained far behind the United States.[86] MHI considered the project sufficiently compelling to devote a large fraction of its investment budget in the late 1980s to refurbishing the Nagoya air works, where much of the development will be done.

The FSX represents a major challenge for MHI and its suppliers. Earlier agreements for coproducing the F-104 and the F-15 entailed comprehensive transfers of manufacturing data from the United States to Japan. The design and development work for modifying major subsystems of the F-16 will be much more complex. There is no guarantee that the FSX will generate direct spinoffs to commercial aerospace, but the project may provide payoffs through other dual-use links. The design and development experience at Nagoya may find subsequent use in MHI's commercial work. In addition, the knowledge of systems integration gained by MHI's workforce through the FSX should have broad, if imperfect, applicability to other projects. Even if the FSX never goes into full-scale production, the program will have demonstrated yet again the determination with which Japanese firms and public agencies seek out opportunities for technological advance, whether for civil or military ends.

WHAT LESSONS?

Each of the four countries compared here relies on distinct institutional mechanisms for pursuing military and civil technology goals. When policymakers seek to guide R&D, they confront widely varying industry structures and competitive configurations. At the level of particular firms, there are organizational differences within as well as across countries that have an important bearing on technology management.

This diversity of policies and outcomes should be a warning to the United States against simplistic efforts to borrow institutions that function well in other societies. There is an important element of asymmetry when these four countries are compared with the United States. None has a defense establishment that approaches that of the United States in size or capability. None of these countries followed the United

---

[86] The importance attributed by MHI to systems integration is discussed in Samuels and Whipple, "Defense Production and Industrial Development: The Case of Japanese Aircraft."

States in pursuing exploratory research across the entire spectrum of potential military technologies. Particularly in the early postwar decades, the U.S. approach helped generate a public reservoir of ideas and technologies, encouraging firms to seek breakthroughs that might confer dominance in commercial markets.[87]

Although these differences prevent controlled comparisons, we can still draw useful insights. Dual use entails a range of mechanisms through which know-how is shared among companies or between divisions within a single company. Such transfers of knowledge occur at the level of specific firms, but they are facilitated or inhibited by external factors.

In France, political institutions and industrial structures combine to inhibit the sharing of technologies within as well as among firms. Rather than promoting cooperative research among firms, French officials have usually funded applied research in dedicated facilities designed to focus resources on centrally defined objectives. In the defense sector, these practices channel research directly from public R&D centers to the prime contractors. Since each of the major suppliers dominates a particular segment, they have little incentive to apply their know-how to other segments where they might provoke retaliation from better established competitors. The concentrated structure of the French defense sector thereby encourages an arsenal strategy with contractors tailoring their capacities to a limited range of systems products.

Although the arsenal strategy worked well in the early postwar decades, the more demanding competitive conditions of recent decades have exposed its limitations. A few firms, exemplified by Aerospatiale, have managed to serve military and civil markets from the same operating divisions. A more typical case, however, seems to be Thomson, where military work has been cordoned off in a special division that deals only with the government agencies responsible for armaments. French policies have reinforced this compartmentalization of defense-related expertise in a limited number of state-oriented companies. Instead of encouraging the diffusion of technological capabilities throughout the economy, the French state has put more emphasis on helping its suppliers gain export orders so that the arsenal strategy can be maintained.

Although Sweden shares with France the objective of military self-sufficiency, its institutional and industrial arrangements encourage

---

[87] Robert Gilpin, *The Political Economy of International Relations* (Princeton: Princeton University Press, 1987), p. 353.

companies to transfer know-how among civil and military divisions. First, the Swedish state does not rely heavily on dedicated R&D facilities like France's *grands organisms*. Instead, Swedish firms must scan outside labs and international markets in search of knowledge to combine with in-house capacities. Second, political pressure for balanced growth means that Sweden's labor-market policies and regional development efforts encourage firms to cultivate multiple competencies so they can respond to shifting markets without sudden reductions in employment or output. High levels of workforce qualification contribute to flexible production strategies in high-value-added product lines. Such strategies allow dual-use gains through shared human resources and infrastructure as illustrated by Saab-Scania's Combitech companies. Third, stringent export restrictions block escape routes for Swedish defense suppliers that cannot adjust or diversify. Except for Bofors, Swedish firms have been dissuaded from concentrating heavily on defense or projecting an arsenal strategy into international markets.

Germany displays political institutions and industrial structures that even more clearly favor intrafirm and especially interfirm transfers of knowledge. The Technology Ministry's civil R&D programs increasingly require firms to collaborate on generic technologies. Much of the defense basic research is performed by Fraunhofer Institutes, which also have independent portfolios of industrial contract R&D. Like Sweden, Germany possesses a highly trained workforce, which is well equipped to apply generic technologies to a range of high-value-added commercial product lines. Germany's mix of large and small firms reinforces opportunities for interfirm technology transfer. There is considerable evidence that the small and medium-size firms comprise a crucial reservoir of specialized know-how that the larger systems companies can tap. One of the central questions is whether the defense sector—kept small in the earlier postwar decades—will push German industry toward a more vertically integrated structure. If so, the still-unanswered question is whether company executives will be able to promote intrafirm knowledge transfers as effectively as public policy and industry structure have promoted interfirm sharing in the past.

Japan provides the best example among the four of both interfirm and intrafirm transfers of know-how. A good deal of civil as well as military research is conducted through mission-oriented projects, as in France. Yet these projects seem to generate more knowledge-sharing. Since the projects typically involve a number of competing firms, the participants themselves steer the R&D agenda away from any single firm's commercial needs and toward more generic know-how. Because the magnitude of Japan's effort in civil technologies has until recently dwarfed its effort in defense systems, it seems that Japanese firms have been able to limit barriers to transfers of know-how across the civil-military boundary. Japanese companies work hard and effectively at technology management—not only in R&D and project-level engineering, but also, for example, through training of production workers. The importance of innovation in Japan's postwar economic development means that firms have had to adapt to an environment of continual technological change.

These cases show that public policies shape the organizational processes on which dual use depends. Public policy cannot dictate how firms will manage R&D or how they will apply it for commercial markets. But policymakers have alternative ways of channelling public funds and know-how to firms, giving them considerable influence over the types of markets to which firms might rationally apply R&D results.

At the same time, national arrangements are not isolated models. Policies for defense R&D and weapons acquisition are defined within each country's larger national security orientation. Thus, in France, the importance of military self-sufficiency opens opportunities for firms to specialize in military markets; as a result, dual use has typically been defined as a problem in spinning off defense technologies to civil markets. Sweden's more restricted objective of armed neutrality has placed the emphasis on identifying and adapting technologies, often from foreign sources. In Germany and Japan, by contrast, U.S. security guarantees have freed resources for commercial technology development. For many firms in these two countries, dual use has meant integrating defense into company strategies aimed primarily at commercial markets.

International factors impinge on neat comparisons in another respect. Cross-border collaborations in defense R&D and production have significantly eroded national procurement boundaries. As a result, national efforts to promote dual use are increasingly linked to international arrangements for military development.[88] The FSX project, for example, provoked vociferous debate in both Japan and the United States. The European Fighter Aircraft (EFA) involves Britain, Germany, Spain, and Italy, while France has tried to go it alone with the Rafale jet fighter. In upstream defense research, several European countries are elaborating a framework called EUCLID (European Cooperative

---

[88] On this topic, see Todd A. Watkins, "Dual Use Technologies: Diffusion Structures and International Policy Convergence," John F. Kennedy School of Government, STPP Discussion Paper 89-04, 1989; and Office of Technology Assessment, *Arming Our Allies: Cooperation and Competition in Defense Technology* (Washington, DC: U.S. GPO, May 1990).

Long-term Initiative in Defense) to parallel the EUREKA program in civil technologies.[89] Such programs raise the possibility that organizational capacities for international collaboration may themselves become an important arena for competition in dual use, as firms devote more resources to the complex tasks of external scanning and monitoring, managing supplier networks, and assimilating foreign-sourced technologies.

Despite the growing import of such international initiatives, the most important factors in dual use are still determined at the nation-state level. Cross-national differences in the institutional landscape for R&D create very different opportunities for firms to appropriate publicly funded know-how. Differences in workforce characteristics—directly influenced by education and labor-market policies—can make it much easier (or harder) for firms to identify and exploit opportunities for dual use. The durability of subcontracting relationships and the quality of component markets can also influence dual use by affecting the general level of technological capability within national borders.

State-led programs to promote technology development comprise only one of many policies that affect dual use. The spinoff model that justified such programs was never the only mechanism for dual use. As the United States has been joined by a group of serious competitors in the world economy, the inadequacy of the spinoff model has become much clearer. The four countries compared in this chapter illustrate a range of alternatives to the U.S. policy of simultaneously supporting all defense-related technologies. No country has the wherewithal or omniscience to realize such a strategy today. While public agencies are likely to remain the key initiators for large-scale technology platforms, their ability to manage such projects will be influenced by a range of other policies aimed at research, education, labor markets, and even regional development. Government's potential as a technology generator will increasingly be conditioned by its capacity to act as a technology facilitator.

---

[89] On EUCLID and other pan-European efforts at defense collaboration, see Martyn Bittleston, "Cooperation or Competition? Defence Procurement Options for the 1990s," Adelphi Paper No. 250 (London: IISS, Spring 1990); and Andrew Moravcsik, "The European Armaments Industry at the Crossroads," *Survival*, vol. 32, no. 1 (January–February 1990), pp. 65–86. On EUREKA, see John Peterson, "Technology Policy in Europe: Explaining the Framework Programme and Eureka in Theory and Practice," *Journal of Common Market Studies*, vol. 29, no. 3 (March 1991), pp. 269–290.

PART THREE
Facets of Dual Use

The chapters in Part III cover three technology families: microelectronics (Chapter 8), computer software (Chapter 9), and manufacturing (Chapter 10). Each illustrates a different facet of dual use (see Table III-1). The semiconductor industry has been hard hit by Japanese competition, making it a center of policy attention. Lacking other mechanisms, the federal government turned to the Department of Defense to support the industry's technology base. Software—even more than microelectronics the paradigm case of knowledge-based high technology—is, like microelectronics, critical for military systems of all types. U.S. software suppliers have so far remained unchallenged competitively. Indeed, the United States seems complacent over its lead in software, although the outlines of a future challenge from Japan are already plain. Manufacturing represents the past as well as the necessary future. Postindustrial economies must make hardware products too, and much of their manufacture is by methods that have changed relatively little since the industrial revolution. Now science and cheap computing power promise radical improvements in manufacturing processes, with payoffs for both DoD and civilian industry.

Compared with the abstract nature of computer software, microelectronics technologies must seem straightforward. Software development is governed by few rules, remains more art than science. In semiconductors, the path of technical progress has been the same for years: ever smaller feature sizes, ever more circuit elements per chip. The problems lie in mastering the tools for fabricating these features, some of them submicroscopic, and in designing chips that can exploit the astonishing capabilities now possible. If manufacturing processes, however specialized, are critical for semiconductors, they are irrelevant for software. Here the challenge lies in improving productivity in program development, a slow and uncertain process, and in managing complex projects requiring the coordinated efforts of hundreds of professionals.

As industries, microelectronics and software differ almost as dramatically as in their technologies. There are a few hundred producers in the U.S. semiconductor industry, including the captive divisions of larger, diversified organizations (some of them defense contractors), only a few dozen of much significance. At least 5,000 American companies develop software. They exhibit a bewildering variety of products

TABLE III-1:
Microelectronics, Software, and Manufacturing Compared

	Industrial Structure	Priorities	Dual-Use Potential	Policy Needs
Microelectronics (Chapter 8)	Mature. Two primary sets of producing firms, independent merchant manufacturers and captive divisions of larger firms. Suppliers of semiconductor or manufacturing equipment also important.	Reestablishing parity with Japan in processing.	Large and likely to grow, driven by DoD needs to make greater use of commercially available chips.	Agreement on legitimate mechanisms for supporting a strategic technology and industry.
Computer Software (Chapter 9)	Large, fragmented, in flux. Many independent firms, along with hardware manufacturers and companies that develop software for internal use.	Improving productivity in software generation; creating sound technological foundations and better tools and methods; managing complex software projects.	Large but underappreciated; awaits recognition by policymakers. Technical and managerial changes necessary to encourage dual use.	Long-term support for better tools, methods, and management practices; DoD emphasis on modular, reusable, dual-use software; human resource development—e.g., mid-career training, university programs in computer engineering.
Manufacturing (Chapter 10)	Very large and diverse.	Improving productivity through better organizational and management practices; exploiting U.S. advantages in science on the shop floor.	Large. Substantial dual use exists in some cases (e.g., aerospace).	Strengthened infrastructure; generic support for process R&D and technology diffusion.

252

and strategies. In contrast to microelectronics, the industry has hardly begun to mature but is already many times larger in terms of revenues. While hardware and software depend on one another, U.S. policymakers have arguably paid too much attention to hardware relative to software.

If the software industry is large and diverse, the manufacturing sector of the U.S. economy is even more so, including some 400,000 companies making everything from shoes to 747s. Chapter 10 focuses on batch manufacturing technologies, typified by firms that produce to order for both civil and defense customers. Manufacturing is mature but hardly static; the products of the microelectronics and software industries have transformed production processes of all types, beginning with numerically controlled machine tools developed in the 1950s under Air Force sponsorship. Defense spending has spawned many related methods and techniques—e.g., computer-aided design and engineering. Ironically, DoD has not been able to take full advantage of these technologies, largely because acquisition policies have not put a premium on cost control and flexibility.

Each of these chapters concludes with an analysis of federal government policies and recommends changes. Each illustrates a different mix of advantages and limitations in dual use.

8

Microelectronics: Two Industries, One Technology

Microelectronics illustrates the dilemmas of dual-use technology policy more clearly than any other case. In effect there have been two microelectronics industries, one military and one civilian, even though much of the technology is common to both. Spokespersons for industry and government have hammered home the notion that semiconductors are "strategic" for national security and civilian competitiveness alike. Despite more than a decade of competitive slippage, however, there have been only a few steps toward consensus on what to do.

Since the beginning of the 1980s, the litany of competitive problems in the U.S. semiconductor industry has been repeated many times over, with seemingly endless variations on a few common themes. In its essentials, the story is a simple one: a once-dominant set of firms, many of them based in California's Silicon Valley, has found itself severely buffeted not only by Japanese competition but by the ups and downs of highly cyclical demand. Eventually, business and government sought a joint response to one part of the industry's dilemma through the R&D consortium Sematech, established in 1987. Funded roughly

Portions of this chapter draw from John A. Alic, Martha Caldwell Harris, and Robert R. Miller, "Microelectronics: Technological and Structural Change," prepared for the conference on *Microelectronics in Transition: Industrial Transformation and Social Change,* University of California, Santa Cruz, May 12–15, 1983; and John A. Alic, "From Weakness or Strength: U.S. Firms and U.S. Policies in a Global Economy," in Lynn Krieger Mytelka, ed., *Strategic Partnerships: States, Firms and International Competition* (London: Pinter, 1991), pp. 149–166.

half and half by the Department of Defense and 14 member firms, Sematech represents a collective effort to regain lost ground in processing know-how.

In two ways, the creation of Sematech—along with the confused debate over high-definition television (HDTV) and high-definition systems during the late 1980s—can be seen as a turning point in U.S. policies. First, an initiative based on technology was put forward, rather than the more comprehensive proposals for sectoral support common during earlier discussions of industrial policy (see Chapter 2). Any linkages with ongoing trade negotiations between the U.S. and Japanese governments remained implicit. Second, Sematech represents an overt subsidy for commercial industry. Many of the advocates of Sematech and of federal support for high-definition systems have seen them as models for the future; success could lead to imitation, failure to a reaction against such policies. The question of follow-ons to Sematech promises to write another chapter in the story of U.S. attempts to reach consensus on policies for bolstering competitiveness.

Regardless of how the Sematech story plays out, policymakers need to absorb the lessons of an earlier undertaking—DoD's Very High Speed Integrated Circuit program, which ran from 1980 to 1990. VHSIC R&D focused on technical problems, notably fine-line lithography, not unlike Sematech. But DoD originally had more ambitious goals for this highly visible effort, managed directly from the Office of the Secretary of Defense (OSD). Pentagon planners sought to speed the incorporation of state-of-the-art integrated circuits in military systems by forging teams that would link major defense contractors with the independent and innovative merchant chip manufacturers. After 10 years and more than $900 million, few if any VHSIC chips had found their way into fielded equipment. Military and commercial microelectronics remained largely isolated from one another, businesses pursued to considerable extent by different sets of companies.

VHSIC had relatively few successes: an R&D program could not change the acquisition practices that delayed the applications of advanced ICs in military systems. Indeed, in the end, VHSIC was captured by the parties whose practices OSD set out to change—the military services and the contractors that shared with them responsibility for design and development decisions on major systems.

When it comes to processing technology, Sematech, with its relatively small budget, could prove too little, too late. In any case, it can do little to overcome the structural handicaps that underlie competitive decline: both the merchant industry and the semiconductor equipment industry—the latter populated by hundreds of small firms supplying production equipment, materials, and supplies—show every sign of remaining fragmented and financially weak.

This chapter does not repeat the full story of U.S.-Japan competition in microelectronics. That story has been told in many places, from many perspectives. Trade, more than technology, has been central, and discussion of such matters as the U.S.-Japan Semiconductor Arrangement would take us too far from the subject of this book. We do summarize the technological dimensions of the competition, beginning with the driving force provided by federal procurements in the early years of the industry, and going on to examine relationships among the merchant manufacturers of semiconductors (firms that sell most of their production on the open market), other chipmakers (including defense contractors), and the suppliers of specialized equipment for making semiconductors.

The policy discussion that occupies the rest of the chapter likewise focuses on technology. Both VHSIC and Sematech have generated controversy. The programs were difficult to design, and consensus was hard to achieve; once the programs were under way, management proved no easier. The chapter suggests that effective dual-use policies will require innovations beyond those yet attempted in the United States. Microelectronics may be the obvious example of dual-use technology, but it does not follow that policymaking will be straightforward.

THE MILITARY MARKET FOR SEMICONDUCTORS: EARLY IMPACT AND DECLINE

During its formative years, the American semiconductor industry depended heavily on sales for defense and space applications. Government procurement was more important than government R&D (Box 8-A). Demand created by the Minuteman II missile and the Apollo project pushed American firms rapidly down design and production learning curves. Indirect demand, through computer purchases by government agencies and defense contractors, also had considerable impact: discrete transistors and then ICs went into machines bought by defense agencies and national laboratories, and by firms in the thriving aerospace industry. Federal R&D contributed primarily through creation of a human resource base—engineers and scientists trained with the help of research contracts from the Pentagon and other agencies, and technical people in industry who got the chance to keep up with the latest developments in solid-state devices and processes.

BOX 8-A: HOW DEFENSE NEEDS INFLUENCED DEVELOPMENT OF THE INTEGRATED CIRCUIT

Conceptually, the idea behind the integrated circuit—putting several transistors on a single "chip" of semiconducting material—was straightforward; it had been publicly discussed as early as 1952. The trick lay in reducing this idea to practice, given the primitive state of solid-state technology. During the mid-1950s, the U.S. Air Force sought denser, more reliable solid-state electronic circuitry from industry under the rubric of an R&D program on three-dimensional "molecular electronics."[1] Westinghouse carried out most of the initial exploration of the ill-defined molecular electronics concept under Air Force contracts. But neither Westinghouse nor other firms were able to meet DoD's requirements. Instead, Jack Kilby, working with internal funds at Texas Instruments (and unaware of the molecular electronics effort), built the first functioning IC. At this point, the Air Force began channeling development money to TI. At about the same time, Robert Noyce was working on the planar process at Fairchild, also without federal funds; Fairchild, in fact, declined DoD R&D support for Noyce's work, which quickly became the preferred approach for fabricating ICs.

With the invention of the IC, it was government's willingness to buy the resulting products—originally signaled by the Air Force's search for molecular electronics modules—that spurred the rapid growth of the emerging merchant semiconductor industry. TI got the lion's share of Minuteman contracts, Fairchild most of the orders from NASA for the Apollo guidance project (some 200,000 chips in 1963). Both companies quickly became industry leaders, with Fairchild the early centerpiece of Silicon Valley.

[1] Norman J. Asher and Leland D. Strom, "The Role of the Department of Defense in the Development of Integrated Circuits," IDA Paper P-1271 (Arlington, VA: Institute for Defense Analyses, May 1977).

U.S. industry led the world both technologically and commercially. But this period did not last long. During the 1960s, commercial markets grew much more rapidly than government markets; by the early 1970s, the U.S. lead in technology began to shrink, and during the 1980s, the U.S. share of the world semiconductor market tumbled.

For a decade after the invention of the transistor in 1947 at AT&T's Bell Laboratories, a few well-known electronics and electrical equipment firms carried out most of the development and production of solid-state devices. These firms—including RCA, General Electric, and Westinghouse, as well as AT&T—supplied components for both military and commercial systems. At the end of the 1950s, this pattern broke down. The first ICs came from outsiders—Texas Instruments (TI) and Fairchild. The structure of the industry began changing rapidly.[1]

As ICs came onto the market alongside discrete transistors, demand took off. U.S. semiconductor sales passed $1 billion in 1965. New firms sprang up, founded and staffed by technical talent from the established leaders. The center of gravity shifted from the semiconductor divisions of a few large, diversified corporations to a varying group of small, entrepreneurial firms competing fiercely with one another to bring new products to market. The merchant industry that these firms created itself began to mature in the mid-1970s.

Prior to the outpouring of new firms and new products, semiconductors were part of the defense industry.[2] By the end of the 1960s, consumer and industrial applications of microelectronics were expanding much more rapidly than DoD purchases. Defense sales grew relatively slowly, and quickly dropped below 20 percent of the total (Table 8-1). Since the late 1970s, the military share of IC sales has remained in the range of 10 percent by value (and only 4–5 percent by number of chips), with rises and falls reflecting not only DoD demand cycles but the tendency of merchant firms to solicit defense orders more aggressively when commercial markets slump.[3]

The most successful companies were those able to quickly negotiate the shift from specialized military requirements to the demands of customers in the civil sector—computer firms and telecommunications suppliers that had begun designing their products around ICs rather than discrete transistors. Some of the chipmakers that fared well in the military market had trouble winning commercial orders, where technical requirements were more diverse and changed more quickly. Moreover, nontechnical factors like price and delivery schedules were far more important to civilian customers than to defense contractors.

With progress toward very large-scale integration (VLSI), the investments in R&D and production equipment that were required to keep up grew rapidly. Companies that could not manage to expand

---

[1] Other forces for structural change included the search for low-cost assembly labor; the first U.S.-owned semiconductor plants in low-wage developing countries were established in 1963. *A Report on the U.S. Semiconductor Industry* (Washington, DC: Department of Commerce, September 1979), p. 84.

[2] Until 1964, procurement for military systems and for the American venture into space accounted for more than 90 percent of U.S. (and world) IC production. John E. Tilton, *International Diffusion of Technology: The Case of Semiconductors* (Washington, DC: Brookings Institution, 1971), p. 91. Until the late 1960s, American firms accounted for nearly all the world's production of ICs.

[3] Integrated Circuit Engineering Corporation, *Mid-Term 1988* (Scottsdale, AZ: Integrated Circuit Engineering Corporation, 1988), pp. 2-15–2-18.

TABLE 8-1:
The Military Market for Integrated Circuits

U.S. Integrated Circuit Sales or Production
(million $; % of total)

Year	Military		Commercial		Total
1965	$61	72%	$24	28%	$85
1970	120	21%	450	79%	570
1975	260	15%	1,450	85%	1,710
1980	400	8.5%	4,300	91.5%	4,700
1985	1,100	12%	8,140	88%	9,240
1990	1,400	8%	16,400	92%	17,800

Sources: 1965, 1970: Norman J. Asher and Leland D. Strom, "The Role of the Department of Defense in the Development of Integrated Circuits," IDA Paper P-1271 (Arlington, VA: Institute for Defense Analyses, May 1977), p. 73.

1975: Estimated, based on *A Report on the U.S. Semiconductor Industry* (Washington, DC: Department of Commerce, September 1979), pp. 39, 44.

1980: *An Assessment of the Impact of the Department of Defense Very High Speed Integrated Circuit Program,* National Materials Advisory Board Report NMAB-382 (Washington, DC: National Research Council, January 1982), p. 64.

1985: *Report of the Defense Science Board on Use of Commercial Components in Military Equipment* (Washington, DC: Office of the Under Secretary of Defense for Acquisition, June 1989), p. A-14.

1990: Estimated, based on figures from Dataquest and the Semiconductor Industry Association.

along with the market remained small and specialized; the more technically stable military market, given acquisition schedules that were much longer than commercial product cycles, provided a refuge from the rigors of the civilian marketplace.

In no small part, technological and commercial competition among American merchant producers drove expansion. There was no meaningful foreign competition until the 1970s, but rival U.S. firms sought to anticipate and understand the needs of customers, and be first to satisfy them. Rapid increases in circuit density—the number of transistors on a chip—followed from improvements in design and fabrication technology. Rising production volumes and the well-known learning curve effect forced prices down. New chip designs and falling prices in turn opened additional markets; like computers, consumer electronic products began to be designed around chips.

The pace in military electronics was more relaxed. Defense continued to attract specialists while losing appeal to the merchant semiconductor firms; today, the military portion of the industry includes several companies that do nothing but produce obsolete parts, discontinued by the original manufacturer but still needed by DoD. During the mid-1980s, although DoD and its contractors were spending more than $1 billion annually for ICs, only about 20 percent of this went for "ruggedized" versions of commercial chips.[4] Most of the rest were specialized, DoD-only parts.

INDUSTRY STRUCTURE AND DYNAMICS

Compared with Japan's microelectronics industry, dominated by a small group of large and diversified companies, American capabilities are fragmented. This structural difference lies at the root of U.S. competitive difficulties. The U.S. merchant firms, focusing on the design of innovative chips, never built a research base in microelectronics. Even more important, they have remained financially weak, a consequence of relatively small size, lack of diversification, and heavy risk exposure (e.g., to Japanese competition—risks appreciated in financial markets better than in the management offices of some semiconductor firms). U.S. suppliers of semiconductor manufacturing equipment have suffered even more severely from these problems. Few large and wealthy American firms have been willing and able to compete for open market sales since the early years, when companies like Philco and RCA tried with little success to keep up with nimbler rivals.

Merchants and Captives

With the invention of the integrated circuit, U.S. microelectronics production split into two parts.[5] The merchant firms—typified not only by the chip pioneer TI, but by such well-known names as Intel and Motorola—sell most of their output on the open market, while still getting perhaps 5–10 percent of their microelectronics revenues from DoD and its contractors. Captive producers, in contrast, consume all or nearly all of their semiconductor output internally, using it in end products ranging from automobiles to computers to defense systems. General Motors makes chips for cars and, with the purchase of Hughes,

---

[4] *Report of the Defense Science Board on Use of Commercial Components in Military Equipment* (Washington, DC: Office of the Under Secretary of Defense for Acquisition, June 1989), p. A-14.

[5] Office of Technology Assessment, *International Competitiveness in Electronics* (Washington, DC: U.S. GPO, November 1983), pp. 132–141.

also for military equipment. Hewlett-Packard's business strategy has been built in part around proprietary chips designed and developed internally; the company buys standard devices on the outside. IBM for many years was the world's largest semiconductor manufacturer, as well as the biggest single customer for chips produced by other firms.

During the 1970s and early 1980s, the ranks of U.S. captives grew, as manufacturers of computers and other electronic systems added semiconductor capability for strategic reasons (e.g., proprietary chips, guaranteed sources of supply). Some of these companies later sold their captive operations, disappointed in the contribution to the firm's business. Ford is probably not that unusual in having been in and out of the semiconductor business at least four times, as owner of Philco in the early years of the industry, with several captive operations in its aerospace divisions, and through purchase of a small merchant firm in the 1970s, later sold; currently, Ford Microelectronics undertakes design and development work, along with prototype manufacturing, but the automaker then contracts for IC production from merchant suppliers.

Such DoD contractors as Rockwell, TRW, and Lockheed also make some of their own chips. Often known as "systems houses" because much of their defense business centers on complex systems incorporating subsystems and components from many sources, these companies, though now less important in microelectronics than they once were, continue to operate captive IC design and production facilities so they can offer specialized capabilities to the Pentagon. Some maintain only the capacity for R&D and pilot production (Bendix, General Dynamics, Northrop). Others produce ICs worth several hundred million dollars annually. Companies like Rockwell and Hughes sell specialized chips on the open market, but for these firms microelectronics is a small part of their overall business, and outside sales tend to be a small part of microelectronics production.[6] Diversified systems firms, including IBM, AT&T, and General Electric, keep their military semiconductor R&D and production separate from commercial operations, for reasons described in Chapter 5.

Captive manufacturers, though less visible than merchant firms, continue to make substantial contributions to U.S. technological capabilities. In the early years, AT&T's Bell Laboratories carried out much

[6]Rockwell's specialized line of chip sets for modems, for example, brings in perhaps $50–$100 million annually, accounting for less than half of the company's IC production, and less than 1 percent of the company's total revenues. Integrated Circuit Engineering Corporation, *Mid-Term 1988*, p. 3-11. Boeing is the only major DoD systems contractor without internal semiconductor capacity.

of the basic research that provided a foundation for the industry. AT&T made the knowledge widely available even before its 1956 consent decree with the Justice Department. More recently, with continued DoD support for R&D on gallium arsenide and other compound semiconductors, much of it performed by defense contractors, the first independent suppliers of gallium arsenide ICs were established as spinoffs from these contractors.[7]

The Equipment Industry

As the merchant industry grew, a quite distinct group of equipment suppliers, also small and entrepreneurial, sprang up alongside it. These companies design, develop, and produce lithographic equipment for creating patterns on chips, equipment for automated assembly and testing, and a wide range of specialized products such as vacuum furnaces. Other firms provide materials and supplies such as ultrapure chemicals. Particularly in the early years, a number of the equipment suppliers grew in part by developing customized hardware for IBM and AT&T, which had large captive operations and abundant expertise. Later, after development and refinement, the equipment firms would introduce similar products on the open market, selling to Japan and European, as well as U.S. manufacturers.

The U.S. equipment industry, always fragmented, is currently populated by about 800 companies, mostly small (i.e., with revenues of less than $10 million annually and fewer than 100 employees).[8] Only a handful have managed to reach levels of $100 million or more in annual sales. Those that did—e.g., GCA and Perkin-Elmer—generally make lithographic equipment, which has sold for steadily higher prices as technical complexity rose to meet the demands of VLSI chip production.

Into the 1980s, equipment firms provided a vital part of the foundation for the U.S. semiconductor industry. But they fell on hard times as their primary customers, the merchant producers, faltered under pressure from Japan. American firms held three-quarters of the world market for semiconductor manufacturing equipment in 1980, but only about 40 percent in 1990.[9] Over this period, Japanese equipment sup-

---

[7] Marc H. Brodsky, "Progress in Gallium Arsenide Semiconductors," *Scientific American* (February 1990), pp. 68–75.

[8] Louise Kehoe, "Strapped in for a Bumpy Ride," *Financial Times*, April 27, 1990, p. 22.

[9] Gary L. Guenther, "U.S. Semiconductor Manufacturing Equipment and Materials Industries: Economic Condition since 1980, Prospects for Future Growth, and Policy Options for Strengthening Their Ability to Compete in Global Markets," Congressional Research Service,

pliers increased their share from 18 percent to about 50 percent. Most dramatic by far has been the decline of the two chief suppliers of lithographic equipment, GCA and Perkin-Elmer. With lithography a core technology in microelectronics, and with both GCA and Perkin-Elmer losing money during the mid-1980s and falling behind technologically, fears over the future of microelectronics in the United States went up another notch.[10]

The equipment suppliers suffer even more than chipmakers from the ups and downs of the market. Companies that buy chips, whether for consumer products or capital goods like computers, may double-order when their own businesses are booming to make sure they will not be caught short. If sales drop, they quickly cancel orders. Moreover, companies like IBM with large captive production facilities add to this volatility by entering the merchant market in a big way in periods of peak demand. By the time these swings in demand reach the equipment suppliers, they are highly amplified, making it very difficult to plan for and finance the development of new products.

Interfirm Relationships and Technology Diffusion

Business relations among the three groups of firms—merchant producers, captives, and equipment suppliers—have generally been at arm's length, in the U.S. tradition. In an environment of fierce cost-based competition, chipmakers continually pressed equipment suppliers for lower prices. Sometimes they got them at the expense of hardware that had not been fully debugged, needing a good deal of trial and error on the factory floor before production yields and quality reached acceptable levels.[11] Merchant firms also kept equipment suppliers at a distance for fear that proprietary information would leak out to their competitors.

While many captives spent a good deal on R&D, including basic research funded by defense agencies, the U.S. merchant manufacturers rarely did much research, devoting their efforts to the never-ending

memorandum, September 26, 1989, pp. 4, 8; *1991 U.S. Industrial Outlook* (Washington, DC: Department of Commerce, January 1991), p. 17-10. In 1980, nine of the ten largest equipment suppliers were American firms; by 1988, the three largest were Japanese.

[10] General Signal bought GCA in 1988; Perkin-Elmer sold its semiconductor division two years later. Office of Technology Assessment, *Making Things Better: Competing in Manufacturing* (Washington, DC: U.S. GPO, March 1990), pp. 138–142; John Markoff, "Perkin Unit to Remain U.S. Owned," *New York Times*, May 16, 1990, p. D1.

[11] *A Competitive Assessment of the U.S. Semiconductor Manufacturing Equipment Industry* (Washington, DC: Department of Commerce, March 1985).

race to be first to market with the latest chip designs. Diffusion of knowledge between captives and merchants was slow and uncertain, the chief exception being research results from AT&T Bell Laboratories in the years before divestiture. Engineers and managers moved from company to company within the merchant industry, but there was far less interchange between merchants and captives.

So long as the merchant suppliers remained a comfortable first in the world in technology and sales, they found these conditions tolerable. Heavily constrained in financing, they focused on chip designs while living off a technology base built in large measure by others—AT&T, DoD, universities, the equipment industry. As the Japanese narrowed the gap and began to pull ahead, particularly in processing know-how, delays in moving technology from laboratory to production took a heavy toll. Box 8-B summarizes the story: the merchant firms fell behind particularly in the high-density memory chips Japanese firms produced not only for open market sales but for use in their own products.

**BOX 8-B: JAPAN'S ENTRY STRATEGY:
HIGH-VOLUME MANUFACTURE OF STANDARD DESIGNS**

The Importance of Industry Structure

The merchant/captive distinction, so useful for understanding the U.S. industry, has little relevance in other parts of the world. Major Japanese producers—NEC, Fujitsu, Toshiba—tend to be large enterprises compared with U.S. merchant firms, more diversified than either the merchants or the end-product manufacturers that operate captive facilities here. With a few exceptions, the same companies in Japan that design and produce semiconductors also design and produce consumer electronics, computers, or telecommunications equipment. They consume much of their semiconductor output internally and sell the rest outside, getting perhaps 5–20 percent of their revenues from microelectronics. In contrast, many of the U.S. merchants depend on semiconductors for 80–90 percent of their sales.

Size and a diverse line of products make it much easier for Japanese firms to finance expansion or weather sudden declines in demand; internal consumption also creates powerful motives for technological improvement, as well as free flows of information between chip designers and users elsewhere in the organization. Japan's major suppliers of semiconductor manufacturing equipment likewise include divisions of such large and powerful corporations as Canon, Nikon, and Hitachi.

Competitive Shifts

Problems in the U.S. semiconductor industry have attracted a great deal of attention.[1] Japan's share of total world semiconductor production roughly doubled during the 1980s. In dynamic random access memory chips (DRAMs), the initial battleground between American and Japanese firms, the U.S. share of open market sales fell from 100 percent in 1974 to less than 20 percent in the mid-1980s, by which point most of the U.S. suppliers had dropped out.

Behind the sales and production figures lay an even more disturbing set of contrasts: Japanese firms were spending substantially more on R&D than the U.S. merchants and investing heavily in new plant and equipment. As a result, the U.S. semiconductor equipment industry also fell on hard times: when their customers ran into trouble and cut back on capital investments, the equipment vendors were caught in a cash squeeze. They had trouble financing development, while Japanese equipment firms were able to increase their sales in the United States as well as other parts of the world.

When it comes to the underlying reasons for decline, the picture begins to blur. While no one denies the technological prowess of the Japanese, many of those close to the U.S. industry place more stress on business and trade practices viewed as unfair—subsidies and cartels allegedly arranged by the Japanese Ministry of International Trade and Industry, dumping by Japanese manufacturers, barriers that hindered sales in Japan by American-owned firms. But at least as important, the structural differences between the U.S. and Japanese industries create fundamental and seemingly permanent handicaps for American firms. In particular, relatively small size made it difficult for them to finance R&D and capacity expansion, given rapidly rising capital intensity for both chip development and production facilities.[2] The much greater financial strength of the large Japanese firms not only allowed them to ride out downturns in sales, but also permitted aggressive forward pricing strategies. The U.S.-based firms that had pioneered forward pricing found another home-grown weapon turned against them.

Design-Based Strategies versus Low-Cost Manufacturing

During the 1970s, market competition centered almost entirely on product innovation.[3] American merchants raced to get new chips to

[1] For example, "A Strategic Industry at Risk: A Report to the President and the Congress From the National Advisory Committee on Semiconductors," November 1989.

[2] Robert R. Miller and John A. Alic, "Financing Expansion in an International Industry: The Case of Electronics," *International Journal of Technology Management*, vol. 1 (1986), pp. 101–117.

[3] OTA, *International Competitiveness in Electronics*, pp. 524–531. Reportedly, no more than 10–15 percent of the R&D expenditures of U.S. merchant semiconductor firms typically go for process development. See Congressional Budget Office, *The Benefits and Risks of Federal Funding for Sematech* (Washington, DC: CBO, September 1987), p. 43.

market, and to establish their devices as de facto standards. With the invention of the DRAM and the microprocessor, competition centered on these two device families. U.S. firms designed DRAMs with different arrangements of input and output pins, developed erasable circuits, and tried to be first with volume production of succeeding generations of memory chips. Microprocessors entered the market at about the same time; each microprocessor that was sold needed memory to go with it. Although the first major DRAM applications had been in mainframe computers, the rapid growth of microprocessor-based systems meant that demand for the two new types of devices grew in parallel. Most of the leading U.S. merchant firms participated in both markets, developing proprietary microprocessor designs or second-sourcing popular parts from Intel, Motorola, or Zilog.

The Japanese strategy had crystallized by 1980, with a commitment to process know-how as a source of competitive advantage. In that year, an artful publicity blitz by the Electronic Industries Association of Japan showed that Japanese firms were supplying U.S. customers with DRAMs exhibiting better quality and reliability than American chipmakers. New entrants like Fujitsu and NEC were able to change the nature of the competition by moving the focus in DRAMs away from design, and forcing American firms to compete on the basis of manufacturing capabilities: the new battlefield was the factory floor. Here the U.S. merchants could not match the Japanese; slow in realizing that the rules had changed, they fell behind in yields, cost, and quality for otherwise similar chips.

The Japanese were able to use their manufacturing capabilities as a strategic weapon for two primary reasons. First, the very success of the microprocessors designed by American firms, which opened huge new markets for DRAMs in embedded applications and small computers, created pressures for standardization in memory circuits. As customers winnowed down the range of design features, settling on those they preferred, DRAMs became commodity-like items. With little scope remaining for competition on the basis of product design, the Japanese could develop chips that suited the needs of American customers nearly as easily as firms based here.

Second, as integrated manufacturers consuming many of their own chips internally, Japanese suppliers had strong incentives to improve quality and reliability. They bore the costs of downstream failures themselves. Finding and replacing a defective chip when a system is undergoing final test, or has reached the customer, costs far more than finding a bad part during fabrication; in-service failures, moreover, harm the reputation of the systems manufacturer more than the chip supplier. The Japanese understood full well the need to minimize the number of bad parts through careful control of the production process.

By the time American companies realized that they would henceforth have to compete on the basis of production costs and quality with financially powerful Japanese rivals—companies that, over many years of experience with other electronics products, had raised manufacturing to a high art—the DRAM market was largely lost to them. Where product design remains central to competitive outcomes—e.g., in microprocessors—American companies have so far been able to hold their own. Now, Sematech has the goal of helping American firms catch back up in manufacturing know-how.

In the diverse array of U.S. companies active in some aspect of microelectronics, IBM for years stood alone in terms of size and technical prowess, isolated except for its working relationships with a few equipment companies, notably Perkin-Elmer. Particularly in the 1970s, IBM chose technical directions quite different from the rest of the industry. Not only was the company secretive, but few employees left; Silicon Valley startups drew people from other firms, but not from IBM. IBM's isolation ended only in the late 1980s, when, concerned for the future of its suppliers, the company took an active role in Sematech, not only joining, but licensing chip designs to the consortium (as did AT&T) and to several merchant firms. Meanwhile, the military systems houses continued to go their own way, despite the Pentagon's efforts to integrate them more effectively into the rest of the industry through the VHSIC program.

DoD'S VERY HIGH SPEED INTEGRATED CIRCUIT PROGRAM

The Department of Defense established the Very High Speed Integrated Circuit program in 1980.[12] Its primary objective was to take advantage

---

[12] On VHSIC, see *An Assessment of the Impact of the Department of Defense Very High Speed Integrated Circuit Program*, National Materials Advisory Board Report NMAB-382 (Washington, DC: National Research Council, January 1982); Glenn R. Fong, "The Potential for Industrial Policy: Lessons from the Very High Speed Integrated Circuit Program," *Journal of Policy Analysis and Management*, vol. 5 (1986), pp. 264–291; Glenn R. Fong, "State Strength, Industry Structure, and Industrial Policy: American and Japanese Experiences in Microelectronics," *Comparative Politics*, vol. 22 (1990), pp. 273–299; *Very High Speed Integrated Circuits (VHSIC)—Final Program Report, 1980–1990* (Washington, DC: DoD, Office of the Under Secretary of Defense for Acquisition, September 30, 1990). For summaries of VHSIC's original objectives, see "The Pentagon's Push for Superfast ICs," *Business Week*, November 27, 1978, pp. 136–142; and William J. Perry and Larry Sumney, "The Very High Speed Integrated Circuit Program," in Kosta Tsipis and Penny Janeway, eds., *Review of U.S. Military Research and Development, 1984* (Washington, DC: Pergamon-Brassey's, 1984), pp. 33–45. On outcomes as the program came to an end, see George Leopold, "Critics Hope Planners Apply Lessons From VHSIC," *Defense News*, April 17, 1989, pp. 14, 23; and Tobias Naegele, "Ten Years and $1 Billion Later, What Did We Get from VHSIC?" *Electronics*, June 1989, pp. 97ff.

of the burgeoning capability in commercial firms by spurring technology transfer to the military side of the industry. The chosen vehicle was OSD R&D contracts that would go to teams including both merchant semiconductor firms and defense contractors. OSD planners hoped to arrest and reverse the growing isolation of military microelectronics. DoD eventually spent five times as much money on VHSIC as originally planned, without achieving this goal. The reasons had more to do with the nature of government-business relations in defense than with the program's research agenda. A centralized, high-visibility R&D effort had little impact on the managers of DoD weapons programs, who continued to make conservative technology choices; they saw no reason to take risks on new and untried chips, promised for delivery midway through their programs. As a result, the microelectronics technology that entered the field at the end of 10–15 year acquisition cycles continued to be old and obsolete by commercial standards.

R&D could not solve a procurement problem. Basic design choices had to be made long before weapons finally emerged; the acquisition process could not accommodate later insertion of new component-level technologies. Perhaps the best that can be said is that if VHSIC had never existed, DoD would now be facing even greater difficulties in exploiting advanced circuitry in its increasingly electronics-intensive systems.

Origins

During the 1960s, DoD requirements drove advances in both chip design and processing technology, but by the mid-1970s commercial applications far outstripped military purchases in volume and in technical sophistication. As a very dynamic merchant industry moved into the VLSI era, Pentagon officials grew increasingly concerned with the slow pace at which advanced chips were finding their way into defense systems. Given 10-year-plus gestation periods, the microelectronics technology showing up in fielded hardware seemed woefully primitive by such criteria as feature size and number of transistors per chip. Even worse, defense planners feared that the Soviets were catching up in the ability to incorporate advanced electronics in weapons systems. In little more than a decade, defense had gone from vanguard to rearguard.

DoD's response—the Very High Speed Integrated Circuit program, initiated in 1980—was intended to push the technological frontier outward, diffusing knowledge rapidly within the United States while keeping it from the Soviets. While the ultimate aim was to get high-performance components into fielded military systems more quickly, the Pentagon sought to accomplish this indirectly, through technology development and demonstration. Once the services saw what VHSIC technology could do for them, so the reasoning went, they would demand the newly available levels of performance. Strengthening the technology base in the commercial industry was viewed in some circles as an additional but clearly secondary benefit. Although outside observers sometimes took VHSIC to be a response to Japan's inroads in microelectronics, that was never DoD's intent. Originally budgeted at $200-plus million over a planned six-year period, VHSIC ultimately ran until 1990, letting contracts worth nearly $900 million.

Planning and Management

The impetus for VHSIC came from the Office of the Secretary of Defense, which housed the program office. This arrangement symbolized the importance attached to the effort by Pentagon management. But it distanced VHSIC from the services, which would be ultimately responsible for incorporating the technical results into weapons programs. And while OSD's goals stressed insertion into fielded systems, VHSIC contracts paid only for R&D.

According to the original plan, VHSIC would first focus on research in order to strengthen technical capabilities for using VLSI in military systems, then move on to applications R&D. This early aim shifted somewhat after meetings of the Advisory Group on Electron Devices (AGED), a panel of industry and university experts, during 1978. AGED warned that VHSIC could only be effective if intended from the beginning to design working chips that would meet the needs of the services. In response, VHSIC managers altered the schedule so that research and applications would proceed in parallel, rather than sequentially.

Even so, as VHSIC planning moved forward, all three military services resisted. They stood to lose control over R&D money that would come from their own budgets; the new schedule, with its more rapid timetable for the design of prototype VHSIC chips, did little to convince the Army, Navy, and Air Force that VHSIC would help them. The Defense Advanced Research Projects Agency also resisted, seeing VHSIC's long-range research goals as more properly a DARPA responsibility. Nonetheless, the OSD program office was able to win DoD-wide responsibility for silicon R&D. DARPA received a handsome budget for alternative or "post-silicon" materials, notably gallium arsenide; indeed, by some accounts, DARPA was spending a good deal more money on gallium arsenide during the early 1980s than warranted by its technical prospects.

As planning ended and VHSIC got under way in 1980, the Pentagon solicited bids from teams that linked established military systems contractors with the merchant firms that had turned at least partially away from the DoD market. The intent was to reinvigorate the former and reattract the latter. OSD also encouraged teams that included semiconductor equipment manufacturers, and, given a still-strong emphasis on research, university groups.

Tension between advocates of research and the user community, consisting of the military systems houses and their customers in the services, continued during the early years. The influence of the services and the systems houses—which in the end won most of the contract dollars—proved decisive by the mid-1980s. A relatively small group of familiar DoD contractors (Honeywell, TRW, IBM) came to dominate the program; these firms, dependent on the services rather than OSD for future business, took care to please the former.

Originally intended to break the mold of business-as-usual in defense electronics, VHSIC by the mid-1980s looked much like other military R&D efforts. The Pentagon had tried to put merchant firms and semiconductor equipment manufacturers together with the systems houses to form technology development teams. Six merchant firms took part as contractors or subcontractors in VHSIC Phase 0, which started in 1980; of these, only Motorola remained during Phase II, beginning in 1984. None of the Phase II teams included an independent equipment manufacturer. Only one—Perkin-Elmer—had made it into Phase I (1981–1984). Although Perkin-Elmer was a member of three different teams bidding for Phase II money, none of those bids proved successful. (Perkin-Elmer did participate in the small Phase III effort.)

Most of the development of VHSIC chips took place in the captive facilities and pilot production lines of the systems houses. The early emphasis on interfirm cooperation and technology transfer faded. VHSIC had no more than limited success in terms of DoD's paramount objective, speeding the incorporation of VLSI components in military equipment. The services remained suspicious; with few incentives for risking commitment to unproven VHSIC components, most program managers favored available technologies. And the chips themselves were slow to materialize. Ten years after the program began, 38 VHSIC chips had been designed; most remained in the prototype stage, with a few entering pilot production during 1990 and 1991. VHSIC technology will make its way into the field eventually. But the first applications of VHSIC chips—gate arrays incorporated in existing systems to emulate obsolete parts that DoD can no longer purchase—illustrate the very problems in weapons acquisition that had slowed military use of VLSI in the first place.

Lessons

The VHSIC program began as an effort to establish new linkages between the merchant semiconductor industry and established military contractors. It ended in capture by the contractors whose practices OSD set out to change. Hindsight shows that AGED gave good advice: the program would probably have had even less impact if it had begun with a stronger focus on research. As it is, VHSIC R&D has strengthened the U.S. technology base in microelectronics primarily by paying for the development of computer-aided tools for IC design. Other spillovers from VHSIC to the commercial industry have so far been small.

What lessons does the VHSIC experience hold for dual use? Most important is the danger of relying on R&D dollars to reach other objectives. The VHSIC program office started with an ambitious plan, ample funds, and the endorsement of high officials in several administrations. By and large, Congress viewed the program with favor (although the House Armed Services Committee did zero the fiscal 1980 VHSIC budget until receiving assurances concerning management and contracting procedures). But OSD was defeated by the ingrown habits and accumulated baggage of an acquisition system seemingly impervious to change. That the Pentagon could not make this flagship program succeed shows the very real difficulties of taking advantage of technologies from the civilian side of the economy to improve national defense.

THE SEMATECH INITIATIVE

There has been one major change since OSD began planning the VHSIC program: the future for U.S. merchant firms looks far less secure than it did in the late 1970s. Thus it came as little surprise in 1987 when a Defense Science Board (DSB) task force recommended a series of steps for strengthening U.S. semiconductor manufacturing.[13] The task force report underlined the dual-use character of integrated circuits, critical components in the smart weapons and other high-technology systems

---

[13] Department of Defense, *Defense Semiconductor Dependency: A Report of the Defense Science Board* (Washington, DC: DoD, Office of the Under Secretary for Acquisition, February 1987).

necessary to counter the numerical advantages of Soviet weaponry. If the U.S. semiconductor industry continued to fall back under the Japanese onslaught, DoD and its contractors might not be able to get the chips needed for future military systems. The DSB task force report did not so much as mention the VHSIC program.

At the same time, leaders of the merchant industry were developing their own plans for what became the Sematech R&D consortium (Box 8-C). With industry promising to invest its own money, the DSB report helped shape consensus in Congress and the administration leading to an implied five-year commitment of federal funds. National security provided no more than a secondary rationale for this decision: Congress, in particular, focused squarely on U.S.-Japan competition for commercial sales.

BOX 8-C: SEMATECH—FEDERAL FUNDING FOR MICROELECTRONICS

Organized in 1987 on a not-for-profit basis with fourteen members, Sematech's objective is to improve U.S. semiconductor manufacturing capabilities.[1] The membership includes not only merchant semiconductor firms, but systems-oriented manufacturers, including IBM, AT&T, and Digital Equipment Corporation, which operate captive facilities. Eight of the ten largest suppliers of chips to the military are Sematech members, but only two—Harris Corporation and Rockwell—get more than 10 or 15 percent of their chip sales from defense. Foreign-owned firms cannot join.

Sematech emerged after many years of discussion among corporate executives and government officials over Japan's inroads into U.S. and world semiconductor markets, and particularly the allegedly unfair practices of Japanese firms. After it had become plain that technology was part of the problem—because, for example, the relatively small U.S. suppliers could not match the investments in R&D and production equipment of much larger Japanese companies—the industry's trade association urged government funding for a joint R&D effort. Congress responded with a $100 million appropriation for Sematech in fiscal

[1] For background, see Congressional Budget Office, *The Benefits and Risks of Federal Funding for Sematech* (Washington, DC: CBO, September 1987); Report of the Advisory Council on Federal Participation in Sematech, "SEMATECH: Progress and Prospects," 1989; Glenn J. McLoughlin, "Semiconductor Manufacturing Technology Proposal: SEMATECH," CRS Issue Brief IB87212 (updated) (Washington, DC: Congressional Research Service, January 19, 1990); General Accounting Office, *Federal Research: SEMATECH's Efforts to Develop and Transfer Manufacturing Technology*, GAO/RCED-91-139FS (Washington, DC: GAO, May 1991); Louise Kehoe, "Loyal Fans But Mixed Reviews," *Financial Times*, May 21, 1991, p. 10.

1988, with the presumption of ongoing contributions through 1992. The sums, channeled through DoD (which in turn assigned oversight responsibility to DARPA), would be roughly matched by the consortium's members, each pledging 1 percent of annual semiconductor revenues (with a minimum of $1 million and a maximum of $15 million).

Although Sematech gets nearly half its annual budget from DARPA, the goals are strictly commercial: to rebuild U.S. capabilities in microelectronics processing so that American merchant firms can compete with their Japanese rivals. Congress channeled the government share of Sematech's budget through DoD largely for lack of alternatives: no other agency had the ability and experience to oversee public spending on a complex technical undertaking of this sort. In addition, the national security link made the arrangement more palatable within the administration and to those in Congress opposed to industrial policy.

The Rationale for Federal Funding

The reasoning advanced in the DSB task force report paralleled that heard from the merchant firms and their trade association: a vital commercial industry was needed to provide technology and a production base adequate for future military needs. Neither group had much to say about the nature of the barriers separating the military and commercial sides of the business. Basically, the argument advanced by both the DSB task force and the industry went as follows:

- A vital commercial semiconductor sector is necessary for DoD to fulfill its overall mission.
- The merchant industry is in serious trouble, in part through loss of its technology edge, and especially in manufacturing.
- The DRAM is the key to manufacturing technology, and, more broadly, the technology driver for microelectronics as a whole. (The initial focus on DRAMs has since diminished greatly.)
- DoD support for an industry consortium focused on processing R&D, together with other, less costly steps (e.g., more money for university research), would help maintain a viable merchant industry and thus ensure DoD's access to advanced microelectronics technologies.

The argument, although both its assumptions and its logic were open to a variety of criticisms, found broad favor in Washington.

Foreign Dependence

The DSB task force skated lightly over such questions as the sources of microelectronic devices actually used in military equipment. Foreign sourcing of chips—whether imported as finished components or assembled domestically by U.S.-owned firms after partial processing overseas—had surfaced as a policy issue in the early 1980s, but had not gained much visibility. VHSIC planners had been concerned primarily with getting VLSI into the field; dependence went unmentioned. While the DSB's report gave the issue a higher profile, the current and potential dimensions of foreign-source dependence remain largely unknown. No databases reveal the foreign electronics content of military hardware, nor do the parts lists for particular systems.[14] With manufacturers free to put their own logos and part numbers on components they purchase, what at first glance might appear to be U.S.-made chips may in fact be imports; with some exceptions, American companies are free to qualify foreign-made devices under the relevant military standard.[15]

In fact, foreign sourcing is bound to grow in the years ahead. Defense systems in production today were designed in the past, when fewer chips of any sort were imported. Systems under development currently will inevitably have greater foreign content. DoD will have to learn to live with dependence, developing policies for ensuring that critical parts will be available when needed (i.e., that vulnerability does not follow from dependence). Supporting the U.S. merchant semicon-

[14] General Accounting Office, *Industrial Base: Significance of DOD's Foreign Dependence*, GAO/NSIAD-91-93 (Washington, DC: GAO, January 1991).

[15] The primary exceptions—JAN (Joint Army-Navy) chips—must be made within the United States in a DoD-certified facility. At present, JAN components account for about 20 percent of the semiconductors in U.S. military systems. In recent years, about four-fifths of the semiconductor content of military systems has evidently stemmed from wafers fabricated domestically by U.S.-owned firms, with the remaining one-fifth split between offshore wafer fabrication by U.S.-owned firms and parts purchased in finished or semifinished form from foreign sources. Domestic wafer fabrication has been most common for linear ICs (93 percent of all linear circuits used in military systems) and discrete transistors (91 percent), least common for memory chips (72 percent from domestic fabrication lines). These figures come from Dataquest, as cited in Glenn O. Ladd, Jr. and John W. Kanz, "Foreign Competition for Integrated Circuits Applications in Military Systems," *1988 Government Microcircuit Applications Conference: Digest of Papers*, vol. 14 (Washington, DC: DoD, Defense Technical Information Center, November 1988), pp. 511–514. All such estimates should be viewed as highly uncertain. For instance, *Report of the Defense Science Board on Use of Commercial Components in Military Equipment*, suggests a rather different picture—that three-quarters of "mil-spec" chips come from overseas plants operated by American firms (53 plants, 15 countries, 10 firms), while a dozen U.S. plants produce the rest (p. A-17).

ductor industry is only one of a number of alternatives for accomplishing this. Indeed, Sematech can at best be no more than a partial response to foreign dependence. The consortium aims to rebuild U.S. manufacturing capabilities for commercial chips, but DoD's requirements differ substantially. Whether they *should* differ as greatly as they do raises another set of issues: critics of DoD acquisition have held that many military standards are unnecessarily restrictive, even counterproductive, and that commercial chips could provide higher quality and reliability at lower cost. These issues have rarely been seriously addressed at the technical level. But only if DoD succeeds in incorporating commercially available ICs in its systems in substantial numbers could Sematech represent a direct response to dependence, as opposed to whatever generalized and indirect effect it may have through improving the competitiveness of U.S. semiconductor firms. Given the global nature of microelectronics production, greater reliance on commercial ICs in U.S. military systems would almost certainly mean greater reliance on foreign sources, no matter how strong the U.S. industry becomes.

The Consortium and Its Activities

Sematech's operating plans emerged after lengthy debate over organizational form and objectives (whether to undertake high-volume production as well as R&D and pilot-scale process development, whether to concentrate on DRAMs). The consortium came in for some early criticism on the premise that, while it was aiming to improve manufacturing technology, the ground rules made membership too expensive for equipment firms. The equipment industry's principal trade association, Semiconductor Equipment and Materials International (SEMI), then established SEMI/Sematech as a mechanism for participation. SEMI/Sematech has about 150 members that account for roughly 90 percent of total sales by U.S.-based equipment, materials, and supply firms; it serves as liaison between member firms and Sematech itself.

During 1990, slightly more than half of Sematech's $224 million budget went for external R&D, mostly contracts with a small group of equipment suppliers; lithography accounted for 55 percent of external R&D. This high level of external contracting represents a substantial shift from the consortium's early planning, which emphasized internal R&D. Much of Sematech's support for equipment firms seems to be going to the GCA company (the consortium does not divulge the dollar value of contracts), evidently for fear the firm might otherwise vanish as a supplier of lithographic equipment. GCA has received not only R&D contracts from Sematech, but orders for 16 wafer steppers to be used for site testing by member firms.[16] Sematech managers hope not only to strengthen equipment firms like GCA technically and financially, but to strengthen the vertical linkages between the equipment industry and the chipmakers—e.g., by creating a forum for discussion of technical requirements, and providing field test sites for equipment debugging and final development, both in Sematech's own facilities and within member companies. As a condition of Sematech's contracts with equipment firms, members of the consortium get first chance at purchases of products developed with Sematech funding.

Sematech had a rocky start, as even the brief account above should suggest, including a long search for a chief executive officer, and the subsequent forced departure of the first chief operating officer. Once operations got well under way, however, most observers have seemed willing to let the organization prove itself. Still, Sematech faces unusual scrutiny, not only because of the visibility of the industry and its problems, but also because the half-billion-dollar federal commitment marks a major shift in U.S. technology policy: a turn toward explicit support for commercially oriented R&D carried out in the private sector.

Sematech has also been highly visible as a much-touted model for new approaches to cooperative R&D; the organization will quickly find itself in the spotlight should new internal frictions surface. This is not that unlikely. Several Sematech members, for example, have ongoing collaborative relationships with Japanese firms—Motorola with Toshiba, TI with Hitachi—in technologies closely related to those on which the consortium is working. Such alliances could lead to conflicts of interest, or to the perception of such conflicts. Finally, the members, and SEMI/Sematech, will have to effectively exploit the results of the consortium's R&D—no easy task. Capitalizing on joint research requires, in addition to effective technology transfer efforts, substantial parallel work inside participating firms so they can absorb the results rapidly. This will be doubly difficult for Sematech, given the need to transfer processing technology both to the equipment industry and to participating chipmakers.

---

[16] Congressional Budget Office, *Using R&D Consortia for Commercial Innovation: Sematech, X-Ray Lithography, and High-Resolution Systems* (Washington, DC: CBO, July 1990), pp. 23–28. Sematech managers view lithography as the primary weakness in the U.S. equipment industry, and have also given R&D contracts to Silicon Valley Group Lithography Systems, which purchased that part of Perkin-Elmer's business in 1990.

Sematech must also guard against unrealistic expectations in Washington. The consortium's R&D centers on fine-line lithography for denser circuits (much as did VHSIC). But parity with Japanese firms in lithography—or even a lead—would not be enough to rebuild U.S. manufacturing capabilities in microelectronics, much less U.S. competitiveness. It was mastery of the systems aspects of production that gave Japanese semiconductor firms their advantages. While lithography has been the central technology in chip production since the exponential increase in IC density began, mastering lithography is not enough; indeed, mastering each of the individual steps in IC fabrication (lithography being one of many) does not lead automatically to efficiency in high-volume production. Sematech's R&D could turn out to be yet another example of a familiar mindset in U.S. industry: the implicit assumption, derived from Frederick Taylor and scientific management, that optimizing the building blocks of a production system ("modules" in Sematech's language) will suffice to optimize the system as a whole. The fallacy is a bit like assuming that sound bricks will more or less automatically yield a sound building. (See discussion in Chapter 10.) Still, without world-class producers of lithographic equipment willing and able to sell to them on a timely basis, American chipmakers risk remaining months or years behind their competitors in Japan, who have close working relationships with Japanese equipment firms.

The Future

When Sematech was established, industry leaders stated their intent to be self-supporting at the end of a five-year period of federal support. Not surprisingly, as 1992 approached, the consortium began signaling that further funding or a follow-on program might be needed, or at the very least, a ramping down rather than the abrupt end of government funding. While many in Washington would no doubt be willing to consider such proposals, some at least will ask for evidence that Sematech could eventually be viable without federal dollars. Sematech's proponents should be able to answer the following sorts of questions:

- What options might government consider for continuing support should informed observers conclude that Sematech has been successful?
- If it were to appear that the consortium has run into serious difficulties, either technically or organizationally, could policymakers afford to back away, believing that Sematech and by

extension the merchant industry had had its chance and failed to take advantage of it?

- Given that the ultimate goal is to rebuild U.S. capabilities in microelectronics processing, which depends in part on a viable semiconductor equipment industry, are there steps government might take to support the equipment industry directly?
- If Congress and the administration agreed that continuing support for the U.S. semiconductor industry were needed, should DoD continue to be the government's agent? If so, might DoD influence gradually turn Sematech's R&D agenda away from the objective of helping the commercial industry? Could some other agency take over?
- Finally, given the outcomes of the VHSIC program, and given that Sematech is not intended to support military microelectronics, what else, if anything, should be done to ensure DoD access to the microelectronics technologies it needs?

POLICY CHOICES

With microelectronics in many respects the prototypical dual-use technology, another question follows. Might this become a laboratory for dual-use policies? In the years ahead, policymakers can either continue to focus on one side or the other of the industry or attempt a more integrated aproach. This section outlines policy alternatives ranging from a straightforward extension of the Sematech model (more money, continuing support past 1992) to steps intended to reduce the barriers between military and civilian microelectronics. Policies in the first group, through primarily directed at the merchant industry, would also provide some support, direct or indirect, for defense microelectronics. The second group is explicitly dual use. (We do not discuss purely defense-oriented policies.)

Policies for Commercial Competitiveness

- *Continued and perhaps increasing federal support for Sematech* (with industry matching funds) after fiscal 1992.

  Merchant firms have been chasing a moving target in manufacturing for years. Regardless of how successful Sematech proves to be, it seems plain that more will need to be done to rebuild

U.S. capabilities in processing. The Japanese will be hard to catch, particularly given their structural advantages as integrated firms with strong incentives for efficiency in manufacturing.

- Cost-shared funding for an *R&D consortium of equipment manufacturers.*

Sematech has given contracts to a relatively small number of equipment firms. Others participate through SEMI/Sematech. Federal support specifically for the equipment industry would complement these efforts. Such a policy would also support defense microelectronics, because captive producers depend on the same group of equipment suppliers as commercial firms.

- Guarantee *large-volume purchases of commercially available, U.S.-made chips.*

Federal contracts could be tied to greater use of off-the-shelf chips in military equipment, or to programs for purchasing end-products such as personal computers for schools or workstations for universities. Given a vast nongovernment market, the impacts of such an approach would necessarily be limited, although an undertaking to build a national information network or "smart highway" system could stimulate demand further. Such policies do have the advantage of relying on market pull rather than technology push.

Dual-Use Policies

- Plan a *dual-use successor to the VHSIC program.*

Such an undertaking would differ from Sematech because technical decisions would be made inside government (with advice from outside), with military requirements given substantial attention.[17] The first step should be unsparing evaluation of the lessons from VHSIC. Subsequent steps might include creation of a military design bureau chartered to keep IC technologies, including processing and packaging, as close to commercial practices as possible.

- *Require use of commercial off-the-shelf (COTS) devices in defense systems* whenever possible.

---

[17] *Report of the Defense Science Board on Use of Commercial Components in Military Equipment* includes a number of specific options.

Some parts of DoD actively encourage military use of COTS now, but implementation of such policies has been generally slow. To accelerate the process, DoD might require that a waiver be requested for use of any IC *except* a domestically produced, commercially available chip. Even then, waiver requests would have to be carefully scrutinized.

• Establish an extensive, structured program of *long-range R&D on next-generation microelectronics and computing technologies,* intended to help the United States keep pace with Japan in devices and their applications to digital systems.

Focusing ample levels of support on long-range goals could reproduce some of the features characterizing the early years of DoD markets for both semiconductors and computers. On the other hand, a highly structured approach could stifle fresh thinking or steer money bureaucratically to alternatives that proved unpromising. As a first and modest step, the federal government could provide added funding, perhaps to Sematech, for next-generation alternatives to optical lithography. Sematech currently spends a small amount on x-ray lithography, as does DARPA. More ambitious steps might include alternatives to x-ray lithography, semiconducting alternatives to silicon (e.g., gallium arsenide), superconducting digital electronics, optoelectronics and integrated optics, optical computing, and neural networks and other massively parallel computer architectures.[18] Work on all these technologies is under way in many laboratories, often with support from DoD and other federal agencies, but funding levels tend to be low, and no coordinated effort exists.

DoD'S ROLE

Policy alternatives that rely on DoD and its needs as a focal point, no matter that the commercial industry might share similar objectives, face an overriding problem in the split between the military and commercial sides of the industry. This split has existed for years. The mili-

---

[18] Alternatives to silicon microelectronics have been of interest to DoD for many years because of the high-speed, high-capacity computing requirements of so many national security systems (not only in weapons, but for code-breaking, processing of satellite images, and other data-intensive tasks).

tary side remains small and specialized, the commercial side under continuing competitive pressure from Japanese firms.

VHSIC aimed to pull merchant firms back into DoD's orbit by creating closer working relationships between the two sets of companies. The carrot for the merchant firms was an injection of research dollars and results. But VHSIC was not a dual-use program: DoD sought to take advantage of commercial technologies for its own purposes. Although Pentagon officials expressed some concern that growing Japanese competition could threaten the long-term viability of the commercial industry, their chief objective was simply to get advanced electronics into the field more quickly. The program did not achieve this, but the reasons had little to do with technology: DoD asked an R&D program to solve a procurement problem.

Sematech is not a dual-use program either. Its purpose is transparently to help commercial firms meet Japanese competition. The money comes from DoD for convenience: DARPA knows how to manage contract R&D. The connection between a world-class domestic manufacturing capability and DoD's mission requirements is at best a loose one. DoD itself continues to rely primarily on defense contractors to fill its needs for chips. These contractors make some of their own ICs and buy others from merchant manufacturers, including Sematech members. Some they buy overseas. While DoD needs continuing access to advanced microelectronics technology, the Pentagon has many possible alternatives for achieving it.

Any substantial shift toward dual use calls for a different approach. Probably the first step should be to create strong incentives for the use of commercially available off-the-shelf chips in DoD systems. Once such incentives began to take hold, dual-use technology development would become a more realistic prospect. Meanwhile, the longer DoD serves as a conduit for funds to Sematech, the greater its influence over the consortium is likely to be. Sematech's management will no doubt continue to be a powerful countervailing influence. Robert Noyce, Sematech's late chief executive officer, was highly visible and highly respected. So is his successor. The consortium's board of directors includes some of the best-known names in the industry. Even so, DoD is not likely to be content as a silent partner. Should federal support be extended past 1992, some arrangement other than channeling funds through the defense budget seems called for.

9

Software: Productivity Puzzles, Policy Challenges
Charles A. Zraket*

Computer software has become one of the nation's largest, fastest growing, and most pivotal industries. Recent estimates place annual revenues at well over $100 billion, much higher than previously thought. Worldwide, American firms furnish 60 percent or more of the dollar value of software in use. Software accounts for half or more of the cost of some defense systems, and with the debate over President Reagan's proposed Strategic Defense System, the public at large began to hear a great deal about its importance for national security.

In the future, software technologies will become still more important for competitiveness and defense. In the computer and telecommunications industries, hardware performance depends on capabilities associated with program design, development, and maintenance. Elsewhere in the economy, companies look to capabilities embodied in software to reduce costs, improve product capabilities and customer service, and manage business operations. Sophisticated computer pro-

*This chapter was prepared with the help of C.E. Kalish, E.L. Lafferty, and R.J. Sylvester. They thank Melvyn P. Galin, David R. Woodward, and Michael V. Vasilik for their survey of the software industry (Table 9-1); John B. Campbell for his analysis of the financial and telecommunications industries; Burton Kreindel for legal expertise on rights-in-data issues; and Margaret S. Jennings for assembling and editing the chapter. The analysis also draws in places on John A. Alic, Jameson R. Miller, and Jeffrey A. Hart, "Computer Software: Strategic Industry," *Technology Analysis and Strategic Management*, vol. 3 (1991), pp. 177–190.

grams have become everyday tools for scientific and engineering tasks ranging from weather prediction to automobile design. Even if these were its only applications, software would be a major element in the economy. However, software has another, perhaps even more critical role, through the built-in or embedded programs that control the functioning of not only consumer products such as microwave ovens, but also industrial and military equipment of all kinds. Examples range from robots and machine tools to guidance systems for the "smart weapons" that performed with such devastating effectiveness during the Gulf War of 1991. Some military aircraft already incorporate software aids that extend the capabilities of both man and machine by helping cope with information overload and flight maneuvers beyond human skill levels.

As the United States continues to move toward a high-technology, knowledge-based economy, software takes on generic significance: the technology cuts across many sectors, with impacts on productivity and competitiveness in industries ranging from retailing to health care to banking. Software itself represents the extreme case of a knowledge-intensive industry. There is hardly any direct production expense other than duplication of programs and documentation; design, development, marketing, and other white-collar tasks account for nearly all labor costs.

This chapter summarizes the workings of the U.S. software industry, and explores the dual-use nature of software technology. A great deal of software is specially tailored for particular applications. Nonetheless, the overriding dynamic in the industry has been to create general-purpose programs in place of custom software. This is one way to deal with a productivity bottleneck resulting from the labor-intensive nature of programming. Software costs have been rising much more rapidly than hardware costs, an unsustainable pattern. A second way of dealing with the bottleneck is to improve the technical tools and methods used in program development—e.g., through computer-aided software engineering (CASE) methods. Finally, by reusing code rather than generating totally new programs for each application, companies (and DoD) can increase productivity, reduce costs, and improve quality (e.g., freedom from programming errors, or bugs).

Software promises more scope for dual-use policies than does microelectronics, an industry that has matured to such an extent that existing institutional patterns, corporate strategies, and government policies would have to be reevaluated and perhaps replaced to improve military/commercial synergies significantly (see Chapter 8). In contrast, software has hardly begun to mature, despite the fact that it is a much larger industry. Flexibility and room for change characterize its industrial structure. Lacking near-term competitive threats, there has been none of the sense of crisis or pressure for immediate policy action seen in microelectronics. This is a disadvantage as well as an advantage. Industry and government suffer from complacency generated by the continuing strong competitive performance of U.S. software firms. Other countries, notably Japan, recognize the need to improve in software and seem bound to do so. But time remains for a measured response, and the policy environment remains plastic, potentially receptive to innovations. Software could become a laboratory for dual use.

MILITARY AND COMMERCIAL SOFTWARE

In the early years of the computer industry, hardware presented the major technological challenge. Since then, hardware and software architectures have become inextricably linked (Box 9-A). As the architecture of a computer system evolves, and its performance improves, the various layers of system software mature. Succeeding generations of hardware typically incorporate much of the old system software, along with upgrades adding new functional layers. In other cases, the architecture will make a sharp break with previous systems because this is the best way to improve performance or take advantage of new capabilities. Dual-use software can be developed only if defense and civil hardware architectures are compatible, while compatibility of operating systems and the software they control are the necessary condition for dual-use hardware. At present, lack of compatibility between defense and civil software/hardware architectures limits dual use. The primary exception is Department of Defense purchases of off-the-shelf systems for general-purpose applications such as accounting.

As we will see, lack of compatibility is not an absolute or permanent bar to dual use. Moreover, DoD continues to play a major role in determining the type and the quality of software developed in the United States. The Pentagon budgeted about $31 billion for software development and maintenance in 1990—more than 10 percent of all defense spending, and more than 20 percent of all U.S. spending on software.[1] As a customer with many special requirements, DoD has

---

[1] D. Hughes, "Computer Experts Discuss Merits of Defense Dept. Software Plan," *Aviation Week and Space Technology*, April 16, 1990, p. 65. The $31 billion estimate seems widely accepted. On the other hand, a 1989 memorandum from the Assistant Secretary of Defense for Command, Control, Communications, and Intelligence (C^3I) stated: "Because of . . . the different offices of primary responsibility . . . there is really *no* accurate data on all DoD software." Charles Condon, Office of the Assistant Secretary of Defense for Command, Control, Communications, and Intelligence, memorandum, August 23, 1989.

BOX 9-A: SOFTWARE AND HARDWARE

Software consists of instructions that the computer translates into the electromechanical actions needed to carry out a defined function or to process information and display it. Software can also simulate physical processes and phenomena in ways that create artificial experiments and experiences. A computer's functionality depends on both software and hardware, but software was recognized only slowly as an important entity in its own right, separate from the machine on which it runs. To change or extend the computer's capabilities, designers must modify the hardware, the software, or both. To modify software, it is necessary only to replace one set of instructions (code) with another; there is no need to alter the physical components of the machine.

Software can implement functions that would be extraordinarily time-consuming and costly in hardware alone. For example, to create a robot arm capable of manipulating a particular set of tools, engineers could build the functionality into the hardware, designing a mechanical system that would operate exactly as intended. But it would operate only in that fashion and for that purpose. To change its operation, new parts would have to be designed, fabricated, and installed. In a factory where requirements shift frequently, or might not have been fully understood at first, this would be a very slow and expensive way to proceed.

An alternative is to design the robot as a universal piece of hardware (indeed, this is the conventional meaning of robot) controlled by a computer program. Coded instructions tell the robot what physical actions to carry out (e.g., the path in space the arm should trace). In this way, robots of the same mechanical design can be used, say, to tighten bolts, to spray paint, or to manipulate a welding torch. Only the software must be changed, along with the tools or devices manipulated by the arm. In some cases, the robot can be programmed to change its own tools. In this way, robots can be quickly switched from one task to another; instead of making mechanical changes, it is necessary only to replace one software program with another.

Another advantage of capturing functionality in software rather than hardware is that the resulting system will often be less cumbersome physically because some of the design complexity resides not in physical components but in computer code. This is critical for many military systems, which must operate in the restricted space inside a submarine, airplane, or missile. In recent years, Reduced Instruction Set Computers (RISC) have been architecturally streamlined by moving even more design features from hardware to software. RISC machines are faster, simpler, easier to manufacture, and hence less costly than conventional computers.

shaped trends in software technology both through its research programs and its operational requirements. The other mission agencies—e.g., the National Aeronautics and Space Administration—have
sometimes had significant impacts, but DoD's influence has been both
broader and deeper.[2] Nonetheless, it is civilian demand rather than
military requirements that increasingly drives the software market, although not to the extent seen in microelectronics.

The Programming Bottleneck

While hardware prices have been dropping for many years, the costs of
software generation—e.g., per line of debugged code—have declined
hardly at all.[3] Highly skilled employees must still write and debug
software on a line-by-line basis; programming remains an extraordinarily labor-intensive activity. Although enhancements of computer
languages and programming aids such as CASE tools have helped,
larger and more complicated programs continue to stretch the capabilities of the best people and the best methods. For such reasons, 80
percent of the life-cycle costs for computer-based systems, on average,
now go for software (including maintenance), with only 20 percent
for hardware.[4]

This bottleneck constrains applications of computing throughout
the economy: software becomes *relatively* more expensive as productivity improves faster in other sectors.[5] It follows that raising productivity

[2] For an extensive treatment of NASA's contributions, see James E. Tomayko, *Computers in Spaceflight: The NASA Experience*, vol. 18, suppl. 3 of Allen Kent and James G. Williams, eds., *Encyclopedia of Computer Science and Technology* (New York and Basel: Marcel Dekker, 1987).

[3] Costs per line remain about the same today as two decades ago, ranging from about $10 for routine programs to as much as $1,000 per line for applications extraordinarily demanding in terms of freedom from errors, reliability, and fault-tolerance (e.g., for the space shuttle). E.J. Joyce, "Is Error-Free Software Achievable?" *Datamation*, February 15, 1989, p. 53.

[4] Charles A. Zraket, "Understanding and Managing the New Benefits and Problems of the Information Society," M82-44 (Bedford, MA: The MITRE Corporation, July 1982). Over a 15-year service lifetime, maintenance and support costs can approach the original software development expenditure.

[5] William J. Baumol, Sue Anne Batey Blackman, and Edward N. Wolff, "Unbalanced Growth Revisited: Asymptotic Stagnancy and New Evidence," *American Economic Review*, vol. 75 (1985), pp. 806–816. The same phenomenon has been driving up costs in education and health care.

in software development, by making better programs available more quickly and cheaply, holds enormous promise for improving efficiency in all industrial economies.[6] There are many reasons that productivity growth in software has been slow; the primary causes lie in lack of unifying technical concepts, effective software engineering tools and methods, and proven management practices, particularly for large-scale systems.

The Dual-Use Issue

Today, both the civil/commercial and the military sectors are voracious consumers of software. Is the relationship between the two sectors essentially positive and reinforcing, or is it divisive? Some critics of DoD's role claim that it distorts the industry by diverting a scarce resource—software professionals—into work that has no application outside of defense, and that at the same time DoD fails to exploit technologies and management practices developed in the commercial sector. On the other hand, through its projects, DoD has financed—and continues to finance—leading-edge software and supporting technologies (e.g., state-of-the-art CASE tools) that have proven valuable in both defense and commercial sectors. Our view is that, on balance, DoD's role remains more supportive than distorting, and that with new policies, defense requirements could become a greater stimulus for commercial software, primarily by strengthening the mutual technology base.

One point emerges above all others: DoD acquisition practices make it difficult for the military to take advantage of commercially available software or, indeed, even of software technologies flowing directly from defense R&D. It takes a decade or more to bring complex military systems from initial conceptual design through full-scale development, a period that far exceeds the life span of most commercial software

---

[6] Despite the unquestioned potential of computers for improving efficiency in both manufacturing and service industries, the productivity statistics show little evidence of such improvements over the first three decades of the "computer revolution." Possible explanations range from poor data, particularly on productivity in the service sector, to long time lags associated with learning to use these new tools effectively. See, for example, Martin Neil Baily and Alok K. Chakrabarti, *Innovation and the Productivity Crisis* (Washington, DC: Brookings Institution, 1988); and Paul A. David, "The Dynamo and the Computer: An Historical Perspective on the Modern Productivity Paradox," *AEA Papers and Proceedings*, vol. 80, no. 2 (May 1990), pp. 355–361.

products (even though software changes less rapidly than hardware). The lengthy time spans for planning, procurement, and deployment of weapons systems mean that software developed for DoD differs greatly from that supplied to other customers. Often, the software will be obsolete by commercial standards when a weapons system first enters the field—at which point a 20-year operational lifetime may lie ahead. Since key defense systems must be functional at all times, modifications during this operational lifetime must be compatible with the rest of the system. This forces suppliers of software upgrades to work with technology that may be 30 years old. Even without the very special requirements of complex weapons systems, DoD's software needs would differ greatly from those in the rest of the economy.

For semiconductors, the military market had become peripheral by the end of the 1960s, falling below 10 percent of industry revenues (though greater for integrated circuits). The situation is somewhat different for software. DoD accounts for a larger fraction of sales—something over 20 percent—and remains a more important source of technology than in microelectronics. In the future, software could go the way of microelectronics, with military software projects isolated. This would deprive the commercial industry of the fruits of DoD's work and deprive DoD of lessons learned from commercial practices. Already, it is industries like banking and insurance—where complex software packages handle record keeping, process transactions rapidly, and manage automatic teller machines and international funds transfers—that drive many of the technological advances in software. Much the same is true of the telecommunications networks that support these industries and others, where highly sophisticated programming integrates switching systems deployed worldwide. And when it comes to embedded software, automobiles and commercial airplanes have become major users, along with military radars, fire control systems, and electronic warfare suites.[7] But divergence and isolation is not a preordained outcome. Later in the chapter, we suggest policies that could benefit both military and civilian sectors.

---

[7] A Boeing 767 carries two dozen digital processors implementing half a million lines of code. This is far less hardware and software than found in a B-2, which carries 200 computers, but substantially more than the older F-16. The F-16A had 7 computer systems, running on 135,000 lines of code; an F-16D has 15 computers and 236,000 lines of code. Operational and support software for the Advanced Tactical Fighter will probably total more than 4 million lines. General Bernard Randolph, "The Importance of Technology Development," *Signal* (August 1989), pp. 35–40.

THE U.S. SOFTWARE INDUSTRY

In reality, there is no single software "industry." Three different types of firms develop computer programs for sale:

1. Most *computer manufacturers* sell systems software for their machines, as well as a broad range of applications packages—for example, for database management. When purchasing computers, customers place considerable weight on the software available for a given platform. Thus hardware manufacturers devote large and growing fractions of their R&D and product development efforts to the software side of the business.

2. *Independent software suppliers* develop and market software on an off-the-shelf basis, offer programming services, or provide customized one-of-a-kind packages. The independent sector includes household names like Microsoft and Lotus Development, known for personal computer (PC) software, along with such organizations as General Motors' EDS subsidiary. The sector includes firms that develop systems software and applications packages for large machines in competition with computer manufacturers (e.g., for cost accounting or inventory control), networking and CASE products, and a wide range of data processing services, including consulting.

3. *Systems developers* specialize in code for embedded applications as part of a larger product or system, not only in defense but in consumer and capital goods industries. Automobile manufacturers, for instance, may work with such firms in developing programs for engine control (fuel injection, ignition timing) or multiplexing.

In addition, companies in many industries develop their own software, for embedded applications in the products they sell or for internal business functions. Proprietary software can provide a competitive edge through management information or factory automation systems, or through direct delivery of service products, as in commercial banking or airline reservations and ticketing.

While all estimates and predictions must be viewed as highly uncertain (see Box 9-B), there is no question that software expenditures are growing much more rapidly than most other economic activities. Such a trend cannot be sustained indefinitely. It is the consequence of growing demand, spurred by ever cheaper hardware, coupled with nearly stagnant productivity in software generation. In the absence of rising

BOX 9-B: HOW BIG IS THE SOFTWARE INDUSTRY?

Past estimates have omitted or underestimated the value of software developed by firms that do not sell computer programs as such. Table 9-1, showing an estimated total of $100 billion for 1988, is based on a survey conducted by The MITRE Corporation that covered all types of software produced by all U.S. organizations, even if software was not their primary product. Including embedded software and application-specific programs developed by companies for internal use gives a truer picture of the significance of software for the U.S. economy.

TABLE 9-1:
U.S. Software Output, 1988

		Estimated Value (billion $)[a]
Packaged		$26
Independent Software Companies	$10	
IBM	8	
Other Hardware Companies	8	
Custom Programming		26
Defense Applications	8	
Commercial Applications	18	
Embedded Software		33
Defense Systems	18	
Commercial	15	
Other (including entertainment, education, and telecommunications)		15
Total		**$100**[b]

Source: Michael V. Vasilik, David R. Woodward, and Melvyn P. Galin, "Survey of the Software Industry," M90-26 (Bedford, MA: The MITRE Corporation, 1990).

[a]Sales for marketed software; estimated development costs for internally developed programs. The estimates given, based on a survey of 47 secondary sources, rely heavily on information from the Massachusetts Computer Software Council (MCSC). The MCSC survey covered some 800 companies, 44 percent of them employing fewer than 10 people. Other major sources included the Air Force Systems Command, Defense Marketing Services, Defense Science Board, National Software Association, Federal Sources, Inc., and the Software Engineering Institute.

[b]Assuming growth since 1988 at 18 percent annually (the average over the period 1965–1985, according to the Commerce Department) would give a 1990 estimated total of $139 billion, rising to $164 billion in 1991.

MITRE's estimates are much larger than the Commerce Department's figure for worldwide software revenues of U.S.-based firms during 1988—given as "more than $30 billion."[1] The MITRE figures are corroborated by others, however. A recent report by Input Inc. placed 1989 software expenditures by U.S. companies at $94 billion.[2] If we add to this purchases of PC software by individuals, estimated at $3.6 billion for 1989,[3] U.S. government direct purchases ($8 billion), and U.S. software exports ($11 billion), the result is a 1989 total of about $117 billion, comparable to MITRE's figures.

All such estimates are subject to substantial uncertainties, for reasons that include:

- There are no consistent and widely accepted definitions for categories of software.
- The proliferation of computers throughout business and government has led to a corresponding increase in the number of consultants and small service firms, the revenues of which are seldom captured by surveys.
- There is little information on expenditures in industries such as telecommunications that develop a great deal of software internally.
- Many costs are hidden; embedded software, in particular, has often been lumped with hardware or services.

[1]*1989 U.S. Industrial Outlook* (Washington, DC: Department of Commerce, January 1989), p. 26-3.
[2]Jeff Moad, "The Software Revolution," *Datamation*, February 15, 1990, p. 22.
[3]*Business Week*, October 2, 1989, p. 98.

productivity, expansion must eventually slow, if for no other reason than eventual shortages of skilled programmers.

The Defense Sector

DoD is the largest single consumer of software in the world, and defense agencies have long been a major source of funding for software R&D, as they have for computing technologies generally (Box 9-C). As defense electronics have become more complex, software requirements have increased exponentially. So has the time required for development and the costs for both development and maintenance.[8] Reviews of major defense systems show a disturbing pattern of cost growth,

[8]Zraket, "Understanding and Managing the New Benefits and Problems of the Information Society"; Randolph, "The Importance of Technology Development"; Defense Science Board, "Report of the Workshop on Military Software," July 1988.

BOX 9-C: EXAMPLES OF PAST FEDERAL SUPPORT FOR SOFTWARE TECHNOLOGIES

Digital computing became a practical tool during World War II. After the war, DoD support continued.[1] Defense agencies paid for development of the Whirlwind computer at MIT beginning in the late 1940s. Whirlwind, often considered the first reliable, real-time digital machine, was originally intended to drive a Navy flight simulator. Once its potential was recognized, DoD funded Whirlwind as a general-purpose computer to handle real-time control problems. Whirlwind became the basis of the SAGE (Semi-Automatic Ground Environment) air defense system. SAGE, in turn, led to SATIN (SAGE Air Traffic Integration), the first automated air traffic control system, supported by the Federal Aviation Administration (FAA). In those years, spinoff was sometimes direct and immediate: Whirlwind and SAGE were developed for specific defense-related purposes but had wide-ranging consequences for the information industry as a whole. Although both projects emphasized hardware, they required radically new approaches to software. Among the outcomes were the first compilers, assemblers, and interpreters.

During the 1960s and 1970s, the (Defense) Advanced Research Projects Agency funded a series of projects that are still spawning technologies found in both military and civilian systems. Examples include computer graphics, artificial intelligence (AI) and, to a lesser extent, avionics. Both local and wide area computer networks utilize technologies stemming from the ARPANET system. Time-sharing began with MIT's Project MAC, funded by DARPA. The same agency supported the evolution of Berkeley Systems Development's popular releases of the UNIX operating system, originally created at Bell Laboratories. Software technologies that have grown out of DARPA contracts exemplify successful dual use at the R&D level.

Other federal agencies have also contributed. Support from the National Institutes of Health, which was interested in AI for medical diagnosis, spurred commercialization of expert systems. Much supercomputer software, particularly in the early years, originated in the Energy Department's weapons laboratories. Highly reliable, fault-tolerant computer systems—now widely used for transaction processing in the civilian economy—owe their birth to NASA, while software for Apollo and the space shuttle has led to many advances in aircraft flight control systems, notably fly-by-wire.

[1] Kent C. Redmond and Thomas M. Smith, *Project Whirlwind: A Case History in Contemporary Technology* (New York: Institute of Electrical and Electronic Engineers, 1975); Special Issue on SAGE, *Annals of the History of Computing*, vol. 5, no. 4 (October 1983); John F. Jacobs, "The SAGE Air Defense System: A Personal History" (Bedford, MA: The MITRE Corporation, 1986); Kenneth Flamm, *Targeting the Computer: Government Support and International Competition* (Washington, DC: Brookings Institution, 1987); Tomayko, *Computers in Spaceflight: The NASA Experience*; Samuel J. Leffler et al., *History of the UNIX System: The Design and Implementation of the 4.3BSD UNIX Operating System* (Reading, MA: Addison-Wesley, 1989); *The Competitive Status of the U.S. Electronics Sector* (Washington, DC: Department of Commerce, April 1990).

schedule slippage, and performance shortfalls.[9] Some of these problems stem from the DoD acquisition process. Others are technological in origin. Mission-critical software, in particular, has become ever more complex as designers build more and more of the functional capability for ground, sea, air, and space systems into the software portion of these systems. Often, the designers have had no alternative—this may be the only way to meet functional requirements—but the resulting development problems frequently lead to inordinately long procurement cycles, cost overruns, and to systems that are unreliable and cannot be maintained.

DoD software purchases fall into two federal budget categories. The first includes software for administrative functions—e.g., accounting and finance, personnel, and similar information system needs. These procurements are governed by the Brooks Act, which sets governmentwide procedures for the purchase of computer systems and software, and requires reporting to Congress.[10] Within DoD, these purchases are reviewed by the Major Automated Information System Review Council. The second category, which includes software for mission-critical functions such as weapons systems, C^3I (command, control, communications, and intelligence), and cryptography, is exempt from the Brooks Act. The Defense Acquisition Board oversees procurements in this category, as it does for all weapons systems.

DoD buys packaged programs for general-purpose applications much like any other customer, but most of DoD's software expenditures take the form of contracts with a few dozen large firms that have many years of experience in developing embedded code and supplying programming services to defense agencies. These contractors maintain substantial in-house programming capabilities, as do a number of smaller companies that specialize in software for particular DoD applications. In the systems-oriented defense contractors, software development is typically spread over several divisions and locations; we have not been able to determine the extent to which these companies seek to combine or to separate software development responsibilities for military and nonmilitary customers. The group of firms qualified to bid on mission-critical projects and experienced in dealing with DoD

---

[9] General Accounting Office, *Schedule Delays and Cost Overruns Plague DoD Systems*, GAO/IMTEC-89-36 (Washington, DC: GAO, May 1989). Each of eight major DoD information systems examined showed significant cost growth, with average overruns of 90 percent. Schedules for seven of the eight systems slipped by three to seven years; two system development efforts were abandoned after millions of dollars had been spent.

[10] P.L. 89-306, October 30, 1965.

acquisition practices is small compared with the overall size of the software industry.[11] This has the effect of limiting DoD's ability to acquire effective systems that incorporate high-quality software.[12]

DUAL-USE POTENTIAL

Quite apart from applications in military systems, maintaining the ability to produce high-quality software at reasonable cost is crucial for economic competitiveness. Many obstacles stand in the way of coordinating national resources so as to make defense and civil activities mutually supportive. Nonetheless, this should be the goal: software resources are limited today, and bound to be more so in the future.

But is it possible that military software is so special as to offer little potential for dual use? Defense systems must meet exceptionally demanding performance requirements—real-time processing, for example. Even within the defense sector, many computer systems and programming languages are incompatible, the legacy of lengthy procurement cycles and esoteric applications. In the remainder of this chapter, we first examine the general characteristics of software development, then turn to the particulars of military software, concluding with a discussion of policies that would permit DoD to take better advantage of commercially driven software technologies. Some of these policies would at the same time result in military software that was more applicable to commercial needs.

The Software Development Process

Software compared with other technologies. In all high-technology industries—jet engines, computers, microelectronics, telecommunications—up-front R&D and product development costs are high rel-

---

[11] The Defense Science Board Task Force on Military Software counted 24 such organizations in 1987, while International Resource Development identified 50 in 1988. *Report of the Defense Science Board Task Force on Military Software* (Washington, DC: DoD, Office of the Under Secretary of Defense for Acquisition, September 1987); International Resource Development, Inc., "Ada Data," Fall 1988. The U.S. software industry as a whole numbers at least 5,000 firms. Department of Commerce, *The Competitive Status of the U.S. Electronics Sector*, p. 6.

[12] Charles A. Zraket, "Four Vital Issues in C³I," Seminar on Command, Control, Communications, and Intelligence, Guest Presentations, Spring 1989 (Cambridge, MA: Harvard University, Program on Information Resources Policy, 1990).

ative to manufacturing costs. This is not to say that production processes or direct labor are unimportant in such industries, simply that the associated costs tend to be smaller as a fraction of the total than in other sectors. In software this tendency is carried to its logical extreme: production costs fall essentially to zero, with hardly any direct labor content. Generation or development of the program, together with ongoing maintenance and updates, comprises the entire technical activity; as in making a painting or a television commercial, creation equals production. These characteristics of software make it particularly appropriate for exploration of dual use. On average, defense is much more knowledge intensive than the civilian economy (see Chapters 4 and 6), and software is more knowledge intensive than any other defense activity.

Software generation differs from other processes of engineering design and development in that the end product is not a tangible physical object like a robot, a bridge, or indeed a computer itself. Nonetheless, software exists independently of the hardware it will control. As Frederick Brooks points out, the essence of software is abstract, and therefore "the hardest single part of building a software system is deciding precisely what to build."[13] The engineers who design robots or bridges think with the aid of schematic representations of physical reality— sketches, drawings, mathematical models. Such tools enable them to predict behavior, such as the inertia forces created by the robot arm's own motion, or the vibratory modes of the bridge. Software cannot be visualized in any simple way (except with flow charts), and because the program is itself a mathematical construct, further modeling is of little use.

In the initial conceptualization stage, software generation resembles other engineering design activities—open-ended, fluid, not heavily constrained. A near-infinite number of alternative solutions exists, and the designer has little beyond intuition, experience, and whatever formal knowledge may be applicable as a guide to meeting the functional requirements. But in the later design and development stages, software differs. Analytical procedures can help in refining the design of a physical system (predicting performance, optimizing parameters), but not a software system. In many cases, it is impossible to evaluate software quality; few quantitative performance metrics exist. Reliability, for instance, one of the traditional measures for comparing alternative physical systems, has a different meaning for software, which does not "wear

---

[13] Frederick P. Brooks, Jr., "No Silver Bullet: Essence and Accidents of Software Engineering," *IEEE Computer* (April 1987), pp. 10–19.

out" (although the physical media carrying it may). Instead, software reliability refers to the probability of undetected errors that might disrupt performance. Even a relatively simple program may have on the order of 10^{20} possible end-to-end execution paths (depending on the number of loops, branches, and subroutines). Because of this, software is by any reasonable measure far more complex than the hardware it controls and correspondingly more difficult to test and verify.

Tools, techniques, and management. The emerging discipline of software engineering is intended to bring order, and a better set of techniques and methods, to the software development process. DoD and the commercial industry alike have strong incentives to seek both particular tools and integrated "environments" that will make it easier to generate, debug, and maintain code, thereby speeding development and reducing errors. Many of these tools are themselves pieces of software. They range from debuggers, compilers, operating systems, and linkers used during programming, to methods for requirements definition, design capture, documentation, and measurement. A good development environment makes a full range of tools available in such a way that individual programmers and project teams can use them conveniently. In such an environment, the programmer might, for example, invoke a debugger while a program is running, in order to step through its execution and spot errors; that is, the program runs under the control of the debugger. The development environment might also include a syntactic editor that catches syntax errors as code is being created. Integrated environments like this can lead to major increases in programming productivity.

Nonetheless, the power of even the newest tools and methods remains limited. After the general "script" for a piece of software has been established, individual programmers must turn that script into working code. Each programmer will express himself or herself differently, no matter the programming language and no matter how detailed the high-level script. Just as no two engineers or architects will design a building in the same way—even if it is to satisfy exactly the same need—no two programmers will design code in exactly the same way. Nor will the same programmer write a program following the same script the same way on two different occasions.

Software development depends more heavily on individual skill endowments than most engineering tasks. Experienced managers know that software projects ultimately stand or fall not on the sophistication and power of the tools used by their development teams, but on the quality of the programmers in those teams. Case histories demonstrate this. Sometimes a company will assign two independent groups to a given project to minimize the chance of failure. Team A may use one set of tools, completing the task 20 percent faster than Team B and producing better code. Then the teams switch tools, but Team A still comes in 20 percent faster and with fewer errors. Most software managers would agree that such experiences reflect the situation today.

Large software development organizations, furthermore, differ greatly in the extent to which they use formal tools and techniques, and follow standardized and more or less rigorous programming practices. Here the scale and complexity of the application makes more difference than whether or not DoD is the customer. A single creative designer can block out the script for a PC application package. But no one person can comprehend a large-scale system, such as an automated air traffic control network or a complex military C^3I project, or supervise its design, construction, integration, and testing; these tasks require large groups of experienced professionals.

Efforts in industry to create better tools and techniques and more effective management practices have their parallels in universities, where courses, curricula, and research in computer science and engineering seek to establish more rigorous foundations for software. Even so, differences among individuals continue to owe more to talent than to training. This has implications for management practices, particularly when it comes to large-scale systems. It could be, for example, that software development does not scale up the way other engineering tasks do, perhaps because we do not yet understand all the relevant technical issues.[14] Managers who fail to appreciate the vagaries of large-scale software development, and try to apply techniques common to other engineering projects, invariably find themselves frustrated and their efforts frequently counterproductive. The reason seems to be that in most engineering work, the "artful" stages are limited primarily to conceptual design, ending relatively early in the project cycle. After this, rigorous, science-based analysis underlies many of the remaining technical decisions. Software development never reaches such a point; it remains art more than science to the end. As a consequence, software projects cannot be managed like other engineering activities.

Military Software

DoD's software needs can be divided into three functional categories: systems operation and support; embedded software; and C^3I. The first

---

[14] Brooks, "No Silver Bullet."

BOX 9-D: SOFTWARE SECURITY

DoD sometimes imposes security requirements on system software—e.g., for encryption—well in excess of what a bank, for example, might demand for safeguarding financial information and ensuring against unauthorized funds transfers. But even in such cases, enhanced versions of commercially developed software usually prove adequate. With continuing demand for secure, reliable, and validated commercial transaction-processing systems, there may be common ground for developing network control and access code that requires little modification to meet defense requirements. In mission-critical defense systems, whether embedded (as in an airplane) or C³I, the most critical pieces of software will normally be protected through encryption, authentication, and code validation, with multilevel and compartmented controls in network environments. Mission-critical applications code will continue to be custom-built. But if validated, reusable code is available—including modules originally developed commercially—that can be augmented with new, application-specific programming, there is no reason that military security requirements cannot be met.

---

category—which includes operating systems, database management, and network management—has broad potential for dual use, while the second has very little. C³I is an intermediate case. Security concerns cut across all three (Box 9-D).

Systems operation and support. In this category, military and commercial needs differ little if at all; in fact, DoD relies almost exclusively on commercially developed programs. There is no reason why dual use should not be the norm. The primary obstacles are institutional, having to do with the timing of military purchases.

Because the procurement cycle and operational life for major military systems spans several decades, software for information management and support often lags far behind commercial practice. Systems code is highly profitable, which increases the gap because software firms keep pushing new products into the civilian marketplace. Today, many of DoD's information processing systems are obsolete by commercial standards in terms of software structure, operating speed, and information display techniques. The Pentagon has recognized the problem and is seeking ways to streamline the procurement cycle for systems operation and support software. Because DoD remains a major customer, policies encouraging dual use could have broad impacts.

Embedded software.   The category of embedded programming includes mission-critical software that functions as an integral part of a larger military system. The very demanding performance requirements, complexity, and specialized nature of embedded code have few parallels in civilian systems, which limits prospects for dual use. Real-time processing—e.g., of radar signals during combat—is a common requirement. While high speed might be desirable in large-scale commercial applications, it is rarely so critical; air traffic controllers have more time to make decisions than fighter pilots. Reliability requirements may also be more demanding: the possible consequences of a worst-case failure in, say, a strategic weapons system dwarf those even for a nuclear power plant.

Embedded military software must handle enormously complicated integration tasks—for example, managing the Aegis radar and fire control system or the many electronics systems in a B-2 bomber. The B-2's software must oversee and coordinate the functioning of avionics, active and passive surveillance, electronic countermeasures, smart weapons, and the intelligence systems on board. Each of these has its own subsystems, and all must function as a harmonious whole, in real time, with extremely high reliability.

The software required by a Strategic Defense System (SDS) illustrates the entire range of technical difficulties found in embedded systems.[15] SDS software would need to be highly distributed, deal with very high throughputs and short response times, and be ultrareliable and secure. Millions of lines of code would need to be developed concurrently with the rest of the system. The software would have to control surveillance, damage assessment, tracking, target discrimination, weapons assignment, weapons control and guidance, network routing and control, security-access control, and system fault tolerance and fail-safe operations. Because an SDS could not be tested in advance of actual combat, design, development, and construction would have to rely on simulations.

SDS software represents the extreme case, but the list of civil-sector applications with requirements even remotely comparable to embedded military systems is a short one: commercial telecommunications, air traffic control, and space information networks such as satellites for communications, weather, and navigation. At component and subsystem levels, code may sometimes be transferable, but software that serves integration functions will normally have little dual-use potential.

---

[15] Charles A. Zraket, "The Challenge for SDS Software," keynote address, Second International Software for Strategic Systems Conference, sponsored by the Army Systems Development Command, Huntsville, AL, October 25–26, 1988.

This is not to say, of course, that tools and techniques for improving the development process would not be applicable to embedded military software.

Command, control, communications, and intelligence. C^3I software, the third and final category, functions as part of mission-critical systems that communicate, correlate, analyze, and interpret information, and provide decision support. These tasks arise in the civil sector too, albeit in other contexts. But if there is potential for dual use, it has not been fulfilled—perhaps because the complexity of particular applications obscures underlying similarities. Indeed, it has been hard for DoD to transfer software and expertise developed for one C^3I program to another. This will be a growing need, as C^3I systems grow more complex, along with the huge data-intensive information systems of the National Security Agency.

If, instead of a unique design for each case, blocks of code could be developed in modular fashion for later reuse, complex information gathering and interpretation systems could be developed far more efficiently. Unfortunately, as we will discuss shortly, reuse of software remains a promise more than a reality. While some work has been done to define modular software packages that could be transferred from one application to another, industry has been slow to recognize the potential of reusable code.[16] Because reuse demands rigorous programming practices, which can mean longer initial development cycles and heavier up-front investments, companies will take this path only if they can expect greater profits. DoD acquisition discourages such investments. For example, government regulations give DoD unlimited rights to software generated under contract. A company that develops modular software stands to lose control of the modules. Finally, cost-based DoD contracts, with profits typically set as a percentage of costs, make it more lucrative to create new software than to reuse existing code.[17]

---

[16] Japan appears to have made progress with their "software factories"—so called because the intent is to put software on an industrial footing, with development becoming routine, hence more easily managed. By specializing in particular applications—e.g., process-control packages for nuclear power plants, or aircraft flight controls—a given organization can more easily reuse blocks of code and train programmers narrowly but deeply. Thirty percent or more of a given package may be recycled from past programs, yielding improvements in both quality and productivity; some years ago Toshiba's Software WorkBench claimed an error rate of 0.3 bugs per thousand lines of code, a factor of 10 below typical U.S. error rates. D. Brandin et al., "JTECH Panel Report on Computer Science in Japan," Science Applications International Corporation, La Jolla, CA, under contract no. TA-83-SAC-02254 from the U.S. Department of Commerce, December 1984.

[17] Office of Technology Assessment, *Holding the Edge: Maintaining the Defense Technology Base, Vol. 2:* Appendices (Washington, DC: U.S. GPO, January 1990), Appendix F.

How Much Dual Use?

In software, as in other technologies, it is easier to envision dual use in a context of design and development tools or components (e.g., program modules) than at the level of end products and systems. Computer hardware, for instance, incorporates many standardized components and subsystems, some of them commodity items (e.g., semiconductor memory chips and disk drives). But software cannot be assembled in this way until modular designs with clean interfaces become the rule. Such modules would have to be standardized, with functions well characterized and documented, to make dual use and reuse simpler and more practical.

Commercial to military.   Military demand is growing for large-scale communication systems, networking, expert systems and other forms of AI, and telerobotics. In all these cases, civilian demand has spurred rapid technological progress. Escalating procurement costs could stimulate the Pentagon to search out commercial software, even though modifications might be required. Software for controlling industrial robots, for example, could offer a baseline for military-qualified systems; possibly a significant portion of code could itself be transferred. If the Army seeks robots for loading ammunition onto trucks, an application much like those of interest in civilian industry, it makes no sense to start from scratch. To promote dual use in such cases, the services must recognize the similarities between commercial applications and military needs; too often, they focus on the differences.

Military to commercial.   Avionics has been the chief success story in transferring military software to the civil sector. Navigation systems, digital engine and flight controls, and communications equipment have all been based on technologies flown first in military aircraft; development and testing practices followed guidelines laid down for military software. Computer-based training also provides examples of successful transfers. Flight simulators depend on software for realism and effectiveness. Honeywell and other companies developed simulators first for training military pilots, later supplying them to airlines and other civilian customers. Much the same has happened with foreign language training. In the future, as training and simulation software spread through the economy, the flow may reverse, with leading-edge products transferred back to the military. Technology transfer also occurs when the same programmers and design personnel work on military and commercial assignments in such companies as IBM and Boeing. IBM, for example, has found that managers with experience in its Federal Sector Division can successfully apply lessons from DoD software contracts to large-scale projects elsewhere in the company (Chapter 6).

Programming languages. At the end of the 1950s, a DoD task force defined the specifications for COBOL, which quickly became the most widely used of all programming languages.[18] But commercial needs have long since taken precedence in language design, while DoD continues to struggle with an inventory of software written in more than 200 languages, many of them obsolete and most of them obscure. In an effort to deal with this problem, DoD has sought to standardize on the Ada language, which originated in a design competition held in the late 1970s. Ada incorporates many structured programming features intended to promote good coding practices. By encouraging software designers to partition their programs into distinct, easily understood, and easily maintained modules, it induces them to pay attention to overall structure and function before starting to write code. The result is software that is easier to debug, easier to modify, and often easier to maintain. Most existing DoD systems have been built with relatively primitive, unstructured languages. By mandating the use of Ada as a standard, DoD has sought and should achieve greater productivity in program development.

Industry as a whole, however, has not accepted the language, although Boeing, for one, has adopted Ada for the avionics systems in its commercial planes. Most companies remain committed to other higher-order languages, seeing more costs than benefits in switching to Ada. There are far more choices today than in the early years of COBOL and FORTRAN, including a number of object-oriented languages, such as C and its variants, that run more efficiently than Ada. The older languages also remain popular, in part because so many people have experience with them. Nonetheless, more civilian than military programmers work in Ada today—a fact that reflects the greater size of the commercial sector, along with the slow pace of weapons systems development.

Software tools and practices. Computer-aided software engineering has been a major focus for both DoD and commercial developers. Much like the computer-assisted tools and techniques used by engineers in other fields, CASE methods provide a framework for program development and automate some design, implementation, configuration management, code documentation, and maintenance tasks. A code generator, for instance, restructures existing software as modularized, reusable

---

[18] Flamm, *Targeting the Computer,* p. 76. The worldwide inventory of programs written in COBOL, still popular for business applications, probably exceeds 75 billion lines of code.

units with clearly defined functions. Automation requires substantial investments in training because the new tools and environments force programmers to follow a prescribed development model, and perhaps to use an unfamiliar language. However, the resulting code is usually much easier to test, and almost always easier to maintain and reuse.

DoD, in its source selections, strongly favors companies that have mastered CASE methods, and have made the necessary investments in hardware, software, and training. Some of today's best CASE tools and development environments are being developed for Ada under DoD contract, suggesting that the military could continue to be a major driving force. NASA has adopted Ada for both ground-based and flight software for the space station. In the years ahead, NASA expects to fund extensive R&D on the CASE tools that will be needed by contractors developing software for the station. The FAA has also chosen Ada for its new automated air traffic control system, the National Airspace System, and will support Ada CASE tools. Taken together, these activities should increase the attractiveness of Ada for commercial applications.

Cost estimation plays an integral part in managing complex software projects. In acquiring software, DoD must constantly balance performance, cost, and risk for systems and their components, keeping the entire life cycle in mind; commercial firms must do the same. Estimating costs for software development and maintenance has been particularly difficult because of the highly variable human factors we have discussed. For many years, DoD has funded work on software cost estimation methods, including assessments of different programming techniques, tools for tracking progress through the various phases of development, and measurements of programmer productivity. Again, industry has benefited: many state-of-the-art cost-estimation models have their roots in projects funded by defense agencies during the 1970s at IBM, TRW, and RCA.[19] Recently, the Software Engineering Institute (SEI) at Carnegie-Mellon University, a DoD federally funded research and development center, has developed an evaluation package for measuring a bidder's level of software process maturity; this can be used by contractors for self-evaluation and by the government in its selection processes.

[19] Barry W. Boehm, *Software Engineering Economics* (Englewood Cliffs, NJ: Prentice-Hall, 1981); F.R. Frieman and R.E. Park, "PRICE Software Model—Version 3: An Overview," *Proceedings, IEEE-PINY Workshop on Quantitative Software Models* (New York: Institute of Electrical and Electronic Engineers, October 1979), pp. 32–41.

Portable software.   Both military and civilian users have recognized the advantages of software that is portable in the sense that it can be used independently of specific hardware platforms or operating system software. Motives have sometimes differed on the two sides of the economy, however. The Federal Acquisition Regulations governing purchases of software and hardware by DoD and civil agencies emphasize competitive procurements, discouraging sole-sourcing and fostering purchases of portable software that can be used with hardware from different vendors. Indeed, this has been one of the reasons for the Pentagon's long-standing sponsorship of Ada. At first, DoD hoped that entire operating systems and database management systems would eventually be written in Ada, breaking the tie between operating systems and hardware, and enabling full competition in hardware procurements. But in part because the commercial sector did not embrace Ada in its early years, no hardware-independent operating system written in the language emerged.

Until recently, many commercial users were in a different situation. Locked into the products of a single hardware manufacturer, they had little reason to care about portability. With the recent wave of mergers and acquisitions in corporate America, many businesses now take a different view. The problems of combining the information systems in newly merged organizations have led to greater awareness of the cost-saving potentials of portable software. Independent software vendors have their own incentives to develop application packages that can operate on more than one hardware platform or more than one operating system. A software firm can increase its potential market with little or no increase in costs if it can avoid the need to maintain separate versions of its products.

While portability has been more common for application programs than for operating system support software, the UNIX operating system (written in the C language) has now been implemented on hundreds of hardware platforms. DoD-sponsored research contributed to several key versions of UNIX, which has become the most common portable operating system, benefiting government and commercial users worldwide. At present, UNIX application programming interfaces have yet to be standardized, although government and industry groups are this time united in pursuing the standardization of UNIX, its divorce from hardware dependencies, and its application-level interfaces. One result of these efforts will be the emergence of application software that exhibits true portability among hardware platforms running standardized components of UNIX.

POLICIES

The federal government faces two sets of policy questions when it comes to software. First, how can DoD more effectively satisfy software needs that already absorb more than $30 billion annually? Second, how can government help ensure a continuing technological and market lead in an industry that is vital for U.S. competitiveness? These questions are related through R&D and acquisition policies. Low productivity in programming affects both military and commercial software; productivity can be improved only through more effective tools and management practices. The human resource base for software also merits attention, along with intellectual property protection. Within DoD, acquisition practices will need continuing review. Many of the policies we are about to discuss would encourage dual use implicitly, but DoD acquisition reform must be the starting point for any explicit dual-use strategy.

R&D and Diffusion

Government, better than industry, can afford to invest in long-term projects, and within government it is DoD that has the greatest need for software. In earlier years, as illustrated in Box 9-C, defense agencies provided liberal funding for computer research and for risky development efforts. But during the 1980s, DoD reoriented much of its software R&D toward short-term mission needs and prototyping. Tellingly, DARPA's Strategic Computing Program, a major undertaking for the agency, has been far less fruitful in yielding widely applicable software technology than earlier, less mission-oriented undertakings.

To help meet U.S. needs in software R&D, the federal government could do the following:

- Direct DARPA and other defense agencies to return to earlier approaches for the support of *long-term, well-funded basic and applied research*. Whether DARPA continues its mission-oriented thrust, a long-range approach is needed to strengthen the foundations for software engineering. Recent DARPA CASE efforts suggest that the agency may be moving in this direction.
- *Fund more software research through nondefense agencies;* for example, work on software engineering metrics (including analysis of what should be measured and when during the course of a project). University groups might conduct controlled experiments to assess the utility of particular metrics, tools, and methods. A

high-priority task appropriate for the Commerce Department would be the establishment of national and international standards for program development and for software interfaces. Widely accepted standards will encourage dual use, reuse, and greater U.S. exports of software products.

- Support *diffusion of information on best practices in software engineering.* The DoD-supported Software Engineering Institute has influenced university curricula throughout the country, as well as continuing education and training for software professionals. SEI is well placed to encourage dual use through technology diffusion, for example, by serving as a clearinghouse for evaluation of emerging CASE tools.

More generally, government must recognize that software engineering is a field just as important as nuclear physics or solid-state electronics and that it deserves continuing support through both research and human resource development.

Human Resources

Despite many predictions of coming shortages of software specialists, serious problems have rarely emerged, here or in other countries. Yet the possibility cannot be ruled out. Software is a relatively new field. It has grown by attracting people with training in other scientific and engineering specialties. But as the disciplinary underpinnings of software engineering become established, lateral movement will grow more difficult and become less satisfactory as a source of skilled professionals.

Over the long term, the much-criticized state of primary and secondary schooling in the United States could pose particularly serious problems for fields like software, where grounding in mathematics is the prerequisite for specialized study. On standardized international tests of mathematics and science, American students compare poorly with their counterparts in other industrialized countries.[20] Relatively few high school graduates enter college with the preparation needed to pursue a technical major such as computer science or engineering. Industry snaps up those who do graduate, offering salaries high enough to divert promising candidates from pursuing advanced degrees. The consequent lack of Ph.D.s contributes to a shortage of faculty, and the

---

[20] A.E. Lapointe, N.A. Mead, and G.W. Phillips, *A World of Differences: An International Assessment of Mathematics and Science* (Princeton, NJ: Educational Testing Service, 1989).

cycle perpetuates itself. Because the greatest demand in industry has been for high-level professionals with advanced degrees—systems analysts and designers, rather than programmers—industry itself suffers from the self-perpetuating cycle (Box 9-E). While improvements in software productivity would help deal with overall shortages of personnel, the most promising CASE tools call for the highest skills.

The first step to increase supplies of software professionals is to strengthen university teaching and research programs. As yet, the field has developed little of the sense of identity that gives focus to established technical disciplines. Software coursework has been spread thinly across departments of electrical engineering, computer science, and mathematics. Universities teach many courses in computer science, but do not offer nearly enough software engineering, particularly at the graduate level. Courses and curricula are needed that focus on the design of complex software packages, on programming aids for improving productivity, and on understanding and satisfying user needs—a perspective closer to engineering design than to the applied mathematics emphasis found in most computer science programs.

Software engineering in the universities will not be enough. Since it began, the field has grown by attracting people trained in other fields—physics, engineering, mathematics. With software projects becoming larger and more complex, software engineering tools becoming more sophisticated, and theoretical foundations for software design beginning to emerge, training must become more systematic, both for people seeking to move laterally into software and for experienced professionals seeking to update their skills. Federal and state governments should also expand support for intermediate-level software training in community colleges, and work with local bodies to improve the quality of primary and secondary schooling.

At the federal level, desirable policy initiatives include the following:

- DoD and the National Science Foundation (NSF) should increase their *support for graduate study and curriculum development in software engineering,* with a particular focus on large-scale systems. Industry partnerships should be encouraged. Under its Engineering Research Centers (ERC) program, NSF already supports several university centers oriented toward industry's software needs. There is room for more such ERCs.
- Federal agencies should fund the preparation of *case histories of large-scale software development projects.* As teaching aids, these could introduce students to the technical and managerial lessons

---

BOX 9-E: SOFTWARE EMPLOYMENT

Projections by the Bureau of Labor Statistics (BLS) suggest that the United States will need about 1.4 million software professionals in the year 2000, an increase of some 400,000 over current levels.[1] In recent years, U.S. employment of systems analysts and designers has grown at about 10 percent annually, five times the rate for programmers.[2]

Software specialists find jobs in many sectors of the economy. In 1988, for instance, an estimated 118,000 software professionals worked in durable goods manufacturing, fewer than half of them for computer manufacturers.[3] Currently about 700,000 Americans hold jobs in the computer and data processing services industry (SIC [Standard Industrial Classification] 737), but only about a quarter of them qualify as software professionals.

Bureau of Labor Statistics projections suggest that employment will grow faster in SIC 737 during the 1990s than in any other major industry, reaching a total of about 1.2 million in 2000 under the bureau's moderate growth scenario. This corresponds to an employment growth rate of 4.9 percent annually. BLS expects output in SIC 737 to increase at only 4.3 percent annually over the period 1988–2000, slower than employment, thus implying negative productivity change. BLS projections show similar behavior for a number of personal and retail service sectors, but for only two minor manufacturing industries. This poor outlook for productivity provides another view of the productivity bottleneck in software.

[1]George Silvestri and John Lukasiewicz, "Projections of Occupational Employment, 1988–2000," *Monthly Labor Review*, November 1989, pp. 42–65.

[2]John Morrocco, "Coming Up Short in Software," *Air Force Magazine*, February 1987, pp. 64–69. Relatively little is known concerning the division of labor in software development, but see Philip Kraft, *Programmers and Managers: The Routinization of Computer Programming in the United States* (New York: Springer-Verlag, 1977).

[3]*Handbook of Labor Statistics*, Bulletin 2340 (Washington, DC: Department of Labor, August 1989), pp. 194–195.

---

of such projects as Honeywell's Multics effort.[21] In addition, careful research based on case histories could help identify, analyze, and disseminate experience from past projects—both successes and failures—concerning tools, techniques, and management methods.

- Government should cooperate with universities, professional societies, and accrediting bodies to encourage *coursework on software development tools and environments* in computer science programs.

[21]As discussed in Frederick P. Brooks, Jr., *The Mythical Man-Month* (Reading, MA: Addison-Wesley, 1975).

- Industry, in cooperation with universities and with financial support from government, should support *internships* for graduate students studying computer science and software engineering. Internships should include participation in large-scale software projects.
- Industry and the universities, in conjunction with professional societies, should expand their *training and retraining* programs. As software tools grow more sophisticated, those already working in the field must update their skills, while software will continue attracting people with degrees in other disciplines, for whom on-the-job training is no longer good enough. Government could not only share in the costs of training, but provide incentives—for example, by ensuring that company tuition payments retain their tax-exempt status.

DoD Acquisition

When practices in military software development lag behind the state of the art, as they often do, the reasons generally lie in the acquisition process. While a host of disincentives stand in the way of commercial off-the-shelf software purchases, the picture has a bright side: substantial improvements are possible if DoD can encourage dual use.

To meet more of its needs through COTS software, DoD will have to permit program managers to balance the costs and functionality of available software against that of specially developed code. At present, it is difficult or impossible to trade off system requirements against one another, which prevents the use of proven COTS software and program modules. In addition, contractors that develop software for DoD must, under current policies, relinquish all "rights in data" to the government. The intent is to promote competition among firms seeking federal business through open access to the results of government-funded development efforts. But without protection for intellectual property rights—which is available, albeit imperfect, for software sold commercially—many firms have been reluctant to bid for DoD software contracts.[22] DoD, for its part, finds itself cut off from available

---

[22] The usual means of protecting intellectual property, copyrights and patents, have limited value for computer programs, particularly in foreign markets where copying of U.S. software has often been rampant. Technical solutions intended to make copying impractical or impossible have not worked; software pirates quickly find ways around new protective schemes, no matter how ingenious, just as thieves manage to keep on stealing automobiles. The

software because companies fear they will lose their ability to protect even code developed with their own funds.[23]

Current DoD acquisition policies likewise fail to encourage the use of proven, up-to-date software engineering methods. Indeed, they perpetuate older, poorer methods. Brooks cites rapid prototyping as one of the most promising methods for more effective software development.[24] However, DoD's software development standards discourage early validation of requirements and design via prototyping followed by evolutionary development—techniques that have proven successful in commercial practice.[25] Contractors also face disincentives for investments in automated CASE tools because DoD regulations make it difficult to recover the up-front costs.

Steps to improve software acquisition policies include the following:

- DoD should establish and follow a *unified software technology plan*. The plan should specify funding levels and other actions, both near and long term, for improving software productivity and quality, and for enhancing dual use. It should emphasize CASE tools, improved programming environments, and reuse of code. DoD should make complementary investments in prototyping technology, distributed and parallel processing software, application generators, object-oriented languages, automated specification aids, trusted (reliable, secure) software, and architectures and tools tailored to each level of software development activity.
- The federal government should review *rights-in-data rules for DoD procurements*, seeking modifications that would permit greater use of commercially developed software. Options include giving

---

problems stem from the fluid and abstract nature of software, which makes multiple solutions to the same set of design requirements easy. Office of Technology Assessment, *International Competition in Services* (Washington, DC: U.S. GPO, July 1987), pp. 318–321. Moreover, U.S. policies for protection of software reflect the implicit assumption that American firms will continue to dominate the industry and technology, rather than the reality of a global software industry that is now emerging. See National Research Council, Steering Committee on Intellectual Property Issues in Software, *Intellectual Property Issues in Software* (Washington, DC: National Academy Press, 1991).

[23] According to the Rights-in-Data Technical Working Group of the Institute for Defense Analyses, DoD's requirements mean that "the government is failing to obtain the most innovative and creative software technology from its software suppliers. Thus, the government has been unable to take full advantage of the significant American lead in software technology for the upgrading of its mission-critical computer resources." Cited in OTA, *Holding the Edge, Vol. 2*, p. 112.

[24] Brooks, "No Silver Bullet."

[25] Defense Science Board, "Report of the Workshop on Military Software."

contracting officers more flexibility to acquire COTS software; ensuring life-cycle support for COTS software through a government-supplier agreement or an optional directed licensing clause in conjunction with an escrow arrangement; and a mixed funding approach for allocating rights to code developed on a cost-shared basis between government agencies and private firms.[26]

- DoD should improve other relevant *acquisition practices and contracting procedures* through such steps as revising DoD Directive 5000.29 (Management of Computer Resources in Major Defense Systems) to streamline project management; reinterpreting and if necessary revising Military Standard 1521 (Technical Reviews and Audits for Systems, Equipments, and Computer Software) to encourage reusable software; and further revising Military Standard 2167A (Defense System Software Development) to promote up-to-date practices such as rapid prototyping and incremental development. DoD should also seek to avoid cost-plus software contracts when these would discourage reuse of code (e.g., in C^3I systems).

THE COMING COMPETITIVE CHALLENGE IN SOFTWARE

In part because of the early impetus provided by defense spending, U.S.-based software firms remain undisputed leaders in world markets. In earlier years, software technologies frequently flowed from military projects to civilian applications, but a more balanced two-way flow is now possible. Greater dual use would contribute to national security while helping preserve U.S. competitiveness in a strategically vital industry. This is important because U.S. dominance in software seems bound to wane over the next several decades, perhaps slowly, but nonetheless inexorably.

Some Americans seem to think that software is special—that countries like Japan will be unable to succeed as they have in other technologies and other industries. This view is wrong. American firms may be the innovators, but software will continue to demand painstaking attention to detail. Japanese firms have amply demonstrated their skills in high-quality, error-free products in other industries; already, they

---

[26] A. Martin and K. Deasy, *Seeking the Balance Between Government and Industry Interests in Software Acquisition, Vol. 1: A Basis for Reconciling DoD and Industry Needs for Rights in Software,* SEI Report CMU-SEI-87-TR-13 (Pittsburgh, PA: Software Engineering Institute, 1987).

have begun applying these skills to software.[27] Many other countries also have policies in place intended to improve their capabilities in software development. In contrast, the United States has so far done little to protect and extend its lead. It should, for software is perhaps the most critical of all so-called critical technologies.

- Not only is software vitally important for improving productivity and competitiveness throughout the economy, but DoD faces urgent needs for controlling software expenditures and improving software quality. Substantially greater dual use is possible and would help both civilian and military users.
- Productivity growth in the generation of software has itself been slow. This results in rapidly rising costs. Software—like other low-productivity-growth sectors—thus accounts for a steadily increasing fraction of the nation's GNP.
- Slow productivity growth in software stems in large part from a lack of unifying technical principles and proven management practices. These should be the subject of focused federal R&D and human resource development programs.

Even if there were no competitive challenges in sight, the United States would have to deal with its domestic software bottleneck. If it does so successfully, American firms will be able to move down the software learning curve ahead of their rivals overseas, and DoD will be able to procure effective systems at more affordable costs.

---

[27] Michael A. Cusumano and Chris F. Kemerer, "A Quantitative Analysis of U.S. and Japanese Practice and Performance in Software Development," *Management Science*, vol. 36 (1990), pp. 1384–1406.

10

Manufacturing: An Agenda
for Competitiveness

Most manufacturing technologies are inherently dual use: defense firms rely on the same machine tools and semiconductor fabrication equipment as commercial enterprises. Materials technologies, likewise, have broad applicability, while new methods of fabrication are needed to take advantage of lightweight fiber-reinforced composites, optical fibers, or heat-resistant ceramics. Defense has often pioneered new materials and new processes alike.

But manufacturing has not been a high priority in the United States. Competitiveness has declined, with American firms falling behind their overseas rivals in product design, production costs and quality, worker skills, and organizational flexibility. Although defense requirements have led to many innovations in manufacturing technology, the Defense Department has failed to reap the full benefits because of its focus on performance to the exclusion of costs. On both sides of the economy, the problems begin with underinvestment in process R&D, and continue with slow diffusion of best practices to industry, particularly to smaller firms and to second- and third-tier contractors.

The generic nature of manufacturing processes makes them a natural candidate for federal support. The last section of this chapter outlines policies for improving U.S. capabilities in both civil and military production. The strategy includes both technology development and investments in production equipment. Explicitly dual use, and capitalizing on U.S. strengths in science, it would focus on technologies

314

suited for quick response to changing requirements through the family of flexible automation technologies known as computer-integrated manufacturing.

MANUFACTURING IN DUAL-USE PERSPECTIVE

The same basic processes are used in making military and civilian products. Functional requirements (e.g., product designs) may differ, along with lot size and length of production run, but manufacturing is a dual-use technology par excellence. The Department of Defense has supported technologies such as numerically controlled (NC) machining that have had broad and deep impacts on manufacturing operations worldwide.[1] This chapter focuses on small- and medium-lot batch production, also drawing on examples from materials technologies to illustrate a closely related aspect of dual use.

Much of the output of the U.S. manufacturing sector is made in lots of 1 to 10,000. This is true for most military components and systems (leaving aside expendables such as munitions), and for a surprisingly large range of commercial products, capital goods in particular (see Box 10-A). Nearly one-quarter of the U.S. manufacturing workforce is employed in the metalworking sector (which includes fabrication of plastic and composite parts), in which batch production accounts for perhaps three-quarters of output by value; three-quarters of that output is made in lot sizes of less than 50.[2] Metalworking technologies are pervasive; in the production of computers, for instance, mechanical components and assembly account for about half of total production costs.[3]

---

[1] We use the term numerical control, or NC, to refer to any and all varieties of production equipment controlled by digitally encoded instructions—whether through a tape reader, as in the early years, or under the direct control of a computer or microprocessor, as has been common since the 1970s. Appendix 10-A summarizes the original development of NC under Air Force sponsorship.

[2] Maryellen R. Kelley and Harvey Brooks, "The State of Computerized Automation in U.S. Manufacturing," John F. Kennedy School of Government, Harvard University, October 1988, p. I-9. The major exceptions to dual-use commonality include mass-produced consumer goods (automobiles, household appliances, some electronics) and continuous process industries such as paper making and commodity chemicals. Military purchases of, say, jet fuel come from the same refineries that supply commercial customers.

[3] H. Barry Bebb, "Quality Design Engineering: The Missing Link in U.S. Competitiveness," paper presented at the NSF Engineering Design Research Conference, University of Massachusetts, Amherst, June 11–14, 1989.

BOX 10-A: BATCH PRODUCTION: EXAMPLES FROM AEROSPACE

The jet engine business provides a classic illustration of small-batch production of precision parts. General Electric and Pratt & Whitney each build 500–1,000 medium and large jet engines annually, in generally similar military and commercial versions. GE's CFM-56 series, for example, which powers such planes as the Boeing 737, uses the F-101 core (hot section) originally developed for the B-1 bomber.

As much as one-third of annual revenues, which run about $5 billion for the engine divisions at GE and P&W each, come from sales of replacement parts.[1] P&W, with some 17,000 engines in service worldwide, supplies parts for 60 different models. In doing so, the company relies heavily on flexible automation to keep its inventories low without forcing customers (whose planes may be out of service) to wait too long for repair parts.

Airframes provide an equally instructive example of dual-use manufacturing. Airplanes are assembled from thousands of individual components—extrusions, castings, and forgings (many of them extensively machined), sheet-metal parts, and a growing volume of composites. Half of all sheet-metal parts for aerospace applications are made in lot sizes of less than 20.[2] Fasteners such as bolts and rivets number in the hundreds of thousands. Some 100,000 holes must be made in producing a single jet engine; many of these holes are now cut with laser beams instead of drill bits. An airplane's total parts count, including propulsion system and avionics, can easily exceed three million. Many of the components and subassemblies come from vendors that specialize in hydraulics or landing gear, supply raw forgings and castings, or engage in subcontract fabrication. Military aircraft push the state of the art in manufacturing, even when the processes are traditional ones; machining titanium, for example, is much more difficult than machining aluminum, while polymer matrix composites require entirely new processes.

Because production quantities are small, aircraft manufacture remains labor-intensive; much of the work is highly skilled—more like making pianos than cars. The vast number of parts and components that must be brought together mean that cost control and on-time delivery depend heavily on production planning and shopfloor management. Batch production, even more than mass production, is a messy business, beset with unpredictable disruptions. At intervals, this stage or that of the production system may threaten to go out of control. The

[1] William M. Carley, "How Pratt & Whitney Lost Jet-Engine Lead to GE After 30 Years," *The Wall Street Journal*, January 27, 1988, p. 1.

[2] Kim A. Stelson, "Sensors for Real Time Control of Sheet Metal Forming," *Department of Defense 1987 Machine Tool/Manufacturing Technology Conference, Vol. 2: Forming*, AFWAL-TR-87-4137 (Wright-Patterson AFB, OH: Air Force Wright Aeronautical Laboratories, June 1987), pp. 219–241.

reasons may be obvious: a bad instruction for an NC machine, or a temperature probe in a vacuum furnace that stops working. In other cases, diagnosis is more difficult. When ultrasonic inspection shows poor bonding of the inner plies of a graphite-epoxy wing skin, this might be a systemic problem or a random, isolated mistake. Perfection in production is impossible, satisfactory performance elusive.

Success in producing high-quality goods at low cost—computers or submarines—stems from highly developed production systems that effectively couple product and process design, work organization, and shopfloor management. To compete effectively on the civilian side, American manufacturers will have to match their Japanese rivals in the design of products and processes for less expensive manufacture, better quality and higher yields, and greater flexibility in adapting to changes in design and output level (Box 10-B). This is a tall order, as illustrated in Chapter 8 for the case of microelectronics.

To maintain adequate military capabilities in the years ahead, the United States will have to design, develop, and produce defense systems with the needed performance at more affordable costs—an equally tall order. Price tags on military equipment—whether computer-based command-and-control systems, or bombers and tanks—have been rising, especially over the past decade, at rates well above that of inflation. Such a trend cannot continue indefinitely. Cost-effective defense requires higher priority for excellence in manufacturing.

From Design to Production

Manufacturing costs depend on the design of the product in question. Good design practices can minimize the total number of parts and components in a product or system, simplifying assembly and saving on both direct and indirect labor. Good design practices also decrease rejects and rework, resulting in better quality and reliability and lower life-cycle costs; experience shows that engineering design and development, rather than manufacturing, is responsible for half or more of all quality and reliability problems.[4] American firms have fallen behind

---

[4] Office of Technology Assessment, *International Competitiveness in Electronics* (Washington, DC: U.S. GPO, November 1983), pp. 222–224; National Research Council, Committee on Engineering Design Theory and Methodology, *Improving Engineering Design: Designing for Competitive Advantage* (Washington, DC: National Academy Press, 1991).

BOX 10-B: JAPANESE ENGINEERING AND MANUFACTURING

As hard as it has been for Americans to accept, the competitive success of Japanese firms stems largely from superior engineering design and manufacturing practices. These in turn reflect differences in corporate management (and, to be fair, financial markets and a host of other variables). Xerox, for example, has found that competing Japanese products have production costs lower by as much as 50 percent, with design responsible for half the difference. Details such as wall thicknesses, tolerances, inserts, and finishing accounted for much of the Japanese cost advantage.[1]

Japanese companies have done a better job of integrating design and production. They bring new products to market more quickly— four years or less for a new automobile model, versus five or six years here—because they manage product/process engineering more tightly.[2] Japanese firms have been quicker to adopt many of the simultaneous engineering techniques discussed later in the chapter. Their design teams also seem to do a better job of capitalizing on lessons learned during past projects.

Manufacturers in Japan grasped the strategic advantages of programmable automation more quickly than companies in either the United States or Europe.[3] Japanese firms have more NC tools and robots in place. They organize production more effectively, making use of automated equipment to run greater varieties of parts in small lots. Equipment breaks down less frequently because they do a better job of process engineering. In many Japanese production systems, quality control is delegated to shopfloor workers, rather than a separate department. The result: higher quality levels from fewer people. Although Japanese firms seem to lag in applications of small computers for production control and scheduling, they have developed remarkably effective solutions for the day-to-day problems of shopfloor organization and management, some devised by production workers themselves. Many American firms, on the other hand, have plugged automated equipment into existing operations, giving little thought to how to take best advantage of the flexibility made possible by the new technology.

An example from microelectronics illustrates another aspect of

[1] Bebb, "Quality Design Engineering."

[2] Kim B. Clark and Takahiro Fujimoto, "Overlapping Problem Solving in Product Development," in Kasra Ferdows, ed., *Managing International Manufacturing* (Amsterdam: North-Holland, 1989), pp. 127–152; Lance Ealey and Leif G. Soderberg, "How Honda Cures 'Design Amnesia'," *McKinsey Quarterly* (Spring 1990), pp. 3–14.

[3] Hiroyuki Yoshikawa, Keith Rathmill, and Jozsef Hatvany, *Computer-Aided Manufacturing: An International Comparison* (Washington, DC: National Academy Press, 1981); Akira Tani, "International Comparisons of Industrial Robot Penetration," *Technological Forecasting and Social Change,* vol. 34 (1989), pp. 191–210; Ramchandran Jaikumar, "Japanese Flexible Manufacturing Systems: Impact on the United States," *Japan and the World Economy,* vol. 1 (1989), pp. 113–143.

product and process technologies. Functionally identical integrated circuits from two manufacturers will differ in a multitude of design details. These details influence costs and yields, quality, and reliability. Memory chips, for instance, are susceptible to "soft" or nonrepeatable errors caused by alpha particles emitted from trace-level impurities in ceramic packaging. While the alpha radiation cannot be eliminated, soft errors can be reduced to tolerable levels through a variety of circuit design techniques. In at least some cases, Japanese companies implemented these techniques before American firms.[4] Design and production technologies frequently enter the competitiveness equation in these complex and subtle ways.

In part because of DoD funding, the United States remains ahead in many of the purely technical aspects of manufacturing, but the resulting tools and methods (e.g., ultrasonic inspection of polymer matrix composites) are much more portable than the managerial and organizational innovations—the "humanware"—that have been such an important source of superior Japanese performance.[5] Thus, while we argue later in the chapter that the United States must take better advantage of its science base to strengthen its manufacturing sector, we must be constantly aware that a lead in hardware alone is not enough to guarantee competitiveness. Japan has proven that the softer, organizational technologies are at least as important.

---

[4] Office of Technology Assessment, *International Competitiveness in Electronics*, p. 249.

[5] Haruo Shimada, "Japanese Management of Auto Production in the United States: An Overview of 'Humanware Technology,' " in Kozo Yamamura, ed., *Japanese Investment in the United States: Should We Be Concerned?* (Seattle: Society for Japanese Studies, 1989), pp. 183–205.

---

many of their foreign competitors in the practice of engineering design. They pay the penalty in the marketplace. Defense contractors design for performance and not for life-cycle costs because DoD wants functional performance at almost any price, and lacks procedures for trading off performance and cost in the acquisition process (and even for minimizing costs when this would entail no sacrifice in function).[5] Taxpayers foot the bill.

---

[5] A review of "design-to-cost"—one of DoD's many efforts over the years to clamp down on overruns for major systems—found that the design-to-cost goals were not established until *after* the services had completed their specification of systems requirements, effectively shortcircuiting the effort. See General Accounting Office, *Impediments to Reducing the Costs of Weapon Systems*, PSAD-80-6 (Washington, DC: GAO, November 8, 1979), pp. 33–35. Integrated circuits meeting military specifications often cost 10 times more than similar commercial devices, take twice as long to develop, and may be less reliable. See Jacques Gansler, *Affording Defense* (Cambridge, MA: MIT Press, 1989), p. 232.

Front-end design tasks—feasibility studies, definition of requirements, concept generation, and preliminary design—largely determine unit production and life-cycle costs. This is true for both military and civilian products. Indeed, fully three-quarters of life-cycle costs may be determined during the critical early design stages (Figure 10-1).[6] Good initial choices contribute to relatively trouble-free engineering development and production. They have especially large impacts on life-cycle costs for military equipment (e.g., logistic support requirements, maintenance, and repair), which can be ten times the original procurement cost for complex systems with long service lifetimes.

Chapter 2 discussed DoD's role in the development of computer-aided engineering methods. These methods have replaced laborious hand calculations, permitting the solution of problems earlier avoided as too time-consuming, and making possible the modeling of phenomena governed by equations that cannot be solved analytically. But computational complexity remains a severe limitation. Computer methods are restricted primarily to the analysis of designs that have already been defined. Modeling involves large investments of time and money, so that the step of assuming a concept or configuration for analysis almost invariably has a lock-in effect. Once the model is up and running, only small variations in design parameters will be considered. Otherwise computing time explodes, even on the fastest supercomputers. As yet, computer-aided engineering analysis offers little help in the early design stages, which still begin with freehand sketching and back-of-the-envelope calculations, liberally supplemented with guesses and intuitions, good and bad.

Until computer-intensive methods can be applied during these preliminary stages, they will not have much impact on overall project success in either civil or military spheres. This is the reason U.S. advantages relative to Japan and other countries in computer-aided engineering have not yet redounded to the competitive advantage of American firms. For instance, a computer model can be used to "tune" the structure of an automobile, reducing noise and vibration caused in driving over rough pavement. But the process of analysis and refinement cannot begin until the auto company has a pretty good idea of what the structure will look like (dimensions, number of doors, whether the roof is slanted). And even with those parameters fixed, there is no way to run the program backward to arrive at a structure

[6]U.S. General Accounting Office, *Effectiveness of U.S. Forces Can Be Increased Through Improved Weapon System Design,* PSAD-81-17 (Washington, DC: GAO, January 29, 1981), pp. 14–18; Committee on Engineering Design Theory and Methodology, *Improving Engineering Design.*

FIGURE 10-1:
Impact of Early Design Decisions on Life-Cycle Costs

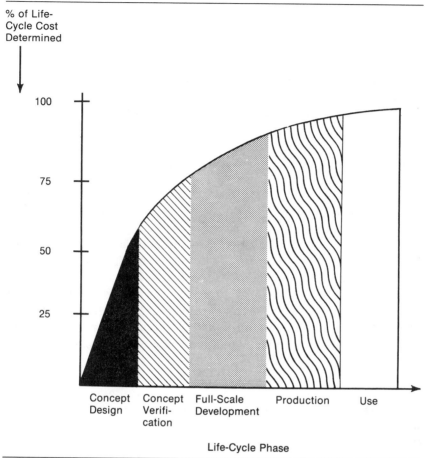

% of Life-
Cycle Cost
Determined

Life-Cycle Phase

Source: Findings of the U.S. Department of Defense Technology Assessment Team on Japanese Manufacturing Technology, Final Report, CSDL-R-2161 (Cambridge, MA: Charles Stark Draper Laboratory, June 1989), p. 19.

that will minimize noise and vibration. It is only possible to start with an assumed structure and calculate its behavior, then change some of the design features (based on judgment) and try again.

The family of computer-aided engineering methods will improve dramatically over the next several decades. Methods useful for preliminary design will emerge, helping both military and commercial firms shorten development cycles, control manufacturing costs, and reduce life-cycle costs. On the military side, rapid, user-friendly techniques for parametric analysis could, at least in theory, help Pentagon decision makers assess trade-offs between cost and performance, and among functional attributes, before making irrevocable commitments to systems requirements. Given the institutional realities, however, such opportunities will probably not be quickly grasped. Since the 1960s at least, DoD has shown little interest in seriously considering trade-offs and design compromises that might curb production costs at the expense of functional performance, even though the performance increments might be marginal, unreachable, or, as a practical matter, irrelevant.[7]

Manufacturing in Systems Perspective

Today, in the plants of U.S. defense contractors and commercial firms alike, products continue to be designed and developed with too little regard for how they can best be made, while high-technology islands of automation can be found embedded in manufacturing systems that function poorly *as systems*. At the same time, many successful firms in the United States and abroad have learned that it makes no sense to automate until the production system as a whole is functioning well and under control. Japanese firms employ such well-known organizational practices as kaizen (continuous improvement) and just-in-time production (JIT or kanban) largely to improve systems-level performance. These practices are much more than technical tools; they help motivate shopfloor workers, integrate them into the production process, and impress upon them the need for continuous and disciplined attention to their tasks. Many American firms have begun to emulate such practices, but others continue to try to integrate people out of the system.[8]

---

[7] *Findings of the U.S. Department of Defense Technology Assessment Team on Japanese Manufacturing Technology*, Final Report, CSDL-R-2161 (Cambridge, MA: Charles Stark Draper Laboratory, June 1989), p. 67. In part as a consequence of DoD's refusal to compromise, the B-1B bomber cannot carry a full load of bombs and a full load of fuel and still maneuver as required for low-altitude penetration. The technical conflicts among these requirements proved insurmountable. Congressional Budget Office, *The B-1B Bomber and Options for Enhancements* (Washington, DC: CBO, August 1988), pp. 15–16. DoD's World Wide Military Command and Control System, to take a different sort of example, reportedly suffered large cost and schedule overruns because its specifications were never clearly defined. Willie Schatz, "The Pentagon's Botched Mission," *Datamation*, September 1, 1989, pp. 22–26.

[8] For revealing comparisons in the case of automobile assembly, see James P. Womack, Daniel T. Jones, and Daniel Roos, *The Machine That Changed the World* (New York: Rawson/Macmillan, 1990). Few such comparisons for small- or medium-lot batch production have been made, but a great deal of case-by-case evidence points to similar conclusions: Japanese

The newer production methods depend heavily on a well-trained labor force. Here, the United States lags considerably behind Germany and Japan.[9] Many American firms have found their employees lacking in basic skills—reading, writing, arithmetic—as well as the social and interpersonal abilities needed to fit into reorganized production systems that demand high levels of worker responsibility. Recently, the poor performance of American schools has excited a great deal of attention, but the spotlight has not yet broadened to include vocational education and workplace training.

TECHNOLOGICAL AND ORGANIZATIONAL CHANGE

There is nothing very new in viewing manufacturing from a systems perspective. What is new is the computer, with its potential for taking over many of the direct tasks of production—guiding machine tools or robots—and also much of the indirect labor that keeps the system functioning smoothly.[10] With NC equipment and other forms of computer automation, firms can, in principle, manufacture small lots to high-quality standards economically (Figure 10-2). Technical developments in advanced materials will also have significant effects on com-

---

firms make better use of their employees, integrating them more effectively into the production system, and relying on them to keep "lean" systems—e.g., those without extensive inventory stocks as buffers against disruptions—running smoothly.

[9] Office of Technology Assessment, *Worker Training: Competing in the New International Economy* (Washington, DC: U.S. GPO, September 1990), pp. 83–95. The German apprenticeship system does a much better job of supplying the mid-level gray-collar skills (e.g., technicians and craft workers) than the fragmented U.S. approach to vocational education and training. Training in Japan contrasts sharply with that in Germany, but yields equally impressive results. Japanese firms rely heavily on instruction delivered by supervisors and managers on the factory floor. Directly involved, managers see the results and place higher value on training than do their U.S. counterparts.

[10] Allen-Bradley's flexible assembly cell for building motor controllers, perhaps the best known of its kind in the country, uses automated equipment to assemble any one of 130-plus different models, each of which has several hundred components. Orders come to a central computer, which schedules production; most shipments go out the same day. Costs have dropped by roughly 40 percent, much of this in overhead categories—e.g., processing the paperwork that once accompanied each order. Douglas R. Sease, "How U.S. Companies Devise Ways to Meet Challenge from Japan," *The Wall Street Journal,* September 16, 1986, p. 1; Barnaby J. Feder, "Allen-Bradley's Stark Vision," *New York Times,* October 6, 1986, p. D2. For an equally striking example of indirect cost reduction, see Lauren A. Perreault and Roger C. Anderson, "IMPCA—Northrop's Paperless Assembly Line," Society of Manufacturing Engineers Technical Paper MM89-749, presented at AUTOFACT '89, Detroit, October 30–November 2, 1989. This computer-based production control system eliminates "over 400,000 active pieces of paper" in building F/A-18s for the Navy.

FIGURE 10-2:
*Flexible Automation Yields Lower Unit Costs
over a Broad Range of Production Volumes*

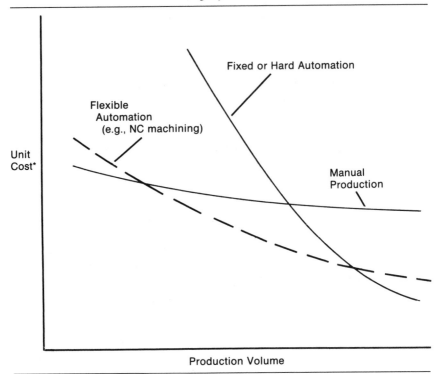

*Including amortization of special purpose equipment, tooling, NC programming, and setup.

mercial and defense manufacturing over the next several decades, as will changes in organizational practices. Since the 1960s, materials scientists and engineers have learned to synthesize polymer matrix composites, plastics, and other materials with properties tailored for specific applications, so that materials (and processes) can be designed along with the rest of the product. Companies around the world, finally, have been reorganizing their production systems, often in conjunction with automation. Many, for instance, have begun to give limited self-management responsibility to multiskilled work groups. Taken together, these three sets of changes suggest a level of technological ferment matching that of the first two decades of the century, when "Fordist" systems of mass production first appeared. This ferment presents both an opportunity and a danger for future American competitiveness—an opportunity because in a period of ferment all traditional sources of competitiveness are up for grabs by companies and nations alert to search out and implement new techniques; a danger if the new paradigms emerge first in the hands of our rivals.

Flexible Automation
and Computer-Integrated Manufacturing

Programmable automation began in the 1950s and 1960s, with NC machining and early computer graphics systems. Since that time, computer control has become cost-effective for replacing manual, mechanical, and electromechanical control systems in ever-wider realms of application (Figure 10-3). Flexible automation has spread from machine tools to robots, automatically guided vehicles and other materials handling equipment, and inspection equipment, including coordinate measuring machines for checking the dimensions of complex parts.[11]

The companies that can make something useful of the following technologies most quickly and most effectively will come out ahead in international competition.

- *CAD, or computer-aided design.* CAD systems not only produce drawings of mechanical parts and components, but can often generate architectural renderings, electrical, piping, and plumbing layouts, and highway routings in 10 or 20 percent of the time once necessary. Such tasks as maintaining databases of part drawings and specifications, and making the changes called for during engineering development (often running into the dozens, if not hundreds, for a single part) have become much easier. Some CAD systems can automatically generate NC part programs.
- *CAM, computer-aided manufacturing.* Descendants of NC machining, CAM installations link together several machines, often with robots, to create automated "cells" and larger flexible manufacturing systems.
- *CAPP, computer-aided process planning.* Many shops schedule jobs

[11] FMC Corporation built the Bradley fighting vehicle using "software gaging" in place of the thousands of hardware gages that would otherwise have been required (at $10,000 to $100,000 each). Software gages, programmed from FMC's computer-aided design database, cut inspection costs by more than 10 times. Early in Bradley production, it took two weeks to prepare for inspection of a moderately complex part, but after rationalization of FMC's design/manufacturing information flow, the work could be done in a day. D.P.S. Charles, "Quality Design Information for Manufacturing," Third International Conference on CAD/CAM, Robotics and Factories of the Future, Southfield, MI, August 14–17, 1988.

FIGURE 10-3:
Choice of Production Equipment Depends on the Number of Design Variants and Production Volume

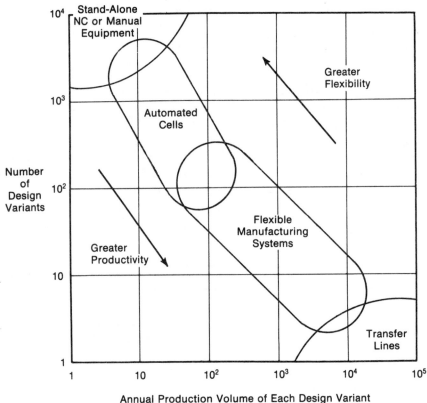

Annual Production Volume of Each Design Variant

Source: Based on G. Spur and K. Mertins, "Flexible Fertigungssysteme. Produktionsanlagen der flexibler Automatisierung," *SWF*, vol. 76, no. 9 (1981), pp. 441–448, as cited in Charles Edquist and Staffan Jackobsson, *Flexible Automation: The Global Diffusion of New Technology in the Engineering Industry* (Oxford, UK: Basil Blackwell, 1988), p. 63.

Note: Automated cells consist of several NC machines, working in coordinated fashion and served by robots or other automated transfer devices. Parts may move from machine to machine in arbitrary sequence.

Flexible manufacturing systems consist of a number of NC machines linked by automated materials-handling equipment. Parts move in fixed sequence from machine to machine.

and manage work-in-process inventories with the aid of small computers and commercially available software packages. Today, managers can run complex CAPP scheduling algorithms on the factory floor.

- *CIM, computer-integrated manufacturing.* CIM implies combining CAD and CAM, and typically CAPP as well. The primary objective: moving from design to production—from CAD to CAM—more or less automatically. This is not yet possible except in a few special cases such as very large-scale integrated circuits and highly automated chemical plants.

- *CAX, computer-assisted everything.* The utopian goal: design + analysis + planning + manufacturing, all automated.

For many years, CAD/CAM developers have worked to erase the slash separating the two acronyms (Box 10-C). The goal has been to move from CAD representation of a part (geometry, hardness, surface condition) to an automated manufacturing process (most simply, an NC part program) without generating any paper, and with little or no human intervention between computer terminal and machine setup. This can only rarely be accomplished today; even for the simplest two-dimensional tasks, integrated CAD/CAM has been elusive.

Process design problems, for example, are ill-defined and open-ended, as are product design problems. Many alternative NC programs can be written, even for a simple part. Lacking unique answers (because, for example, of a lack of firm and unambiguous criteria for optimization), automation of both product and process design will remain a matter of support for people—e.g., through expert systems—rather than replacement of people. Planning and setup for small lots can easily take twice the time of machining itself. Clamping and fixtures have to be chosen, along with feeds, speeds, and cutting fluids. NC programs must be verified. At present, it may be necessary to make half a dozen parts just to prove out tooling and programming; for cost-effective production of very small lots, the first parts must be good.[12] But despite the technical difficulties, enthusiasts foresee a time when CAD/CAM will be economical for lot sizes down to one, so that parts can be produced on demand. The implications of these technologies for small-lot defense production are plain, while in some sectors of commercial manufacturing the effects have already been to reduce economies of scale and hence optimum plant sizes.

Advanced Materials: Driving Force for Process Technology

The Wright brothers were able to get off the ground because they had a relatively light power plant. Today, thrust-to-weight ratio remains a central design goal for jet engines. The limiting factor is high-

---

[12] Paul Kenneth Wright and David Alan Bourne, *Manufacturing Intelligence* (Reading, MA: Addison-Wesley, 1988).

BOX 10-C: NC EQUIPMENT AND CIM-RELATED TECHNOLOGIES

If some of the past efforts to implement CIM now seem overambitious, that should be no surprise. Technological innovation of any stripe brings with it unanticipated difficulties more frequently than unexpected serendipities. The great difference between adopting computer-based systems for factory automation and computer-based systems for, say, aircraft flight control is simply that American companies typically put their best people to work on projects like flight controls, and give them ample budgets, while too often they leave manufacturing systems to less capable people with less than ample budgets.

Early generations of NC technology, like the first CAD systems, depended on skilled technicians and engineers to keep the equipment running and to write programs. The paper tapes that guided the machines had to be prepared using specialized and complex computer languages. The programmers who prepared these tapes needed design skills as well as knowledge of machining practices.

Today, numerous vendors supply simple and reliable CAD and NC equipment. Nonetheless, recent surveys indicate that no more than 10 to 12 percent of installed machine tools in the United States have NC capability.[1] More than 30 percent of those NC machines are 10 years old or more; nearly 40 percent are simple models that can read instructions but do not incorporate computer controls—equipment that has been available for more than 25 years.

This relatively slow penetration should not be too surprising. Machine tools have useful lives measured in decades—indeed, the average age of the equipment in the U.S. machine tool base is nearly 30 years (in part because companies retain older models as backups, even if they rarely use them). On the other hand, investment in CIM-related equipment (including CAD, CAPP, programmable controllers, and local area networks, as well as NC machines) grew at about 15 percent annually between 1983 and 1989—a respectably high rate. Two-thirds of U.S. manufacturing establishments have implemented at least one such technology. Nonetheless, relatively few firms, mostly large, account for most of the investment. Ninety-four percent of U.S. manufacturing establishments employing 500 or more people have invested in at least one type of computer-assisted technology, compared with 67 percent of firms with fewer than 500 employees. More companies have invested in CAD and CAPP than in NC machinery, because the costs are lower—starting simply with a PC and an off-the-shelf software package.

The aerospace industry—home to only 5 percent of U.S. machine tools—accounts for some 10 percent of all NC equipment. Companies

[1]Kelley and Brooks, "The State of Computerized Automation in U.S. Manufacturing"; OTA, *Worker Training*, Appendix 4A, pp. 119–123; *Current Industrial Reports: Manufacturing Technology 1988* (Washington, DC: Department of Commerce, Bureau of the Census, May 1989).

that do most of their business with DoD make greater use of programmable automation than those selling primarily to the private sector (Figure 10-4), with prime contractors and subcontractors showing broadly similar patterns of adoption. Smaller American firms serving commercial markets have been particularly slow to adopt advanced technologies, compared not only with large U.S. firms, but with small firms abroad.[2]

The overall picture for the United States is one in which computer-based manufacturing technologies seem to be diffusing at a pace about like that for comparable technological innovations in the past. Vexing problems in practical applications have tended to counterbalance the economic driving forces. At the same time, many smaller U.S. companies have plainly failed to grasp the logic of programmable automation and have not made investments that would be cost-effective in business terms.

[2]Louis G. Tornatsky and Daniel Luria, "Technology Policies and Programs in Manufacturing: Toward Coherence and Impact," in *Strengthening Corporate and National Competence through Technology* (special issue of *International Journal of Technology Management*, to appear).

FIGURE 10-4:
Penetration of NC and CAD among Defense and Nondefense Firms

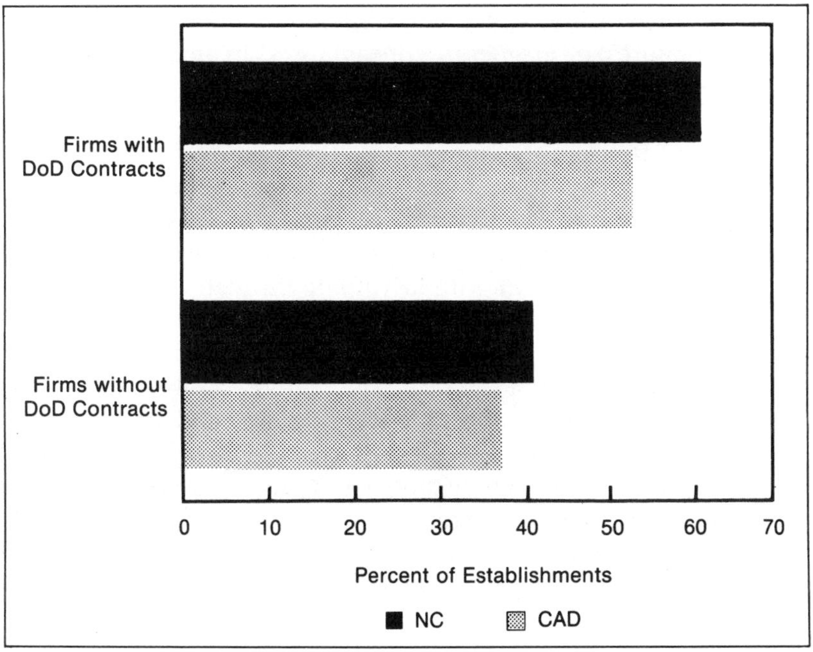

Percent of Establishments

■ NC   ▨ CAD

Source: Current Industrial Reports: Manufacturing Technology 1988 (Washington, DC: Department of Commerce, Bureau of the Census, May 1989), pp. 31, 35.

temperature materials for turbine blades and other critical engine components. In airframes, which took on their present form in the 1930s, aluminum construction remains the norm, with titanium alloys and ultra-high strength steels in some components. Polymer matrix composites (PMCs) found early applications in rocket motor casings for the military and have been heavily used in helicopters. The Air Force's Advanced Tactical Fighter, now under development, will be about 40 percent PMC by weight. In addition to their greater strength-to-weight and stiffness-to-weight ratios, the properties of PMCs can be tailored by varying the orientation of fiber plies in the layup. Composites not only reduce weight, they reduce assembly costs. A composite structure will have fewer components, which can be joined by adhesive bonding, avoiding the labor-intensive work of drilling holes and setting rivets.[13] The family of design methods, fabrication processes, and inspection techniques required for PMCs—ranging from computer codes that can optimize their directional properties to automated equipment for filament winding—represents a shift as great as that faced in earlier years by the electronics industry in moving from vacuum tubes to transistors.

At first, new materials are produced in small volumes for applications where their advantages in properties justify high costs. These advantages may include strength and stiffness, resistance to wear or to high temperatures, or magnetic permeability, as in amorphous (glassy) metals used in transformer cores. Material properties depend on processing; for instance, the only way to make an amorphous metal is to cool it so rapidly from the liquid phase that it cannot crystallize. The development of practical low-temperature superconductors illustrates the importance of processing; it took years after useful values of critical current density had been demonstrated in the laboratory to learn to fabricate niobium-titanium wire in volume through carefully tailored sequences of wire drawing and heat treatment. In recent years, DoD has been putting a good deal of effort into automated fabrication of PMC structural components, which now require extensive hand labor. Because the fiber plies are limp, hard to grasp, and wrinkle easily, automation has been quite difficult (just as it is in the apparel industry).

In many cases, early production for defense, where high performance outweighs high cost, creates a base of production and service experience that serves as a springboard for commercial applications. This was the case with graphite-epoxy composites, titanium alloys,

---

[13]There are 2,600 graphite-epoxy structural members in the Beech Starship (a business plane), compared with 10,000 metal parts in a comparable aluminum airframe. Eric Weiner, "Innovative Plane Making Its Debut," *New York Times*, June 5, 1989, p. D1.

and gallium arsenide semiconductors. With scale and learning pushing costs down, materials and processes filter into the civil sector. Even when the materials are familiar, military demand for the ultimate in performance often means that process innovations get their start in defense plants, just as interchangeable parts did more than a century ago. This was the case with NC machining, which grew out of the Air Force's desire to carve wing skins from single plates of aluminum alloy (Appendix 10-A).

Organization and Management

New forms of work organization have begun to reshape production in the United States, Europe, and Japan. While few of the trends outlined below have penetrated very deeply in American industry, their potential for improving efficiency in both civil and defense production has already been amply demonstrated.[14]

The traditional approach.   When design and production were craft skills, product technology, manufacturing equipment, and work organization evolved together. Mass production in the twentieth century brought not only separation, but hierarchy. The design of the product set the parameters for processing and fabrication: the job of manufacturing was to produce whatever product planning, engineering, and marketing called for. The equipment on the factory floor in turn governed the organization of work: people adapted to machines, as exemplified by the scientific management paradigm associated with Frederick Taylor.[15]

The underlying premise of scientific management is straightforward: for maximum efficiency, the manufacturing system should be broken down into small and discrete tasks. Each of these should be optimized, under the assumption that optimizing the individual tasks will result in optimization of the system as a whole. It was a sensible procedure half a century ago, but today we can see that this approach is too great a simplification: in fact, optimizing individual tasks—or optimizing subsystems in isolation—does not necessarily result in an optimum for the system as a whole.

---

[14]Thirteen percent of American firms surveyed in early 1990 were using some form of self-managed work group, with another 4 percent reporting plans to move in this direction. "Workforce 2000—Competing in a Seller's Market: Is Corporate America Prepared? A Survey Report on Corporate Responses to Demographic and Labor Force Trends," Towers Perrin, Inc. and Hudson Institute, July 1990, p. 27.

[15]John A. Alic, "Who Designs Work? Organizing Production in an Age of High Technology," *Technology in Society*, vol. 12 (1990), pp. 301–317.

Scientific management, moreover, sees people as unpredictable—as sources of errors, mistakes, and uncertainty—rather than as solvers of problems. If such a view could ever be justified, it can be no longer. If anything, people become more important as automation spreads; the greater the degree of automation, the more the system as a whole depends on the judgment, skill, and experience of the workers who remain. There are also many tasks that machines cannot yet perform. People will keep on doing a great deal of assembly work, even in high-volume production, as well as machine setup, inspection, and maintenance and repair. At the same time, new technologies can help avoid the human errors that can never be totally eliminated.[16] Automated inspection methods, for example, can locate one-in-a-million defects impossible for human operators to spot.

Reorganized production systems. Short production runs and design variety require work reorganization as well as programmable automation. At the most general level, the overall shift can be described as one from Fordist mass production to more flexible organizational structures.[17] There is no need to accept the theorizing that goes with many discussions of both Fordism and flexible specialization to sketch out the implications. Table 10-1 traces the shift by contrasting two ideal types: an older model characteristic of U.S. manufacturing in the 1950s and 1960s, and a new pattern encompassing the primary features found, singly and in various combinations, in reorganized production systems. Rarely does any one company take all the steps listed on the right-hand side of Table 10-1. Partial, halting, and piecemeal implementation has been the rule. But many companies, whether they produce semiconductor chips or aircraft engines, are experimenting with at least some of these steps.

Production planning and control. With falling lot sizes and scale economies, planning, scheduling, and coordination become ever more important for efficiency. In batch machining, indirect costs often exceed the direct costs of operating equipment and making products. Many of these indirect expenses are associated with managing the flow of production: ensuring that equipment is available and in good repair when scheduled for use; getting information, including NC part programs, to shopfloor personnel when they need it; coordinating flows of materials, parts, tools, and supplies. In mass production too, as auto-

---

[16]Robert U. Ayres, "Complexity, Reliability, and Design: Manufacturing Implications," *Manufacturing Review*, vol. 1 (1988), pp. 26–35.

[17]Michael J. Piore and Charles F. Sabel, *The Second Industrial Divide* (New York: Basic Books, 1984); Richard Florida and Martin Kenney, "High Technology Restructuring in the USA and Japan," *Environment and Planning A*, vol. 22 (1990), pp. 233–252.

TABLE 10-1:
Old and New Patterns in U.S. Manufacturing

Old Model: Traditional Manufacturing, 1950s and 1960s		New Model: Flexible Decentralization, 1980s and Beyond
Low cost through vertical integration, scale economies, long production runs	OVERALL STRATEGY	Low cost with no sacrifice of quality, coupled with substantial flexibility, through partial vertical disintegration, greater reliance on purchased components and services
Centralized corporate planning; rigid managerial hierarchies		Decentralization of decision making; flatter hierarchies
Internal and hierarchical; in the extreme, a linear pipeline from central corporate research laboratory to development to manufacturing engineering	PRODUCT DESIGN AND DEVELOPMENT	Decentralized, with carefully managed division of responsibility among R&D and engineering groups; simultaneous product and process development
Breakthrough innovation the ideal goal		Incremental innovation and continuous improvement valued
Fixed or hard automation where possible, but much hand work in low-volume defense contractors	PRODUCTION	Flexible automation
Cost control focused on direct labor		With direct costs low, reductions of indirect cost become critical
Off-line or end-of-line quality control		Real-time, on-line quality control
Fragmentation of individual tasks, each specified in detail		Selective use of work groups; multiskilling; job rotation
Employees mostly full time; heavy reliance on semiskilled workers in mass production industries	HIRING AND HUMAN RESOURCE PRACTICES	Smaller core of full-time employees, supplemented with contingent workers (part time, temporary, and contract) as a source of flexibility

TABLE 10-1 *(continued)*:

Old Model: Traditional Manufacturing, 1950s and 1960s		New Model: Flexible Decentralization, 1980s and Beyond
Layoffs and turnover a primary source of flexibility; workers, in the extreme, viewed as a variable cost		Core workforce viewed as an investment; management attention to quality-of-working-life as a means of reducing turnover
Specialized training (including apprenticeships) for gray-collar, craft and technical workers. Minimal training for production workers, except for informal on-the-job guidance		Broader skills sought for both blue- and gray-collar workers
Supervisors as policemen, organization as army	GOVERNING METAPHORS	Supervisors as coaches or trainers, organization as athletic team (the Japanese metaphor: organization as family)

Source: Office of Technology Assessment, *Worker Training: Competing in the New International Economy* (Washington, DC: U.S. GPO, September 1990), p. 115.

mation drives down direct labor content, indirect costs become relatively more important.[18]

Production control remains highly labor-intensive, involving stock clerks and toolroom attendants, fork-lift operators and expediters, foremen and middle managers. Jobs enter the batch production system in the form of orders characterized by specifications, drawings, and bills of materials. At any point in time, a shop will have a certain number of jobs in process. Lot sizes will differ, as will material requirements. The shop will have a certain stock of production equipment, more than likely with overlapping capabilities. Two machines that can, in principle, be assigned a given job may have different costs of operation (perhaps in terms of setup time versus run time) and produce parts

---

[18] At Tandy Corporation's Fort Worth, Texas, plant, direct labor accounts for less than 2 percent of the cost of each personal computer, the remainder consisting of purchased inputs (which, of course, have their own labor content) and overhead. "North American Profiles," *Datamation,* June 15, 1990, p. 67. In automobile production, direct labor now comes to about 10 percent of total costs, with indirect labor adding another 15 percent. Bruce Beier and Mary Gearhart, "Productivity vs. Profit Sharing," *Automotive Industries* (April 1990), pp. 53–56.

that differ qualitatively (e.g., a milled surface versus one produced by a shaper or a surface grinder). Some machines will need more highly skilled operators than others. The shop can subcontract part of its work, and may have to for specialized operations (e.g., plating, electric-discharge machining). Added to these factors will be such imponderables as equipment breakdowns and late deliveries of materials and supplies.

This complexity is a major reason behind the traditional approach of breaking the factory system down into small and simple pieces in the hope that they can be handled more easily—the genesis not only of scientific management, but of functional departments for turning and grinding, materials handling, and inspection. Organizational approaches such as group technology and cellular manufacturing represent attempts to improve efficiency through factory layouts that minimize production cycle times and work-in-process inventories. Indeed, the entire agenda for change summarized in Table 10-1 can be viewed as a response to complexity.

POLICY

The defense sector may have led in NC and in computer-assisted engineering methods, but it has lagged behind commercial industry in shopfloor reorganization. In some cases, defense has lagged even in effective use of manufacturing technologies originally developed with DoD dollars. The reasons lie partly in a DoD procurement system that creates few incentives and many disincentives for contractors to invest in modern manufacturing equipment and advanced organizational practices.[19] Unless a bidder can expect immediate reductions in production costs, there is little reason to invest in new equipment or new technology, and every reason to avoid risks. If costs do come down, DoD will seek to adjust the contract and recoup the savings. Excess capacity in the defense sector makes new investments look even riskier.

Whether acquisition practices change, the imbalance that sees DoD underinvesting in process technologies relative to product technologies should be rectified. In fact, because of its broad strength in high technology, the United States has a unique opportunity to bring theoretically based technical knowledge to the factory floor through R&D. Support for generic processing technologies, by strengthening the U.S. infrastructure for manufacturing and expertise in production engi-

---

[19]"U.S. Manufacturing: Problems and Opportunities in Defense and Commercial Industries," Office of Technology Assessment staff paper, Washington, DC, June 1990.

neering and management, would also help with a second critical problem: slow diffusion of best practices to industry.

R&D Needs: Strengthening Manufacturing Technology

The United States has one major source of advantage in seeking to improve its manufacturing capabilities: an unmatched science base. So far, this source of advantage remains largely untapped. But in the years ahead, many of the traditional processes of manufacturing (casting, forging, machining) will finally be put on a rational footing (Figure 10-5). Today, new processes such as electron-beam welding, or fabrication methods for advanced materials, originate in organized research rather than craft-based empiricism. U.S. leadership in the social and behavioral sciences might also contribute. Taking advantage of American science to improve American manufacturing will require both business and government to reshuffle their priorities and policies.

Both new and old processes involve poorly understood physical phenomena. Examples include fiber-matrix adhesion and resin curing in PMCs; flow of matrix and fibers during forging of metal-matrix composites (candidate materials for high-performance aerospace appli-

FIGURE 10-5:
Hardware Technologies That Affect Efficiency in Manufacturing

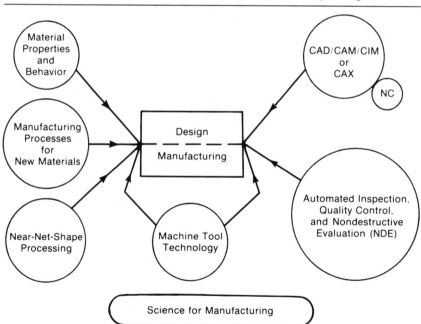

cations); densification during sintering of powder-metal compacts; nondestructive evaluation (NDE) for finding micron-sized flaws in ceramics; catalyzed chemical reactions (largely a mystery from the viewpoint of science). Analyzing almost any such process in useful detail poses great difficulties. Mathematical modeling of machining has yet to provide much help in improving productivity on the shopfloor, although the incentives are huge: annual costs simply for cutting metal probably exceed $125 billion in the United States.[20] At the same time, many of these long-standing technical problems are now ripe for attack. Finite element methods (FEM), for instance, make it possible to simulate the flow of metal in a forging die, or of polymers during injection molding. Among the results may be shorter design cycles, with less need for trial-and-error to prove out tooling; more nearly optimized processes, with less material waste and energy consumption; higher yields, with fewer bad parts to be scrapped or reworked. Rarely, however, do such problems get much R&D support. Lack of funding means that research in manufacturing seldom attracts the most talented engineers and scientists.

With cheap computing power and ever more sophisticated modeling techniques, substantial opportunities now exist for improving productivity and efficiency in U.S. manufacturing through focused R&D. The work is unlikely to attract extensive funding from industry because it is widely applicable, thus unlikely to be a source of proprietary advantage. But civilian firms and defense alike would benefit from expanded R&D on such topics as

- the engineering and science base for metalworking;
- practical feedback control systems for shopfloor processing;
- integrated materials and processes R&D;
- automated methods for nondestructive evaluation, inspection, and quality control;
- better techniques for planning, scheduling, and production control; and
- computer automation in support of simultaneous product and process design.

Metalworking. More or less conventional casting, forming, machining, joining, heat treating, and finishing remain foundations for much of U.S. industry. But despite the vast experience base accumulated over the years, few of these processes are well understood on a

---

[20]William P. Koster, "Machining," *Advanced Materials & Processes* (January 1990), pp. 67ff. Perhaps 10 percent of total production of primary metals ends up simply as chips.

fundamental level. They take place at extremes of heat and frictional force (machining, wire drawing), involve very large plastic deformations (both machining and forming), steep thermal gradients and rapid cooling (casting, welding, grinding). Cutting tools reach red heat, chatter and break; chips clog; cutting fluids decompose. Steady-state conditions are rare, highly rate-dependent phenomena common. Nucleation and growth of solid from the liquid phase during casting, for example, takes place far from equilibrium, so that phase diagrams provide little help with practical problems.

In the past, when predictive models could be formulated at all, they were empirical—often crudely so. Extrapolation to new conditions or generalization to other materials was problematic. Yet progress is possible; indeed, the pace has been accelerating. New surface coatings can reduce cutting-tool wear by factors or ten or more. Closed-loop control has been applied to arc welding. Great scope remains for continuing improvements and cost savings; for example, a seemingly trivial step such as deburring (more properly, final finishing) still accounts for about 15 percent of the machining costs of a jet engine, and up to 30 percent for other products.[21]

Feedback control.   Current NC machine tools rely on feedback from sensors indicating the positions of cutting tools, rather than the state of the part being machined. Part dimensions and surface condition can be checked only after stopping the machine. Continuous measurements reflecting the state of the part and the process (tool wear, conditions at the chip-tool interface) demand inexpensive sensors that will operate reliably under shopfloor conditions. Optical devices that work in the laboratory have short life expectancies in a factory. Simple transducers, such as accelerometers used to count acoustic emission pulses, can be used only after extensive database development, and even then may need to be tuned to a particular machine. Control systems are even more of a limitation for robots, and a major reason why they have diffused more slowly than expected 10 or 15 years ago. Most robots still rely on relatively primitive open-loop controls.

Integrated material/process R&D.   When DoD requirements drive materials R&D, processing normally takes a back seat. High-cost pro-

---

[21] Lawrence J. Rhoades, "Machine Tool Industry Perspective," *Department of Defense 1987 Machine Tool/Manufacturing Technology Conference: Executive Summary, vol. 1,* AFWAL-TR-87-4137 (Wright-Patterson AFB, OH: Air Force Wright Aeronautical Laboratories, June 1987), pp. 28–52; Frederick M. Proctor and Karl N. Murphy, "Advanced Deburring System Technology," Winter Annual Meeting, American Society of Mechanical Engineers, San Francisco, December 1989.

cesses may then limit civil applications. Titanium components were first used in military aircraft where aerodynamic heating raised temperatures above the limits for aluminum alloys. Costs were high and production rates low because titanium is so difficult to machine. Even today, DoD's contractors continue working to overcome these disadvantages, in part through diffusion bonding and superplastic forming techniques that reduce the need for machining. By integrating R&D on new materials and on production processes suited to them—e.g., net-shape processing (minimizing material removal by forming parts as closely as possible to their final dimensions)—DoD could save money. At the same time, new materials would find their way into commercial markets sooner.

Automated inspection and evaluation.   Practical applications of materials as different as ceramics, superalloys, and PMCs depend on nondestructive evaluation. Advanced ceramics such as silicon nitride have very attractive properties—high hardness and abrasion resistance, excellent temperature resistance, chemical stability. They have been of interest to DoD for years and would find applications even at very high cost. But these materials are extremely brittle and will never fill critical load-carrying applications without failsafe methods for finding the very small flaws that can lead to catastrophic breakage.

More broadly, inspection is one of the places where it is appropriate to design people out of the system if possible: people make mistakes, and the goal in inspection and quality control is first to avoid mistakes, and second to detect and correct those that do occur. Still, human operators are needed where judgment and experience come into play: although a computerized vision system can spot solder runs on a printed circuit board, it cannot interpret x-rays of pipeline welds; people must still do the latter.

Planning, scheduling, and production control.   In batch production, expensive machine tools may sit idle 95 percent of the time because there is no work for them to do. A single part may move 50 or more times from machine to machine—several miles within the confines of a 10,000 square-foot building—while undergoing processing only 1–2 percent of the time over a period of months. In general, these and other shopfloor management problems have no optimum solution; CAPP algorithms aim at satisfactory outcomes as aids to decision making. Managers must still rely heavily on their own judgment in making day-to-day scheduling decisions. As CAPP methods improve, computers may reduce indirect labor costs as much or more than automation of production reduces direct labor costs. In addition, effective use of flexible automation in direct production requires better scheduling methods because flexibility, by opening up more alternatives, makes planning even more complex.[22]

Simultaneous engineering.   Computer automation in support of engineering design has proven more difficult than the development of techniques for analysis. Similar limitations have slowed the development of simultaneous or concurrent engineering—parallel product and process design—particularly for mechanical systems.[23] Simultaneous engineering is partly a matter of technical tools and partly a matter of management. From the management standpoint, the goal is to break down the organizational barriers separating product and process engineering, making sure that system designers respond to manufacturing constraints, and permitting process designers to influence product features during the early design stages. Managers may also assign employees from marketing, purchasing, and after-sales service groups to project teams. Some companies involve their suppliers as well.

From a technical standpoint, simultaneous engineering is difficult because of the fluid nature of design. With many design features unresolved—a nearly blank slate—the cascading consequences of alternative choices on either the product or the process side quickly overwhelm the capabilities of even the fastest algorithms and most powerful computers. These purely technical problems will gradually yield to R&D. When they do, it will be possible to give the designer not only an automated sketchpad, but also quick and easy access to a menu of analytical procedures that provide feedback in a few seconds on a wide range of questions—for example, the suitability of plastic parts of a given shape for injection molding. A sufficiently smart system might tell a designer her part could not be molded, or suggest ways that it could be molded with smaller radii or better surface quality. So far, however, the computer codes are too slow for routine interactive use.

To take another example, design groups now spend endless hours verifying that the components of, say, a transmission for a truck or a tank will fit together. Today no computer can answer this question. Determining assembly sequences is still more difficult—that the parts fit together geometrically does not mean they can be physically assembled. The usual way of attacking such problems is to begin with a set

---

[22] David A. Bourne and Mark S. Fox, "Autonomous Manufacturing: Automating the Job-Shop," *IEEE Computer* (September 1984), pp. 76–86.

[23] In defense acquisition, "concurrency" has another meaning, referring to programs in which production begins before the design has been frozen; thus production overlaps engineering development and testing. We use "simultaneous engineering" to avoid ambiguity when referring to overlapping product and process engineering.

of drawings representing the assembled design, then work backward to determine if disassembly is possible. (If so, assembly is possible by reversing the sequence.) More difficult questions may follow. If hundreds of assembly sequences are possible, which is best?[24] Nor is this the only question. Simple maintenance procedures are needed for military equipment, but a design suited for factory assembly may not be best for disassembly and repair in the field.

Computer aids for such problems are many years away. Satisfactory operating performance in complex mechanical assemblies, as well as reliability and maintainability, continue to depend on expensive prototype development and testing. In electronics, however, simultaneous engineering has advanced more rapidly. Computers and chips are designed in parallel because decisions on, say, processor architecture depend on the components available. The chips themselves are developed and verified on pilot fabrication lines that can be fine-tuned to raise yields before transfer to a production setting.

Support for Process R&D

The United States underinvests in a broad range of generic industrial technologies, with manufacturing a prime case. U.S. firms typically split their R&D roughly 2:1, with product development getting twice as much money as process engineering. Japanese firms reverse these proportions, spending more on facilities, tooling, and special manufacturing equipment (Table 10-2). On the evidence of competitive outcomes, American firms invest too little in process R&D. All the indications are that DoD also underinvests on the process side.

Table 10-3 summarizes federal spending for manufacturing-related R&D. DoD and the Department of Energy account for the great majority of funding, with most of DOE's manufacturing R&D related to nuclear weapons programs. DoD's manufacturing R&D, together with the defense-related share of the DOE spending, accounts for some 80 percent of the federal total, but only about 2 percent of all defense-related R&D.

Defense department manufacturing programs.   Since the 1950s and its sponsorship of NC development, the Air Force has been the leader within DoD in supporting manufacturing technologies. Far more

[24] Daniel E. Whitney et al., "Tools for Strategic Product Design," *Preprints: NSF Engineering Design Research Conference,* June 11–14, 1989 (Amherst, MA: University of Massachusetts, 1989), pp. 581–595. The authors note that a particular automobile rear axle they analyzed could be assembled following any of 938 different sequences.

TABLE 10-2:
Product and Process Development Expenditures by U.S. and Japanese Companies[a]
(percentage of total project cost)

	U.S. Companies	Japanese Companies
Research, development, and design	26%	21%
Prototype or pilot plant[b]	17%	16%
Tooling and equipment	23%	44%
Manufacturing startup[c]	17%	10%
Marketing startup	17%	8%

Source: Edwin Mansfield, "Industrial Innovation in Japan and the United States," *Science,* vol. 241 (September 30, 1988), p. 1770.

[a] Survey figures from 1985 for 50 matched pairs of U.S. and Japanese firms. The total of 100 included 36 chemical companies, 30 machinery, 20 electrical and electronics, and 14 from the rubber and metals industries.

[b] For cases of product development, the costs are for prototyping; for process development, they include investments in pilot plants.

[c] Because Japanese companies spend more in preparing for production, their startup costs are less. Adding the percentages for tooling and equipment to those for manufacturing startup gives a total of 40 percent for the U.S. companies, 54 percent for the Japanese. The greater proportion of total project expenses for tooling and equipment reflects the higher priority managers in Japan place on manufacturing as an element in competitive strategy.

than the other services, its primary missions—aircraft, missiles, and space—demand advanced materials and exotic processing. The Air Force's Integrated Computer-Aided Manufacturing (ICAM) program, begun in 1978 and funded for a number of years with ManTech (Manufacturing Technology) program dollars pursued work including fabrication methods for composites, system architectures for CIM, and robotics. Defense agencies have also supported robotics research because of the potential direct applications in military systems; robots might, for instance, be used to clear minefields. With the exception of a few programs like ICAM, however, most of DoD's manufacturing R&D has been fragmented and short term in orientation.

Much of DoD's manufacturing-related R&D has been associated with particular weapons systems programs. Most of the rest flows through the ManTech program (Table 10-4). ManTech funds go primarily to defense contractors for applied R&D expected to yield near-term savings. As part of their technology base funding, the services and DARPA also provide some money for generic processing research. These sums are relatively small—a few tens of millions of dollars annually.

TABLE 10-3:
Federal Funding for Manufacturing R&D[a]

	Fiscal Year Budget Authority (million $)		
	1990	1991	1992[b]
Department of Energy	$406	$ 498	$ 614
—defense-related	369	436	525
—nondefense-related	37	62	89
Department of Defense	368	536	372
National Science Foundation	56	58	64
Department of Commerce	54	60	68
National Aeronautics and Space Administration	31	43	43
Totals	$914	$1,194	$1,160

Source: Office of Management and Budget, February 1991.
[a] Excludes Department of Agriculture, which spends about $60 million annually on R&D related to processing of food and biological materials.
[b] Budget request.
Note: Totals do not add because of rounding.

In one form or another, ManTech can be traced back at least as far as 1960, but the program has never enjoyed the stability and acceptance that long life might suggest. Appropriations have come under repeated threats, with budgets (each service runs its own program) fluctuating widely from year to year (Table 10-4), sometimes by as much as 50 percent. ManTech has been reviewed several times by outside groups, with generally consistent findings: the program is too important to terminate, but has not been well or consistently managed.[25]

DoD also provides support for manufacturing technology through the Industrial Modernization Incentives Program (IMIP, formerly

[25] U.S. General Accounting Office, *DoD Manufacturing Technology Program—Management Is Improving But Benefits Hard to Measure*, GAO/NSIAD-85-5 (Washington, DC: GAO, November 30, 1985); *The Role of the Department of Defense in Supporting Manufacturing Technology Development* (Washington, DC: National Academy Press, 1986); *Manufacturing Technology: Cornerstone of a Renewed Defense Industrial Base* (Washington, DC: National Academy Press, 1987). For an overview of DoD's manufacturing R&D, including ManTech, see Office of Technology Assessment, *Computerized Manufacturing Automation: Employment, Education, and the Workplace* (Washington, DC: U.S. GPO, April 1984), Chapter 8, pp. 305–334.

TABLE 10-4:
DoD Manufacturing Technology (ManTech) Program Funding
(million $)

Service or Agency	FY1990	FY1991[a]	FY1992[b]
Army	$ 24.6	$ 31.7	$21.1
Navy	48.4	109.7	25.3
Air Force	85.0	109.2	50.5
Defense Logistics Agency	15.0	10.9	0.0
Office of the Secretary of Defense	0.0	50.0	0.0
Totals	$173.0	$311.5	$96.9

Source: Department of Defense, June 1991.
[a]The FY1991 figures are so large because of congressional additions to DoD's ManTech request.
[b]Budget request.

known as TechMod, for Technology Modernization). IMIP funds pay for implementation of new manufacturing technologies in defense plants (in addition to contract funds tied to specific weapons system programs). ManTech and IMIP dollars go almost exclusively to prime contractors, not to second- and third-tier firms. With reductions in defense spending, money for process R&D and implementation of new manufacturing technology will probably be among the first items to disappear, unless manufacturing gets special policy attention.

Civilian agencies. As Table 10-3 showed, other agencies spend little on manufacturing R&D compared to DoD and DOE. The National Aeronautics and Space Administration funds modest levels of process R&D. Most of this work is mission-related, like that of DoD and DOE. While, for instance, there has been a good deal of effort on robots for use in space, it is hard to find examples of transfers to industry. NASA has also funded modest levels of generic CAD/CAM R&D related to aircraft and space vehicle design.

Manufacturing research in the universities, funded by the National Science Foundation, tends to be more generic. But few schools undertake research and teaching relevant to manufacturing, in part because there has been so little money for individual investigator research. The foundation does support several engineering research centers that emphasize manufacturing, including Purdue's Center for Research on Intelligent Manufacturing Systems, and Ohio State's Near Net-Shape Manufacturing Center. ERCs are expected eventually to become self-supporting, but it will be surprising if more than a few large U.S.

firms—the IBMs and Xeroxes—prove willing and able to provide continuing support for manufacturing R&D.

Department of Commerce support for manufacturing takes place almost entirely through the National Institute of Standards and Technology (NIST). The institute's Center for Manufacturing Engineering, funded at $8.3 million in fiscal 1991 (with another $18.3 million in reimbursable work paid for by other government agencies and nongovernment users), concentrates on standards-related matters such as computer interfaces. The Advanced Manufacturing Research Facility, part of the center, serves as a testbed for new equipment and for standards development.

The Omnibus Trade and Competitiveness Act of 1988 gave Commerce a larger role in manufacturing technology. By 1991, NIST had established five new manufacturing technology centers (two just authorized) intended to help diffuse best-practice know-how to industry. But the centers' appropriation had reached only $11.9 million by fiscal 1991. NIST's activities come closer to providing a focal point for manufacturing than exists elsewhere in government, but limited funding remains a severe constraint.

A U.S. Manufacturing Strategy

Not only do both the public and private sectors spend too little on process R&D, but the United States also lacks an infrastructure with more than a few centers of expertise in manufacturing. As a result, there is little sense of technical community. While NIST and DOE's Lawrence Livermore National Laboratory have excellent programs in manufacturing technology, both are small. Beyond Purdue's ERC, Carnegie-Mellon's Robotics Research Institute, and a very few other university groups, research in manufacturing is scattered; few schools have more than one or two faculty members of recognized accomplishments, and many have none. Lack of faculty makes it hard to offer manufacturing engineering courses even at the undergraduate level, much less train graduate students and offer postdoctoral research opportunities. Creating a manufacturing sciences directorate within NSF, as we suggest in the subsequent list of recommendations, would be a big step toward higher priority and visibility for manufacturing within the university community.

In industry, the National Center for Manufacturing Sciences (NCMS)—a nonprofit R&D consortium based in Michigan—began operations in 1987. The impetus grew out of concern over the decline of the U.S. machine tool industry, and the implications for the defense industrial base. Nearly 100 companies have joined NCMS, representing a wide range of industries. The consortium gets about half of its budget (which totals about $10 million) from DoD, the rest from its members; most of the money goes back out in the form of R&D contracts, with a next-generation NC controller the major current project.

The United States also suffers from a lack of testbeds for the development of generic processing technologies. There are few opportunities to gather experience in pilot facilities and in scale-up. With manufacturing specialists isolated from the mainstream of science and technology, new knowledge moves slowly from R&D laboratory to factory. For example, computational methods used in process analysis often lag well behind the state of the art. These infrastructure problems must be addressed if the United States is to take advantage of the new technological opportunities in manufacturing.

We recommend the following as first steps toward a U.S. strategy for manufacturing technology development.

1. *DoD funding for manufacturing R&D should be increased, and DoD should improve program management.* The Pentagon should channel more money to process R&D, both in the early stages of the development cycles of new weapons and for generic work. In the past, process engineering has been left to the end of weapons systems programs; generic process R&D has never received much money.[26] The primary reason seems to be that DoD managers, particularly at the upper levels, have not realized that more efficient production could save billions of dollars annually—money that could go toward the purchase of more ships, planes, and tanks (or that could be shaved from the defense budget).

2. *DoD could also begin a 10-year program of purchases of advanced manufacturing equipment* for meeting future defense needs. The objective would be to create a modern, flexible defense industrial base, able to respond rapidly to new matériel requirements. Funded in the range of $300–$500 million annually (about 10

---

[26] A 1989 study team recommended that "DoD should establish an explicit RDT&E . . . design and manufacturing research budget line . . . which is a significant percentage of funds allocated for device research and weapons system development. Similar line items should be created in the service RDT&E budgets." It also recommended that "consideration should be given to increasing the IR&D . . . ceiling and requiring that the increase be spent on design and manufacturing research." *Findings of the U.S. Department of Defense Technology Assessment Team on Japanese Manufacturing Technology,* pp. xxxv, xxxvi.

percent of current U.S. machine tool output), plus R&D as needed, such a program would help stabilize the market for U.S. toolbuilders. Half or more of this government-owned equipment could be placed in plants operated by second- and third-tier firms (as opposed to prime contractors) that produce for both defense and civilian markets. These companies would be permitted to use the equipment for filling commercial orders on the condition that they train their employees on it, provide maintenance, and give priority to defense work when called upon.

3. *The Commerce Department should begin a capital equipment leasing program* to accelerate diffusion of innovative manufacturing technologies to industry.

    Under such a program, qualifying companies could lease advanced production equipment (including computer software), learn to use it, and later convert the lease to a purchase agreement if they chose to do so. Companies would qualify by demonstrating potential cost savings or productivity/quality improvements and showing that they could not finance purchases from current revenues.

    American industry has been relatively slow in adopting even well-proven technologies such as NC. There are many reasons, some of them targeted by NIST and state-level factory extension efforts—for example, lack of standards for compatible interfaces, lack of understanding of the economic benefits, lack of skills in the blue- and gray-collar labor force. For smaller manufacturing firms, however, financing is the most serious obstacle. Under a federally financed leasing program, companies able to generate enough cash flow to cover lease payments, but not the upfront costs of immediate purchase, would have access to technology otherwise unavailable to them. While tax credits or loan guarantees might seem more consistent with the normal U.S. policy approach, a leasing program would enable DOC to target process technologies developed with federal support (including DoD funds), thereby encouraging innovation rather than simply subsidizing safe purchases of conventional equipment.

4. The administration should begin a comprehensive program for strengthening the *manufacturing technology infrastructure*, with the ultimate objective of a network of several hundred centers for manufacturing R&D and industrial extension.

    Possible early steps include the following:

- Support for more industry consortia following the Sematech/NCMS model.
- Substantial new funding for university programs in manufacturing through DoD and/or NSF.
- Establishment of a manufacturing sciences directorate in NSF, standing alongside the existing directorates for science and for engineering, and supplementing but not duplicating the activities of the latter. New money for NSF is the best way to create centers of excellence in manufacturing in the universities; one-third or more of the new money should be set aside for R&D that would bring expertise from the social and behavioral sciences to bear on problems of shopfloor management and organization.
- Expansion of DOC's existing program of manufacturing technology centers, and the addition of extension centers at laboratories funded by DoD and DOE with expertise in manufacturing. An appropriate goal would be creation of three dozen centers by 1999 (five exist now), with federal funding totaling at least $100 million annually. Workforce training should be integrated into the activities of some or all of these centers.
- Federal cost sharing for state industrial extension programs that have a manufacturing technology component.
- Pilot plant/demonstration facilities serving small and medium-size firms. These could be operated by NIST, each specializing in a particular set of manufacturing processes (e.g., fabrication of PMCs, nonconventional machining).
- Rapid expansion of NIST's Center for Manufacturing Engineering to the $50 million annual level. NIST is the logical candidate for a federal government center of excellence in manufacturing. As such, it needs to be able to sustain the center's activities with its own funds, independent of support now received from other federal agencies. Budget growth would permit NIST to complement the science-based thrust of its Advanced Manufacturing Research Facility with work of a more down-to-earth character.
- Joint establishment by the Department of Labor with the Department of Commerce and/or DoD of one or more university-based institutes for research and demonstration programs on workplace organization and management, including training and retraining. NSF might also participate,

particularly in R&D and demonstration efforts involving technologies and skills associated with new processes and/ or advanced materials.

Federal policies for supporting manufacturing need both focus and money; the first depends on the second.

NEEDED: NEW PRIORITIES

American manufacturing firms, regardless of whether they sell to the Pentagon, to original equipment manufacturers, or to consumers, share a common set of process technologies. Major defense systems are made in small lots, as are many of the products of civilian industry. Federal initiatives should raise priorities for manufacturing, encourage process R&D and workplace training, and stimulate capital investment. This would help American industry meet foreign competition that every year becomes more intense. Such policies would also help DoD procure military systems at prices the nation can afford.

It would be a mistake to rely too heavily on DoD, however. Despite many expressions of anxiety over weaknesses in the industrial base, the Pentagon remains largely unconcerned with process R&D. Product performance, rather than process improvement, drives the design and development of military systems. Competitive pressures have already forced many commercial firms to change, but no such pressures exist in defense. This is a major reason why DoD's manufacturing programs, whether aimed at R&D or at implementation and diffusion, have had limited impact in the past.

More than 98 percent of federal R&D goes for product-related work. Greater support for process R&D could help put manufacturing engineering on a footing comparable to other technical disciplines and strengthen the research base for manufacturing (including research in the social sciences). With much of American industry lagging behind Japanese firms in manufacturing expertise, often emulating Japanese practices without fully understanding them, the United States needs to find sources of advantage of its own. There is only one candidate: the application of science to manufacturing. Many empirically based production methods are ripe for rationalization through high technology and ever more powerful computational techniques. The United States should turn its unmatched scientific expertise to manufacturing problems.

APPENDIX 10-A:

The Air Force and Numerical Control

By 1952, when the MIT Servomechanism Laboratory converted a three-axis tracer mill to numerical control under Air Force contract, the pattern had been set for the development of NC machine tools and the future direction of much R&D in flexible automation. Digital code replaced analog template; a punched tape, prepared by computer, guided the mill in machining helicopter blades.[1]

Over the next two decades, NC equipment was installed primarily to automate long and complex machining operations. The Air Force and its contractors had needs even more demanding than helicopter blades—large and intricately machined spars, for example, and integrally stiffened wing skins. Rather than making wings by riveting aluminum sheet to a skeleton of ribs and stringers, aircraft designers hoped to start with a single large plate, machining most of it away to leave a smoothly curved aerodynamic surface backed by a skein of stiffeners—all in one piece. Ninety percent or more of the metal would have to be removed, much of it by milling pockets on the back surface.[2] One mistake hundreds of hours into the job could mean an expensive piece of scrap. Hence the desire for automation: perfection—the elimination of irreducible human error through automatic control—not flexibility, was the goal.

Diffusion and the U.S. Machine Tool Industry

Satisfactory NC performance, much less perfection, was slow in coming.[3] During the late 1950s, when manufacturers declined to invest their own money, the Air Force purchased more than 100 five-axis

---

[1] The Air Force began supporting automation research during the late 1940s. The NC story has been told in a number of places, most comprehensively and most polemically by David F. Noble, *Forces of Production: A Social History of Industrial Automation* (New York: Oxford University Press, 1986).

[2] With the shift to integrally machined structural components, machining rose from about 15 percent of aircraft manufacturing cost to 40 percent. M.G. Dronsek, "The Development of Aircraft Production—Computerized Integrated and Automated Manufacturing at the Augsburg Plant of Messerschmitt-Bolkow-Blohm GmbH," *Advances in Computer Technology—1980*, vol. 2 (New York: American Society of Mechanical Engineers, 1980), pp. 351–361.

[3] Anderson Ashburn, "The Machine Tool Industry: The Crumbling Foundation," in Donald A. Hicks, ed., *Is New Technology Enough? Making and Remaking U.S. Basic Industries* (Washington, DC: American Enterprise Institute, 1988), pp. 19–85; Artemis March, "The U.S. Machine Tool Industry and Its Foreign Competitors," *The Working Papers of the MIT Commission on Industrial Productivity*, vol. 2 (Cambridge, MA: MIT Press, 1989), especially pp. 20–27.

milling machines and placed them in contractor plants. The factory floor proved a harsh environment for early NC equipment, which incorporated fragile vacuum-tube controllers and error-prone tape readers, while program preparation required costly mainframe computers. Although inexpensive NC-controlled drilling machines soon followed, in 1960 no more than a thousand NC machines were at work in the United States.

During the early 1960s, American machine tool companies began producing versatile NC machining centers featuring automatic tool changing. These showed that NC could mean flexibility as well as precise control over lengthy machining sequences. As computers themselves dropped in price, tape readers gave way to direct control, first by minicomputers, and during the 1970s by microprocessors. Banishing tape readers and off-line editing opened new avenues for flexible utilization of NC equipment.

But already the U.S. machine tool industry had come under severe competitive pressure from lower-priced imports. Japanese companies had been quick to license NC technology. As in so many cases, their export strategies began with straightforward, soundly engineered equipment produced in volume to standardized designs. Firms like Yamazaki and Makino built for stock, not to order; when U.S. backlogs grew during upswings in this highly cyclical business, Japanese firms were able to deliver immediately. Japan's Ministry of International Trade and Industry encouraged Fanuc, a Fujitsu subsidiary, to develop standardized controllers. American toolbuilders, meanwhile, were slow in adopting microprocessor-based controls. By the mid-1970s, more than half the output of Japan's machine tool industry was sold abroad, mostly in the United States.

Imports accounted for 12 percent of U.S. machine tool sales in 1975, reached 21 percent in 1980, and have remained in the range of 30–50 percent since the mid-1980s; the highest percentages are for NC equipment.[4] Japanese firms account for about half the imports. Since 1987, with negotiated Voluntary Restraint Agreements limiting shipments from Japan and a number of other countries, Japanese toolbuilders have increased their investments in the United States, putting up new plants and buying U.S. firms.

---

[4] Gary Guenther, "Machine Tools: Imports and the U.S. Industry, Economy, and Defense Industrial Base," report 86-762E, Library of Congress, Congressional Research Service, Washington, DC, July 17, 1986, p. 15; U.S. General Accounting Office, *International Trade: Revitalizing the U.S. Machine Tool Industry*, GAO/NSIAD-90-182 (Washington, DC: GAO, July 1990); Joseph Jablonski, "World Machine-Tool Output Survey: Japan, Europe Boost Consumption," *American Machinist* (February 1991), pp. 35–39.

Military Influence: Technological Blind Alley
or Positive Boost?

Some observers have blamed the troubles of the U.S. machine tool industry on the Air Force and its contractors, which together set the original technical directions for NC technology. At least by implication, some of these same observers continue to blame defense agencies, with their very demanding production requirements and R&D agendas.[5] With hindsight, it is easy to criticize the path of development during the 1950s and 1960s. In those early years, Air Force desire for high-technology automation was more than matched by the willingness of MIT's Servomechanism Laboratory to provide it. The first generation of equipment had little application beyond the specialized tasks of aircraft production; few firms in other industries needed the five-axis milling machines originally built for sculpting wing skins. In the view of the critics, the United States developed gold-plated equipment, more complicated and expensive than needed for most machining jobs in civilian industry, while the Japanese produced less costly, easier-to-use tools that they could sell around the world.

But blaming defense for the decline of the U.S. machine tool industry is too simple. In particular, it ascribes too much significance to the legacy of the 1950s. If the initial NC programming language, APT (Automatically Programmed Tooling), developed at MIT, was less than user-friendly, so were all the computer languages of that era. APT had many and simpler descendants, while a variety of competing NC languages emerged independent of military influence. Today, much NC programming can be done on CAD terminals. The operator selects options, responds to queries and prompts, and calls up help screens as needed, much as with the software packages used in office automation. Because the equipment is straightforward, semiskilled shopfloor workers can do a good deal of programming themselves, limited not by computer skills but by their knowledge of machining. It is hard to see a legacy of gold-plated military requirements in the turnkey systems widely available today.

---

[5] Noble, *Forces of Production;* Anthony di Filippo, *Military Spending and Industrial Decline: A Study of the American Machine Tool Industry* (New York: Greenwood Press, 1986); David J. Collis, "The Machine Tool Industry and Industrial Policy, 1955–82," in A. Michael Spence and Heather A. Hazard, eds., *International Competitiveness* (Cambridge, MA: Ballinger, 1988), pp. 75–114, especially p. 108.

Corporate Strategies

In fact, nothing in the early history seems as significant for the decline of the U.S. machine tool industry as the lag in adoption of microprocessor-based controllers, a mistake of the 1970s—not the 1950s or 1960s—and one for which the companies themselves must take responsibility. During the early years of NC, few of the 1,000-plus toolbuilders then in the industry (many have since disappeared) had any connection with the Air Force or its contractors. Admittedly, most of these companies were small, with limited technical resources and financing, while the highly cyclical nature of demand made it difficult to plan for and invest in innovation. But they were free to go their own way; there was no lack of ideas; and at least some of the larger firms were in a position to adapt and improve upon the NC technology emerging from the Air Force/MIT work.

By and large, U.S. toolbuilders, conservatively managed, did none of these things: stagnation and lack of innovation, not technological excess, have been hallmarks of the industry. American firms were positioned to lead the world in NC at a time when the United States still enjoyed a dominant position in computer and microelectronics technologies, but they failed to take advantage of the opportunity. They did not adopt aggressive export strategies, even in Europe, where demand was strong and import barriers much lower than in Japan. They failed to understand the competitive threat posed by Japanese firms until it was too late (like the semiconductor firms discussed in Chapter 8), while their major customers, accustomed to long production runs for automobiles and other consumer durables, saw little need for "mechatronics" in their factories and did not press for new capabilities.

Japanese toolbuilders were able to undersell American firms for reasons that ranged from lower labor costs (in the 1970s) to an overvalued dollar (in the first half of the 1980s). Perhaps most important, Japanese firms were able to use their own products—machine tools—to achieve productivity levels substantially higher than those of American toolbuilders.[6] Most were divisions or affiliates of large firms with internal needs for production equipment; they could test ideas at home before launching them in external markets. Japan's rapid economic growth meant strong demand, while keiretsu (industrial group) linkages provided access to capital for R&D and new capacity and helped

[6]U.S. International Trade Commission, *Competitive Assessment of the U.S. Metalworking Machine Tool Industry*, USITC Publication 1428 (Washington, DC: U.S. ITC, September 1983).

Japanese toolbuilders ride out downcycles in demand, even as they continued to expand and improve their product lines.

The competitiveness problems of the U.S. machine tool industry have had more to do with industrial structure and corporate strategy than with Air Force sponsorship of early developments in NC. The course of decline resembles that of the steel or consumer electronics sector; it is wrong in this case to see negative spillovers from defense spending. Indeed, without Air Force spending for R&D and procurement, decline might have been faster and more U.S. toolbuilders might have disappeared. DoD pioneered NC technology and pushed military contractors and subcontractors to adopt it. The U.S. industry lost competitiveness despite the positive influence of DoD. "Trickle down" worked well for the United States in computers and semiconductors. It failed in machine tools.

PART FOUR

Toward a New National Technology Policy

Part IV summarizes our conclusions, which are of two types. First, in Chapter 11, we assess the prospects for synergies between defense and commerce, and whether Defense Department policies can be modified to achieve them. We conclude that it is possible—and important—for DoD to strengthen linkages with the commercial sector, but that this will not suffice as a national technology policy. In an open international economy, in which U.S.-based firms face rivals every bit as capable as they are, government policies rooted in defense simply cannot address the needs.

Second, therefore, in Chapter 12 we explore the types of civil technology policies the nation should adopt, proposing new policies to enhance the generation—and probably more important, the diffusion—of commercially relevant technology. New policies are required when the intent is to boost U.S. competitiveness rather than to generate knowledge or purchase systems and equipment that the government itself needs.

Much of what Tokyo can do, Washington cannot. What is legitimate in Brussels may be illegitimate in the United States. American politics places fundamental limits on the policies and processes of government—limits different from those in other countries. But while political traditions and institutional history will condition our future technology policies, they do not preclude them. Recognizing the potential pitfalls ahead, Chapter 12 seeks to outline policies that capitalize on U.S. strengths.

On the one hand, our system—by design, highly decentralized—will make it difficult to define and pursue consistent technology policy. In the absence of overriding imperatives such as those the cold war created, different interests will jockey for position, some winning and some losing. American industries compete among themselves for resources at the same time they compete with international rivals. But the openness of our political structure, the many centers of power and authority, also bring opportunities for innovation. Flexibility and creativity have been among our strengths—strengths we should count on for implementation of measures such as those suggested in Chapter 12.

11

Dual-Use Technology: The Search for Synergy

The preceding chapters describe how the American defense effort after World War II shaped industrial technology and stimulated the rise of high-tech industry here and around the world. U.S. defense policy, in turn, depended on the nation's position of world technological preeminence to compensate for the Soviet Union's quantitative advantage in armaments and its strategic position adjacent to Europe. Thus did the defense and commercial technology bases reinforce each other, both in actual fact and, through the prevalent spinoff paradigm, in policy lore.

Today, we find that spinoff alone is not up to the job of supporting American industry in international competition. Technical breakthroughs from defense cannot compensate for industry's failure to control costs and quality, and to respond quickly to market signals. In the past twenty years, German, Japanese, Korean, and other foreign firms have exploited the weaknesses created by excessive American reliance on the "pipeline" view of the relationship between new science and commercial success and on the spinoff view of public policy for technology.

Simply enhancing the synergy between defense and commercial technology bases is not the answer to America's competitiveness challenge. Other public policy responses are necessary, and these are detailed in Chapter 12. Yet enhancing dual-use synergy is still an important objective. First, government-sponsored projects (largely defense and space) still account for a large fraction of technology spending in the United States, employ a significant fraction of scientists and engineers, and influence national attitudes toward technology and engineering culture. Second, while dual-use synergy by itself cannot be regarded as an adequate solution to strengthening competitiveness, it will be increasingly critical to the future of national defense. The Defense Department must draw on the global commercial technology base if America is to continue to have a defense strategy that relies on U.S. forces having equal or better technology than any opponent's. In this chapter we address the ways in which synergy between defense and commercial technologies can be enhanced. In the final chapter, we address the broader menu of technology policies that are needed to enhance American economic performance, where dual use is not enough.

The Dual-Use Relationship

The preceding chapters identified the key features of the existing military-commercial relationship, which can be summarized as follows:

- The United States spends as much on R&D as its major industrialized competitors combined, and devotes roughly the same fraction of its GNP to R&D. But a third of U.S. R&D is sponsored by the Pentagon, whereas military sponsorship accounts for a very small fraction of the innovative effort in other Western nations (Chapter 4). From these crude spending measures, one can argue that either the United States must derive substantial civil benefit from its defense technology spending, or it must dramatically increase its commercial R&D spending.
- Defense R&D, though still large, has been declining in relation to commercial R&D spending for three decades, and the declining Pentagon budgets projected for the next decade will hasten this trend.
- Most defense R&D dollars go for development rather than for basic or applied research. Defense support for basic and applied research is considerably smaller now than 25 years ago, both in absolute terms and as a share of defense R&D (Chapter 4).
- Defense is many times more R&D-intensive than commercial industry. For defense, R&D is 31 percent of the total acquisition (procurement plus R&D) budget, whereas in civil manufacturing

as a whole the corresponding ratio of R&D to sales is less than 5 percent (Chapter 6).

- Defense R&D is concentrated in a few industries. Seventy percent of all federal R&D funds spent in industry are concentrated in the aeronautics and electronics industries.

- The business units that serve the defense market typically have few or no commercial markets. On the other hand, these defense-dominated business units are almost without exception embedded in much larger firms dominated by commercial markets. Civil sales account for 83 percent of the output of the top 50 defense manufacturing sectors, and 91 percent of the business of the largest defense firms (Chapter 6). Thus one sees in industry structure both the substantial segregation of defense, and the substantial potential for dual-use synergy.

- Dual-use synergy is not the normal state of affairs: high barriers separate defense and commercial business (Chapter 5). Some of these barriers reflect DoD's special contracting requirements, others the contrasting business cultures and requirements for success in civil and military innovation. Most important are the legal requirements and regulatory practices, intended to ensure fairness and accountability (among other goals), that are placed on all government procurements. Waived in wartime, but often reimposed with a vengeance in peacetime, these requirements define a highly specialized environment for doing business with the Pentagon. Other barriers are a consequence of incompatibility between military and commercial technologies; for example, there are few civil uses of "stealth" technology or of nuclear weapons, and few military applications of agricultural technologies. Finally, export controls place explicit constraints on diffusion of technology. With the changing East-West security situation, controls are being reevaluated; however, with new attention being given to the proliferation of high-technology weaponry in the third world, controls will remain an important tool of U.S. policy.

The Importance of Dual Use to Defense: Ensuring Military Access to Commercial Technology

Innovations important to defense increasingly will originate outside the Pentagon's R&D system. Even at today's level of defense expenditures, DoD's share of the output of key high-technology sectors (e.g., computers, microelectronics, and aircraft) is small. At the same time, with increasing frequency commercial firms are declining to bid on defense procurements; as the money shrinks, more companies will shun defense work.

Thus the Pentagon will need to learn to work with and borrow from the world of commercial technology in a manner completely foreign to its cold war practice. Moreover, this adjustment will need to be made in an era in which procurement budgets are declining and when, as a consequence, DoD R&D will less and less often lead directly to production. Thus the cultural gulf between defense's practices and the commercial world's tight linkage of R&D to manufacturing will probably grow. Finally, since much crucial technology will originate outside U.S. borders, defense planners will have to view foreign technology as an asset to be acquired and exploited. Problems raised by dependence on foreign suppliers will have to be managed by diversification of sources, rather than wished away in the hope of somehow reestablishing the American technological hegemony of decades past.

During the upcoming period of retrenchment, the temptation to slash technology investments may be irresistible, particularly if Congress resists other forms of savings such as closing defense bases and plants. But a strong case can be made that greater selectivity in weapons systems development and production should be accompanied by an increase in long-range R&D—a "technology reserve."[1] A small percentage of the savings from halting unneeded weapons development and acquisition programs, and closing redundant facilities, would be enough to support a high-quality research and technology activity. Unless the relative decline in the defense technology base during the Reagan and Bush administrations is reversed, the United States may find itself unable to rebuild an effective and affordable defense in the future, should the need arise. But if the Pentagon does move to bolster technology base investments, even a leaner Defense Department could become a valued partner in national technology effort.

Coupling defense more closely to the commercial sector will prove challenging, with fiscal pressure only one of the difficulties. DoD will have to bring something of value to the table: an aggressive investment strategy aimed at generic technology that the Pentagon is willing to share, or a liberalized and expanded IR&D program. Dual-use technol-

---

[1] See discussion in Chapter 5, and the report of the Carnegie Commission on Science, Technology and Government, *New Thinking and American Defense Technology* (New York: Carnegie Commission, August 1990), p. 23.

ogy development, in this view, would be seen by defense as advantageous for both sectors, rather than as a threat to unilateral control, a conversion of public resources to private gain, or a source of technology leakage. Indeed, it has been suggested that DARPA's role in defense-justified dual-use technology be emphasized by changing its name to the National Advanced Research Projects Agency (NARPA), with a policy board comprised of the secretaries of Defense and Commerce and the director of the White House Office of Science and Technology Policy.[2] This idea was advanced as a mechanism for promoting a more collaborative relationship with commercial industry, perhaps using cooperative agreements more often than contracts as the vehicle. Without changes in the U.S. political environment, however, the perceived and the actual hazards of mixing government and private interests might prevent this approach from being tried.

Should defense prove unable to work with the commercial sector, it would be forced to retreat into a specialized ghetto, with military technology focused on specialized requirements pursued by a few private firms or by dedicated government arsenals. With little access to commercial technologies, components, and subsystems, costs would rise and capabilities shrink. In the extreme, this scenario ends in collapse—a debilitated and isolated enclave. The only solution is to tap the commercial sector. To do so, Pentagon managers will have to sponsor dual-use technologies of their own, not just to attract commercial industry, but to remain aware of and to take advantage of market-driven advances in commercial technology worldwide.

Over the long term, the Pentagon must find a way to reduce the disincentives that drive commercial firms to wall themselves off from defense. Reforming defense acquisition has proven difficult for many reasons, but that does not make it impossible. The following steps should come first:

- Replace inflexible and inflated defense system requirements with cost/performance criteria that encourage trade-offs. Doing so will require DoD and its contractors to make realistic cost projections rather than "buying in" with artificially low estimates. At the

---

[2] Carnegie Commission on Science, Technology, and Government, *Technology and Economic Performance: Organizing the Executive Branch for a Stronger National Technology Base* (New York: Carnegie Commission, September 1991). The commission suggested that under this management structure NARPA could manage dual-use projects funded jointly by Defense and one or more civil agencies.

same time, Congress must be willing to make binding commitments and to allow DoD to retain some savings to be reinvested elsewhere.

• Provide incentives for simultaneous engineering (Chapter 10) and in other ways give more emphasis to manufacturability and field maintainability during the early stages of design and development. This will be hard to do if Congress continues to insist on separate competitions for R&D and production contracts; it will be especially so if production contracts are not in sight at the time of development.

• Change the accounting and procedural requirements that keep firms from integrating their commercial and government businesses. Doing so will be difficult in an atmosphere of long-standing public suspicion and distrust of the "military-industrial complex," but it is a necessary part of any dual-use strategy.

• Bring military standards and specifications where possible into agreement with best-practice industrial counterparts, a necessary step to encourage military use of commercial technologies. The burden of proof in acquisition decisions should be shifted so that it is the use of government-developed technology, rather than commercial, for which justification is required.

The long history of frustrated attempts at acquisition reform may lead to skepticism. However, with the right political strategy and a national perception of crisis, sweeping policy changes involving gains and losses for large numbers of stakeholders can be legislated. The combination of crisis in U.S. economic competitiveness, the virtual certainty of a drastic reorientation in American defense strategy, and severely constrained DoD budgets could create the conditions for the sweeping reforms of defense acquisition that have been recommended for so long with so little result.

Other policy changes do not depend on acquisition reform. Defense policymakers should take the following steps:

• Increase support for DoD's science and technology base (budget categories 6.1, 6.2, 6.3A) to an even larger fraction of the total than it represented in 1980. Because reduced levels of procurement and downstream development will contribute less spillover to the technology base, explicit technology base investments must be expanded just to hold the effective level steady.

• Encourage DoD/DOE labs to use their existing authority to cooperate with private firms in the development of emerging technologies at much earlier stages, so that commercial considerations will have more influence. The labs should view such collaboration as a two-way street—to pull in commercial technology and at the same time contribute to it.

- Encourage the Defense Advanced Research Projects Agency to take a broad view of its role as a source of new and promising technologies, even though defense applications may not be in view. This role for DARPA was envisioned forty years ago by Vannevar Bush, whose wartime experience as head of the Office of Scientific Research and Development convinced him of the need for farsighted technical exploration, independent of the services. In his report to President Truman entitled *Science, The Endless Frontier*, Bush envisioned a National Research Foundation (created in 1950 as the NSF) that included a division of defense research devoted to long-range, high-risk projects. Not a part of NSF as established, such a function was created later with the establishment of the Advanced Research Projects Agency (ARPA, later DARPA) within DoD. A smaller Defense Department will need visionary R&D more than ever. This is not the time to divert DARPA to near-term tasks.

Adjustment

Decline in the defense share of the nation's technological activity will have both short- and long-term ramifications. In the short run, the United States faces the problem of shifting resources out of defense—technically trained people, research laboratories, and production facilities. This process is sometimes called "conversion" (from military to civil activities), but the implication that defense firms can easily put their people and facilities to work on commercial product lines is unjustified. Minimizing the social costs of adjustment provides additional motivation for policies that reduce barriers between military and civil businesses.

Presumably, airframe and jet engine manufacturers can shift their people and capital into commercial work, provided demand exists. But most corporations with defense and civil subsidiaries will find the barriers between them too great to surmount. They will have to cut jobs in defense and redeploy as many of their people as they can into their commercial divisions. Government has a role in cushioning the shocks that will hit defense workers and defense-dependent communities. But it would be a mistake to assume that other national problems are as amenable as defense to a centralized, government-orchestrated, and in considerable measure government-executed technology strategy.

The Contribution of Dual-Use Technology to Competitiveness

Looking back over 50 years of DoD investments, one can only be impressed with the diversity of government contributions to the growth and technological sophistication of industry. To be sure, many of the images Americans hold are misleading: product spinoffs are rare, the bundling of technology projects in billion-dollar packages is often more politically attractive than technically sensible, and the coupling between defense technology development and the commercial sector tends to be indirect, slow, and inefficient. But those connections certainly exist. Government, and especially defense, played—and still plays—a central role in the development of American technology.

Defense agencies pioneered in supporting basic science in U.S. universities. The National Science Foundation's policies were patterned after those of the Office of Naval Research. Defense agencies led in supporting many fields of science and tools of engineering—in electronics, materials science, lasers, computers, and feedback control systems, to name a few. High-technology industry would not look the same without DoD's past investments.

Government procurement of sophisticated weapons systems challenged the ingenuity of American engineers; they created tools, materials, and design methods that pervade commercial practice worldwide and are now taken for granted. Second-tier manufacturers enjoyed improved economies of scale and of scope from serving both defense prime contractors and commercial customers.

The most significant investments from a commercial standpoint often seemed the riskiest at the time. Indeed, it may be defense's willingness to gamble on expensive and highly uncertain undertakings that will be most sorely missed in the long run. Justified on security grounds, such programs were not subjected to the cost-effectiveness tests, market analyses, and public scrutiny that would face a program that took such risks to explore commercially relevant technologies.

But defense programs, important as they once were, are a poor match, in magnitude and in kind to the needs of commercial competition. There is a limit to the contribution dual use can make to the competitiveness dilemma, as we have stressed repeatedly in the preceding chapters. First, DoD's relative level of effort is shrinking. Second, defense's engineering culture is poorly matched to the requirements of the commercial world. True, some of defense's peculiar features, such as its emphasis on performance regardless of cost, can and should be eliminated in favor of practices more like those in successful commercial practice. But other features of defense, including the politics of public spending, will always set DoD apart.

Third, whereas DoD's programs emphasize knowledge generation, especially high-risk exploratory technology, commercial manufacturers have other, sometimes quite mundane, needs than dramatic new technologies spun off from defense. While some of these needs can and should be met by government, they are unlikely to be met by DoD in the pursuit of its national security mission. They will require new federal technology policies.

Thus dual use is an ingredient, but not the most important ingredient, of a technology policy to address the international competitiveness of American industry. We discuss the issue of overall federal technology policy in the final chapter.

Enhancing Defense's Contribution to Commercial Competitiveness

While they are not sufficient by themselves, DoD has played the following roles in the past and they should be continued:

- Exploring risky new technologies;
- Supporting the instrumentation, test methods, technical standards, and other underpinnings of a modern economy;
- Using government procurement (even at a reduced level) in imaginative ways to stimulate commercial vendors and underwrite some of the costs of early technical learning; and
- Supporting high-quality technical education to expose engineers to a broad range of challenges.

New tasks for DoD include:

- Bringing second-tier manufacturers and small firms generally into better contact with appropriate technologies and practices for enhancing productivity and innovation;
- Increasing the productivity of R&D itself, and of downstream technical activity;
- Ensuring that scientific and technical information services pro-

vide efficient and economical access to useful and accurate knowledge; and

- Encouraging the services and suppliers ("complementary assets") important for competitive success.

Defense agencies should contribute to these objectives, not by abandoning national security priorities, but by pursuing those priorities in ways that enhance support for and access to dual-use technologies. However, this will not do the job alone. The country also needs new policies that are independent of defense—the subject of our concluding chapter.

12

Beyond Spinoff: Mapping Out a New Federal Technology Policy

The preceding chapters show that defense spending is still a major contributor to the technology base upon which U.S. economic performance depends. But they also show that defense contributions are growing smaller and more specialized relative to the efforts of the private sector, here and overseas. Moreover, changing times have revealed gaps in the nation's technology policies that defense never filled, such as insufficient attention to manufacturing, slow diffusion of best-practice know-how through the economy, and the needs of the many industries that benefit little if at all from DoD's R&D dollars. If the U.S. government is to play as significant a role in the national technology base in coming decades as it has played in the past half century, it will have to devise a new approach. The nation has relied for too long on the indirect effects of defense spending.

The American public has begun to ask government to add economic performance to its list of explicit objectives alongside defense, energy security, health, and environmental protection. Yet few federal institutions or programs now seek to foster the technological competence of industry. As long as the spinoff paradigm held sway, policymakers saw little need. Reliance on spinoff admitted only an inadvertent, indirect role for government, denying legitimacy to more direct contributions by federal agencies.

But as we have seen, spinoff cannot serve as a realistic basis for technology policy. We propose an alternative, focusing on technology to support the nation's commercial sector. Much of this necessarily involves a gray area lying between basic research and product development, falling outside the traditional domain of federal government policy yet not intruding on the domain of private investors. Policies dealing with this gray area will have to avoid dangers associated with public efforts to influence the private sector. In particular, such policies must:

- Meet the widely varied technological needs of the largest, most diverse economy the world has yet seen;
- Ensure that the benefits of federal investments are primarily retained by Americans and their institutions;
- Minimize the fact and the appearance of diverting public funds for private benefit;
- Prevent vested interests from capturing programs and perpetuating unsuccessful efforts; and
- Ensure that government officials are equipped to make competent decisions.

We examine four new modes of technology policy that we believe can avoid these pitfalls if properly designed and implemented. None of the four is entirely new to Washington; examples of past success can be cited for each. What is new is a clearer justification for federal investments and for the basic principles that define the risks, benefits, and criteria for successful implementation.

One new mode emphasizes the *diffusion* of nonproprietary technical information. The other three address categories of federal investment:

- *pathbreaking* technology, emphasizing technical challenge;
- *infrastructural* technology, emphasizing breadth of application; and
- *strategic* technology, emphasizing the criticality to the nation of the industries to which the technology applies.

These three investment categories have different rationales. Not only do criteria for investment differ, but this framework allows us to isolate the most problematic—and thus legitimately controversial—policy choices. Of the three, strategic technology investments will raise the most questions.

Our discussion revolves around the process by which particular projects of these four types can be selected for public support, with proper attention to the risks of waste and pork-barrel programs and to the inherent limits on government understanding of the marketplace. We seek to keep one eye on market failure, the other on government failure.

We deliberately eschew analysis of particular policy initiatives that dominate the current debate, such as semiconductor manufacturing or aerospace planes. Controversy swirls about such proposals in Washington. We believe that these controversies cannot be resolved without a framework, which we describe, into which specific proposals can be placed. In the current policy vacuum, initiatives emerge by political happenstance rather than on their merits, and it follows that controversy frequently attaches to particular projects that are not good exemplars of what is in fact a valid category of public policy.

Neither do we believe it useful to prescribe policy by selecting among or prioritizing technologies themselves. Today's ubiquitous lists of "critical technologies" say little about what the government should do, or indeed whether more government support is necessary at all.[1] The fact that a given technology shows up on one or more such lists is a measure of its recognized importance, which may in fact indicate that it is already attracting sufficient public and private support. Even if a given technology is receiving inadequate attention, simply singling it out does not indicate whether or how government can contribute.

Finally, we discuss allocation of responsibility among federal agencies for implementing these new technology policies. None has the job today. The major R&D agencies—the Departments of Defense and Energy, the National Aeronautics and Space Administration, the National Institutes of Health, and the National Science Foundation—have preexisting missions. The National Institute of Standards and Technology, in the Department of Commerce, comes closest to having the job of enhancing industry's technological competence, but its budget, experience in contract R&D, and political clout are minuscule in comparison to the other agencies and to the scope of the problem at hand. In fact, there is more than enough for all the agencies to do, but their missions will have to shift, and their institutional structures adapt, to encompass their new roles.

---

[1] Examples of such lists include *The Department of Defense Critical Technologies Plan,* AD-A219300, March 15, 1990; Department of Commerce, *Emerging Technologies: A Survey of Technical and Economic Opportunities* (Washington, DC: Technology Administration, Department of Commerce, Spring 1990); Office of Science and Technology Policy, *Report of the National Critical Technologies Panel,* March 1991; Council on Competitiveness, *Gaining New Ground: Technology Priorities for America's Future* (Washington, DC: Council on Competitiveness, March 1991).

Washington innovated very profoundly after World War II, and again when Sputnik threatened America's sense of well-being. These spurts of bureaucratic creativity were in response to political shocks. Competitive slippage and the four revolutions described in Chapter 1 portend not crisis, but "death by a thousand cuts," as Lester Thurow called it. In the absence of some dramatic shock, it is unlikely that new federal agencies will be created to strengthen economic performance through technology. Thus we will probably be left with what we have: a mix of mission agencies, likely to change slowly at best, a National Science Foundation devoted to basic science, and an infant Commerce Department effort. Only the latter has the mandate to carry out the policies described in this chapter. Collectively, however, the agencies possess an enormous range of expertise and experience. Thus we emphasize making use of existing agencies rather than creating new ones.

AN ERA OF TRANSITION

If, as noted in Chapter 1, national governments and national economies are losing their ability to influence and control technology—a trend that will increasingly constrain any nation's technology policy— why do we call for such a policy here? We have two reasons. First, a truly integrated international economy remains in the distant future. For now, the United States must try to interpolate between the policies and institutions suited to the postwar world of national economies and U.S. technological hegemony, and the yet-unborn policies and institutions appropriate to a borderless world. The United States is still the dominant innovator, unrivaled in science, and the world's largest market for high-technology goods. Washington's policies still influence innovation within U.S. borders and the prosperity of those living there. National technology policies still have validity and force.

Second, a great deal of technology does not and will never move freely from nation to nation like disembodied capital or goods: it is in the heads of scientists and engineers who do not or cannot write it down, and it is embedded in the organizations that design, manufacture, and market goods and services of all kinds. Skilled people and organizational know-how constitute the technological competence of the nation. In a knowledge-based economy, comparative advantage comes from human and institutional capital. Americans should expect their government to help them develop technological competence in cases where the private sector cannot: where substantial investments are required whose benefits, although they might also be very large, cannot be appropriated by private investors. The United States also needs to raise its priorities for the diffusion of knowledge without lowering its priorities for generation of new knowledge: we need to raise average levels of technological competence, at the same time extending the frontiers. Hence we prescribe public policies for both *generation* and *diffusion* of technology.

A CAUTIONARY NOTE

Private Sector Responsibilities and Public Policy

In the traditional U.S. view, government technical activity is appropriate in two cases: in upstream basic research, where specific applications are remote and results are published for all to read and exploit, or when the objective is an accepted public good such as a weapon system or medical procedure. In the second case, public support is acceptable in all pipeline categories from basic research through product development. In this view, government has little business funding technical effort whose returns could be captured by a firm, or that leads to products that the market could provide.

We do not challenge this traditional assignment of public and private roles. But a wide gray area lies between them (see Figure 12-1). We have seen in earlier chapters that government influences the capabilities of U.S. firms in ways that are much more diverse and complex than usually appreciated, and that taken together constitute a de facto technology policy. Nevertheless, many of the economy's needs will be addressed haphazardly, if at all, without mechanisms for addressing them directly.

The burden for those proposing government investments in technology is to demonstrate that such investments create "public goods" benefiting many citizens in the long term; that private actors have insufficient incentive to invest; and that government can and will do so without distorting the market or displacing scarce technical talent and resources that would otherwise be invested more wisely by the private sector.

We do not suggest that government bear *primary* responsibility for the technological competence or competitive ability of U.S. firms. All the government-funded technology in the world will not help a com-

FIGURE 12-1:
Public and Private Roles in Technological Innovation

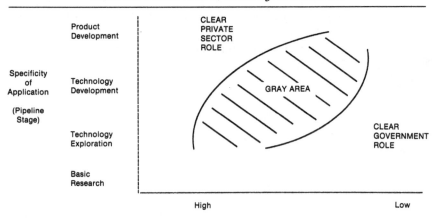

pany that treats shopfloor production as a low-level endeavor, fails to listen to its customers, or ignores the strategies of its competitors.

Pitfalls

Before exploring the new gray-area technology policies, it is worth noting some of the difficulties ahead. If government is to support R&D in the broad public interest, it will need to find ways of identifying what that public interest is, and convincing potential critics that it has done so fairly and objectively.

Officials and agencies responsible for a U.S. technology policy attuned to the needs of civilian industry will have to satisfy a range of constituencies with different and sometimes conflicting interests.[2] Money for industrial technologies will inevitably bring forth politically effective interest groups ready to make the case for small-business set-asides, declining sectors, and distressed regions. Perceived inequities between competing firms and industries would be difficult to avoid; money for R&D, even for long-term projects, means winners and losers. Industry, moreover, will want immediate help, rather than medium- to long-term R&D.

In such an environment, government must demonstrate prudence

[2]The following paragraphs draw from John A. Alic and Dorothy Robyn, "Designing a Civilian DARPA," *Optics & Photonics News* (May 1990), pp. 17–22.

in spending public funds, fairness in dealing with the private sector, and some degree of balance with respect to regional interests. At times, inevitably, these demands will clash with those of sound R&D management and economic efficiency. The risks are real ones, more so in the United States than in most countries, given our decentralized government with many points of entry for interested parties.

Government decision making: winners and losers.   All proposals for federal financing of civil technologies in the name of competitiveness face a common problem: without a well-understood and widely accepted mission (like those of DoD, NASA, or NIH) to discipline the process of setting priorities and making funding decisions, agencies can easily end up subsidizing marginal undertakings. This has been a common fate of government support for commercial R&D in Europe, as well as of past transportation and energy projects in the United States.

Nurturing technologies that are intended to succeed in the marketplace is more complex and exacting than supporting technologies for which government is the customer. In the latter case, the negotiation over means and ends is bilateral, between the R&D performer and the government as customer/user. But if the end user is a customer in the market economy, the negotiation is indirect, taking place between the R&D performer and the government as surrogate for the eventual customer. No matter how competent and motivated they are, there is every reason to fear that government officials, left to themselves, will be too far removed from technical substance and market feedback to make sound decisions.

The harm can extend beyond wasting taxpayers' money: government actions can distort private sector decisions. Investors may be discouraged from pursuing alternatives if government loads the dice in favor of a chosen approach. Because of learning-curve effects and economies of scale, the government choice may become so entrenched that potentially superior alternatives have little or no chance of overtaking it.[3]

The limits on government's ability to select projects suggest two necessary criteria for policy design. First, a well-defined *process* for allo-

---

[3] A prime example is the advantage given to light-water reactors in competition with high-temperature, gas-cooled reactor technology by early Atomic Energy Commission support. Would the gas reactor have won out in the end on a level playing field? Government support ensured that this question would never be answered. See Harvey Brooks, "The Typology of Surprises in Technology, Institutions, and Development," in William C. Clark and R.E. Munn, eds., *Sustainable Development of the Biosphere* (New York: Cambridge University Press, 1986), pp. 325–350.

cating support should be established, requiring competitive evaluation of proposals by experts from both government and industry. Criteria for project selection should be clear in advance, though they will differ greatly among the three types of federal investment (pathbreaking, infrastructural, strategic) that we will describe. Indeed, we propose these three categories in part to force advocates to declare which category their proposal fits into, and thereby to invoke the appropriate rules for evaluation. Second, in some cases—particularly strategic technology projects—government should put the private sector to the test by requiring *cost-sharing* in proportion to the privately appropriable benefits expected.

It is not U.S. industry collectively, but U.S. corporations individually, that compete in world markets. Even if government picked only winners, there would be limits to the public's tolerance for the use of tax revenues to benefit particular firms. Projects focused on technologies that are crucial to particular industries should be jointly funded by government and industry, and the vehicle in these cases should normally be a consortium of firms. The broader the applicability of the investment, or the harder it is to identify beneficiaries, the less will be the perception of an inappropriate diversion of public funds for private gain.

Losers picking governments. Every program that transfers public assets to private hands, or that creates a new bureaucracy, generates an instant constituency. As Sylvia Ostry has observed, "Governments may fail to pick the right winners, but losers can be very skillful at picking governments."[4] Bureaucracies created to foster an "infant technology" tend to become solutions looking for problems, persisting after the technology matures to the point where it should be ready to survive on its own.[5] Local and regional self-interest reinforces agency self-interest. This is an especially acute problem in the United States, where reelection to Congress may depend on the ability to bring home federal funds.

Not all technology development efforts will be successful—nor should they be. One of government's jobs is to pursue projects too risky for the private sector to fund alone. But the public deserves assurance that unsuccessful efforts will be halted before too much money is wasted. Measures of progress and of ultimate success, and criteria for

[4] Sylvia Ostry, *Governments and Corporations in a Shrinking World: Trade and Innovation Policies in the United States, Europe, and Japan* (New York: Council on Foreign Relations, 1990), p. 58.

[5] This point is discussed in the chapter entitled "Science and the Allocation of Resources" in Harvey Brooks, *The Government of Science* (Cambridge, MA: MIT Press, 1968).

a clear end to the project, should be identified from the beginning. Projects should either be handed off to industry and government support ended; returned to the status of basic research, if the technology cannot be developed on the current base of scientific knowledge; or terminated, if assessment indicates that prospects for ultimate commercial viability are too dim.

Every industry and every technology is different.   Industries differ in their structure and in their markets, making it unwise to generalize. Seemingly minor differences can alter the appropriate policy recipe. Technologies draw in different measure on the fields of science and engineering—some remain largely empirical, others rest on a rigorous theoretical foundation. Each progresses through a life cycle.[6] Americans enjoy a commanding lead in some technologies, lag seriously in others. Some technologies are highly proprietary; others draw on knowledge widely dispersed among actual or potential competitors. This variety would be challenge enough for government agencies and industrial firms even if technologies changed slowly, but many are changing rapidly. Those that change the fastest (biotechnology and microelectronics, for example) are among the most important, often for exactly that reason. When government seeks to establish technology policy, it must be prepared to specialize in order to meet the needs of specific technologies at particular stages of their life cycles.

The variety and complexity of commercial technology place a heavy burden on government decision makers. Indeed, government cannot make decisions alone; industry must be closely involved. Cooperation will not be easy for either party. Misunderstandings and mistrust characterize the government-business relationship in the United States. Business leaders have preferred to keep Washington at arm's length, which is why the Commerce Department is a weak sister among federal agencies. Business may be reluctant to engage the technical agencies of government in dialogue. Many government officials, for their part, when seeking the advice of technical experts from industry will be nervous—often with good reason—about the danger of bias because of economic, organizational, or professional self-interest.

''Who is us?''   The underlying purpose of U.S. technology policy should be to strengthen the technological competence of American workers and organizations. Yet it is increasingly difficult to help Ameri-

---

[6] For a good explanation of the technological life cycle, see James M. Utterback, ''Innovation and Industrial Evolution in Manufacturing Industries,'' in Bruce R. Guile and Harvey Brooks, eds., *Technology and Global Industry: Companies and Nations in the World Economy* (Washington, DC: National Academy Press, 1987), pp. 16–64.

cans without also improving the capabilities of foreign individuals and organizations. A nominally "American" firm owned by Americans and headquartered in the United States might do a good deal of its manufacturing and even R&D in another country, with the citizens of that country benefiting in large measure from the firm's success. On the other hand, a foreign-owned company might locate many of its operations in the United States, where American workers, scientists, and executives draw salaries and share in its prosperity. In a world of global technology, transnational business, and highly mobile capital, national governments must ask themselves, in Robert Reich's words, "Who is us?"[7]

The simplest answer is for government investments to focus on those parts of the U.S. science and technology system that in aggregate are the least mobile—its infrastructure and its workforce.[8] Improving these relatively "fixed" assets will induce firms, both foreign and domestic, to invest in and conduct high-value-added activity in the United States. Thus investments by U.S. taxpayers in their technological future, through government, will redound to their own benefit.

In practice, however, no simple guidelines will suffice. Politicians, nervous at the prospect of public funds flowing to U.S. subsidiaries of overseas competitors, are likely to attach tests of "American-ness" to technology policies. They may require, rightly, that firms manufacture or even perform R&D in the United States, and that other governments accord equivalent treatment to American subsidiaries. But foreign subsidiaries are not the only problem. Increasing numbers of U.S.-based companies are entering into strategic alliances with foreign firms that involve technical collaboration and the sharing of embedded or tacit know-how. If government programs are closed to such firms, they may end up open only to an uncompetitive enclave of inward-looking firms or captive suppliers to federal agencies.

Avoiding the Pitfalls

Two lessons emerge from the preceding discussion. First, considerable knowledge of technical details, and of market and industrial structure, will be needed to put government policy on a sound footing. This means active, ongoing cooperation between government and industry. Current levels of ignorance and even hostility on both sides exact costs

[7]Robert Reich, "Who Is Us?," *Harvard Business Review* (January–February 1990), p. 53.
[8]Lewis M. Branscomb, "Toward a U.S. Technology Policy," *Issues in Science and Technology*, vol. 7, no. 4 (Summer 1991), pp. 50–55.

to the nation that far exceed the limits set by reasonable principles of democratic governance.

Second, federal officials must have the discretionary authority to make case-by-case decisions tailored to individual circumstances. By their very nature, government investments will go to projects where no clear and convincing arguments can be made for private investment. Input from industry—validated by the willingness to share costs—will be important for good decisions. But this input cannot provide certain guidance. In at least some cases, public officials will have to base their actions on subjective judgment. Without the freedom to fail, they cannot succeed.

The dialogue between government and industry must be broad-based, not restricted to a few large, favored firms, as has been the case in such European countries as France (see Chapter 7). Moreover, Sweden and Germany have shown that involving labor can broaden the base of expertise and advice, enhancing both the political legitimacy and the quality of decision making.

Technology Policy Assessment

Given the magnitude of federal R&D spending and the many existing policies with direct and indirect impacts on productivity and competitiveness, government should continually assess the impacts of its actions. The Commerce Department and many other agencies conduct economic analyses, but the depth and sophistication of their technology assessment leaves much to be desired. Defense and intelligence agencies engage regularly in net technical assessments, but usually from a narrowly military perspective. For ongoing assessment to have value for public policy, it must be the result of joint industry and government effort by experienced professionals. To ensure access to corporate proprietary data, such an assessment may need the protection of a level of confidentiality like that afforded much of the information provided for the Census of Manufactures or to the U.S. International Trade Commission: raw data are accessible only to government employees sworn to protect their confidentiality, and only tabulations that do not reveal proprietary data are publicly released. Such an assessment should have two primary objectives:

1. *Assessing the performance of U.S. industry in international competition.* The first goal of such an evaluation should be to identify technology needs that could enhance U.S. competitiveness. The second and related aspect is competitive benchmarking—the comparison of U.S.

industrial performance with foreign performance. Policymakers need to know what is due primarily to technology and what is due to factors such as management, industrial structure, financial market conditions, workforce capabilities, regulations, and so on.

2. *Determining the technical value and long-run costs of "big science" and "big engineering" projects.*   Policymakers need better understanding of both the benefits and the forgone opportunity costs of large-scale government projects with life-cycle expenditures of $1 billion or more.[9] To what extent can projects like the Superconducting Super Collider and NASA's proposed space station be justified by claims of spinoff? Might analysis show that greater value to the economy could be obtained at lower cost by a more modest approach?

NEW MODES OF TECHNOLOGY POLICY

The United States needs technology policies of two types. The first type fosters the *diffusion* of best-practice technologies to American firms that can use them to improve their competitive performance. The second supports the *generation* of technology through public investments in the vast gray area between basic research and commercial product process development. Within this second category, as noted earlier, we distinguish three types of public investment in the generation of commercially useful technology—*pathbreaking*, emphasizing technical challenge; *infrastructural*, emphasizing productivity improvement and breadth of application; and *strategic*, emphasizing the importance to the nation of the industries to which the technology applies. All three aim at the creation of technical knowledge that is more directly applicable to commercial products than basic research, yet useful to a wider segment of industry than the applied product and process development routinely conducted by private firms. There is abundant precedent in the postwar American experience, as well as in the practices of other industrialized nations, for all three. What is controversial in the United States is the notion of competitiveness as a policy objective on the same footing as defense or health.

---

[9]Linda R. Cohen and Roger Noll, in *The Technology Pork Barrel* (Washington, DC: Brookings Institution, 1991) analyze six cases of large-scale commercialization, concluding that such projects often build constituencies that prevent the redirection or termination of the project if the initial objectives prove unattainable.

Clear evidence that Washington is not comfortable with the gray area between science policy and industrial policy is the linguistic shrubbery that litters this landscape—labels including "enabling," "precompetitive," "emerging," "critical," "generic," and "strategic." New words are invented in the hope that the confusion engendered by earlier labels can be avoided by starting afresh. Because these words blur together into an indistinguishable thicket, we too are forced to attach our own specific meanings to "pathbreaking," "infrastructural," and "strategic."

These three categories are distinguished from one another by their time horizons, by their mix of business and technical risks and rewards, and perhaps most significantly by their beneficiaries. The technical content of projects in any of the three categories could range from laboratory research to production engineering and prototype testing. Results could be made public when appropriate, or kept proprietary when implementation—and the ensuing public benefits—depend on maintaining exclusive access to the technical knowledge that the projects create. All projects, however, should have the features we identified earlier as necessary:

1. There should be a well-defined *process* for project selection involving business and technical experts from outside government.
2. Industry commitment should be demonstrated by *willingness to share costs* in proportion to privately appropriable benefits.
3. The project should have *broad, long-term economic benefits* substantially exceeding the benefits to particular firms in the short term.
4. Criteria for a *clear end* to the project, or to government support thereof, should be identified from the beginning.
5. The main beneficiaries of the project should be *American workers, scientists, and engineers,* and the organizations that employ them.

Table 12-1 summarizes the three categories and compares them with the more familiar categories of basic research, mission-oriented R&D, and commercial product development. In the following sections, each of these gray-area categories is described in greater depth and illustrated with examples of past and possible future government involvement. For each, we offer a proposed method of selecting R&D projects for federal support, with the sections on pathbreaking and

TABLE 12-1:
Policy Categories for Federal Investments in the National Technology Base

Characteristics	Traditional Public Sector Role		Traditional Private Sector Role	Proposed New Public/Private Modes		
	Basic Research	Technology Development for Public Missions	Commercial Development	Pathbreaking Technology	Infrastructural Technology	Strategic Technology
Time horizon	Up to 30 years or more	About 10 years	2 to 5 years	10 years or more	Continuous payoffs	About 5 years
Risks/rewards	High risk; few direct rewards expected	Moderate to low risk; reward is accomplishment of public mission	Calculable risks; planned profit	High risk; potential new industry	Low risk; incremental improvements in productivity or quality	Business risks exceed technical risks; high rewards through economic leverage
Beneficiaries	Other scientists in the short term; long-term beneficiaries unidentified	The public, through accomplishment of mission	Single firm or joint venture	Unidentified or entirely new industries	Many firms in many industries	Identified group of firms
Main performers	Universities; national labs	Government contractors, national labs, and think tanks	Single firm, product/process development team	Universities and national labs; industrial labs	Universities and national labs; industrial labs; consortia	Industry-centered consortia

Criteria						
Appropriate sources of advice and evaluation	Peer evaluation on a technical basis	Experts on mission requirements; public and elected officials	Firm management	Scientists and technical experts; primarily technical judgment	Applied scientists and engineers; managers; mix of technical and business judgment	Business executives and technical experts; primarily business judgment
Appropriate criteria for project selection	Scientific value; quality of research	Relevance to public mission	Business case based on rate of return	Technical promise; entrepreneurial vision	Importance to productivity; technical quality	Competitive impacts
Examples						
Technical subject	Mechanisms of conduction in crystals; stellar evolution; structure of biological molecules	Engineering "stealth" aircraft; manufacturing techniques for composite materials	Commercial HDTV receiver; flexible manufacturing system for a specific product line	Parallel processing computing; fusion energy; National Aerospace Plane	Solid-state chemistry; mechanics of metal forming; software engineering	Semiconductor process tools; digital video
Sponsoring agency	NSF, NIH	DoD, NIH, NASA, DOE, EPA, etc.	None	DARPA, DOE, NASA	All agencies; NIST	Previously DARPA; in the future, Dept. of Commerce?

strategic technologies giving sample evaluation criteria. Infrastructural technologies investments, however, have a wide range of motivations and cannot easily be described by a single set of criteria in the manner of pathbreaking and strategic technology investments.

Creating New Industries through Pathbreaking Technology

Pathbreaking technologies, usually arising from new science, are those that might evolve into a significant new industry or transform existing industries. They are characterized by high technical risk and by uncertain and possibly long-delayed practical payoffs. Industry avoids pathbreaking projects, not only because the rewards are uncertain and far in the future, but because it is rarely evident who will benefit. Biotechnology, for example, has potential impacts on many industries; but without the stimulus of medical R&D, investments would have lagged because the payoffs were too uncertain both in their probability and in their market character for any one company to determine that it stood to gain. When pathbreaking technologies do appear to have relevance to existing industries, large firms in those industries may see the potential for dramatic new capabilities as a threat to current products and installed production capacity, preferring to invest defensively in improvement of existing product lines and only enough in the new technology to monitor the work of others. Government support for the infant technology may then be essential for development.

The government has played a pivotal role in the development of pathbreaking technologies. Federal agencies often funded generously and imaginatively the science out of which such opportunities arose— sometimes with conscious intent, but in other cases with no predetermined application in mind. The early U.S. lead in biotechnology, for example, is almost entirely due to research in fundamental molecular biology and biochemistry, primarily supported by NIH. Not only was NIH responsible for the underlying science, but its mission of improving the nation's health through advances in medical science stimulated early biotechnology applications. Soon after the discovery of recombinant DNA in the early 1970s, NIH altered its own patent policies to encourage some of the best NIH-supported scientists to participate in the commercial exploitation of their discoveries by forming startup firms. Yet NIH has never considered itself as having an explicit mission to encourage a biotechnology industry.

The Defense Department does not have the mission of fostering new commercial industries either. Nevertheless, its cold war imperative of advancing a broad front of technologies that might have military application gave DoD license to pursue pathbreaking technologies, many of which have had important civilian applications. Earlier chapters illustrated the contributions made by defense to the computer, microelectronics, and aviation industries.

Within DoD, DARPA has been an especially productive source of pathbreaking technology, willing to invest a decade or more before widespread applications could reasonably be expected. A recent example: the investigation of alternative computer architectures that involve interconnecting many hundreds or even thousands of identical processing elements. Such "massively parallel" computers, made possible with the invention of the microprocessor, were pioneered at universities. DARPA explored a wide variety of options without premature commitment to a single path of development. In the almost 20 years since Carnegie-Mellon University's DARPA-supported C.MMP machine first demonstrated the feasibility of the idea, a number of new firms have been launched to exploit massively parallel processing. Some have disappeared, others have brought out products, and the mainstream computer industry now takes a strong interest in the technology's potential. However, complementary inventions are still required, especially in the software that controls the processors and tries to keep them equally busy, before these architectures will represent major commercial opportunities.

Both NASA and the Atomic Energy Commission (now DOE) took direct responsibility, unlike DoD and NIH, for development of nascent industries resulting from their pathbreaking investments. The difference is that NASA and the AEC had missions defined by the multipurpose exploitation of the technologies of spaceflight and atomic energy, respectively, rather than by achievement of a well-defined social or political goal. Both were established as dual-use agencies, created to emphasize and make highly visible the peaceful applications of technologies which had originated in the military sector. Each emphasized the value of technology spinoff to the commercial sector or to nonmilitary public services. But each was established following a perceived threat to U.S. security, and each retained a significant responsibility for the security aspects of its technology. The result was that, although both agencies could claim successes, both tended to keep their technology closely held and to resist the kind of decentralized partnership with industry that would have been appropriate if diffusion of technology to the private sector had been a principal objective.[10]

[10] See Brooks, *The Government of Science*, especially the chapter titled "Science and the Allocation of Resources."

Based on NASA investments, the satellite communication industry is one of the purest examples of the government's ability to generate pathbreaking technology for commercial purposes. The federal role in this case went well beyond technology development to include the creation of a legal and institutional structure that made transition to a self-sustaining industry possible. The institutional innovations include COMSAT, the quasi-public corporation established to promote satellite communications in the United States, and INTELSAT, the international consortium responsible for worldwide satellite telecommunications. Also necessary was the existence of a huge space launch and ground support infrastructure, built by government for its own purposes but made available, at marginal cost, to the nascent satellite communications industry.

The Atomic Energy Commission's development of nuclear medicine—the production and use of radioactive materials for biomedical research, medical diagnosis, and treatment—is arguably a successful example of a civilian industry developed out of a technology—atomic energy—originally developed for military purposes. However, the nuclear medicine industry is relatively small in scale and scope compared to the nuclear electric industry, where the story has not been fully played out. Widespread public dismay over perceived hazards resulted in an ever more unfavorable regulatory environment, escalating capital costs, and investment uncertainties. No new reactors have been ordered in the United States since the mid-1970s. On the other hand, nuclear power already accounts for more than 20 percent of U.S. electricity generation, and with the completion of previously ordered plants has been by far the fastest growing source of electricity in the United States. Nuclear power is also expanding worldwide, with several countries, including France and Japan, generating a higher fraction of their electric power from nuclear reactors than does the United States. Nevertheless, there is public opposition even in these countries, and the long-term future of the industry remains far from clear. While final judgment on the effectiveness of past policies would be premature, it is unquestionable that the United States has lost its leading position in a technology in which it was, until recently, overwhelmingly dominant.

These examples indicate the major role that government has played in promoting pathbreaking technologies. Yet there has been no general mechanism or policy by which such projects are selected, evaluated, or compared with one another. We propose that a more systematic and transparent process be established for judging proposals according to a consistent set of criteria, in which alternative technical approaches would be compared to one another before final decisions (see Box 12-A). In this approach, advocates identifying their proposals as falling into the category of pathbreaking technology would submit themselves primarily to technical, not business, judgment. Economic considerations cannot be entirely ignored, however. Viability must not be based on totally unrealistic economic assumptions (as in the cases of the Concorde and the proposed U.S. supersonic transport of the 1960s). Technical evaluation must guard against white-elephant projects and provide ways to manage down the budgets of programs that fail to meet their initial technical vision. The closer a project moves toward technical success, the more important it will be to reexamine the economic and social assumptions on which viability was originally predicated. At the same time, long-term funding commitments are needed to avoid scenarios in which advocates concoct artificial interim breakthroughs to maintain political support.

We assume that proposals for pathbreaking technologies would be submitted by groups, inside or outside government, that seek federal subsidies for R&D, and perhaps other forms of support such as regulatory relief, tax benefits, institutional innovations such as COMSAT and INTELSAT, or infrastructure investments (like nuclear waste disposal sites or space launch services). In each case, the question would be whether the federal government should fund or facilitate the project up to the stage of proof-of-principle—the fabrication and reasonably realistic testing and evaluation of prototypes or plausible scale models. The goal of government support would be to establish technical feasibility sufficiently to attract private investors, or to justify public procurement for operational use by agencies delivering services to the public.

Projects selected for approval could at first be funded by government, with industry contributions increasing as technical risks were reduced. If successful, the project would be handed off to industry and public funding phased out; if unsuccessful—whether because of initially overoptimistic technical projections or of changed economic and social assumptions—the project would be returned to the level of basic research or terminated outright.

Establishment of a government-industry partnership is vital from the earliest stages, even though a long-term, high-potential program rests on technical vision with only a dim outline of the subsequent business opportunities. Existing federally funded consortia such as collaboration in high-temperature superconductivity are in this spirit. It is essential to involve industry groups at the earliest possible stage

BOX 12-A: SAMPLE CRITERIA FOR PATHBREAKING TECHNOLOGY INVESTMENTS

Here, the primary criterion for selection is technical promise. Yet not every technically promising idea can be supported, and choices must be made among them. Certainly one cannot expect detailed technology assessments or benefit/cost analyses at the outset of a long-term, high-risk project. Yet advocates should still be expected to do the best they can to provide answers to the questions set forth below. Even on the basis of incomplete and uncertain information, the answers provide a tool for establishing priorities among technical problems to be investigated. The criteria that follow are not necessarily weighted equally, and the relative importance of a question may change as the project evolves.

1. What are the principal technical goals or milestones against which success or failure is to be judged? If economic or cost-effectiveness criteria are important, what are they?[1]
2. If the technical goals of the project can be achieved, what are the potential benefits to society?
3. Given the potential social or commercial benefits of the project, what alternative technical approaches could lead to the same or similar benefits, and how do they compare with the proposed approach in terms of technical risk and economic cost? Are any of these alternates sufficiently promising to be pursued in parallel with the suggested approach until sufficient information is accumulated to permit a plausible choice?[2]
4. What are the potential "show-stopper" technical questions that, if not resolved favorably, might make the goals of the project unattainable? To what extent should the level of commitment to the project—and the development of ancillary technologies that would eventually be needed—be held back pending favorable resolution of these key technical questions?[3]
5. What is the qualitative appraisal of the social benefits versus costs, assuming favorable technical outcomes?

---

[1] An example might be the reduced launch cost criteria that were originally advanced to justify the Space Shuttle and the operational assumptions on which they were based.

[2] This point is important primarily to avoid premature commitment to a single approach, and to avoid discouraging private investment in alternatives through giving the government-chosen approach an early competitive advantage that later proves unjustified.

[3] This is not a trivial issue, because waiting to resolve key questions could significantly delay the schedule of the project as a whole and increase its total cost even if those questions were eventually resolved favorably. On the other hand, if the project proceeds too far before resolution, much money and time would be wasted if the answers proved unfavorable, and in the meantime the project would have accumulated vested interests that would make it much more difficult to stop or redirect.

6. What is the sensitivity of the potential social benefit/cost ratio to unexpectedly favorable or unfavorable technical outcomes?

7. Who are the possible or likely winners and losers if the technical goals are achieved and the results are implemented on a significant scale? Have all the potential stakeholders been properly identified, particularly those that might be affected by externalities or spillovers if the technology should be deployed on a large scale?

8. Assuming favorable technical outcomes, to what extent can the benefits be captured by the private sector? To what extent are the benefits "public goods" (nonappropriable to the innovating organization), and hence eligible for federal sharing of the costs of implementation and deployment beyond the original development and proof-of-principle?

9. Who should be involved in judging the feasibility and desirability of the proposed project: technical experts, business managers and market experts, potential users, public officials, consumers, environmental impact experts? Who should represent the possible stakeholders (including future generations)? At what stage should the various stakeholders become involved?

10. Who should be involved in the decision regarding whether or when technical progress warrants transition to implementation or application? If the ultimate application has positive externalities or public-good aspects, how should costs be shared between public and private sectors over time?[4]

11. Assuming technical success, what kinds of political, legal or institutional, and infrastructural changes would be needed to encourage commercial implementation? To what extent can federal policies bring about these changes? If it is politically unlikely that these changes can be brought about, how should this affect the desirability of public funding of the precommercial phase of the project?[5]

12. What are the potential cost savings or synergistic benefits of undertaking the project on an international basis? How would international planning and funding be likely to affect the later competitive position of U.S. firms if the technical goals are realized?

---

[4]Some public seed money may be needed to encourage commitment by private investors in the early stages of implementation, but this could be phased out as technical and market risks are further reduced. This was the case for COMSAT, where the federal subsidy was soon eliminated.

[5]One has to be cautious about putting too much weight on this criterion too early. Demonstrated success in meeting precommercial goals may drastically alter the political climate in which policies to facilitate implementation are considered. This would be especially so if the potential social or economic benefits are perceived as large.

so that they can influence the evolution of the technology before its characteristics become frozen. Industry technical experts are more likely to understand the importance of technological infrastructure—tools, processes, supporting component vendors—than are their academic or government colleagues. Industry experts are also more likely to face up to evidence that a program has reached the point of diminishing returns and should be curtailed. Government officials, on the other hand, may be more farsighted in addressing potential social costs.[11]

Since the rewards of successful commercialization are so far off, pathbreaking programs may be suitable for international collaboration. The worldwide effort in fusion research is one of the more successful examples of extensive international collaboration in a major scientific or technical undertaking. This attests not only to the common worldwide interest in attaining the benefits of fusion power, but also to the decades that will pass before the technology matures sufficiently for international competition to arise.

Improving Productivity through Infrastructural Technology

Infrastructural technology consists of research, development, and institutional and technical support that enhances the performance of a broad spectrum of firms in the near to mid-term, but whose benefits cannot be predominantly captured by any one firm.[12] Like basic research, infrastructural technology has many of the characteristics of a public good. Unlike basic research, its principal goal is not to advance the frontiers of knowledge but to make the design and development of products and processes more efficient. Although the benefits from infrastructural technology investments may be modest in the short run, they can cumulate to generate a significant long-term payoff. Like highways and other public infrastructure, infrastructural technology investments benefit many at modest cost and low risk, yet no party would invest in them alone. Therefore society must invest collectively.

---

[11] Strong advocates of a new technology, whatever their institutional base, are inherently unlikely to be receptive to evidence suggesting potential drawbacks. See Cohen and Noll, *The Technology Pork Barrel.*

[12] What we have called "infrastructural" often carries such labels as "generic," "precompetitive," "enabling," or "public-good" technology, although the meanings attached to these words may not all coincide.

Infrastructural support emphasizes low-risk incremental investments in technical activity such as development of engineering methods, measurement tools, development and characterization of materials, compilation and validation of technical data, refinement of manufacturing processes, and instrumentation. It is especially likely to contribute to the production or design of products that are already close to or on the market. National facilities such as wind tunnels, synchrotron light sources for the study of materials, and supercomputer centers fall into the category of infrastructural technology. So do facilities and services that enhance the productivity of the technical community such as the national computer networks that interconnect universities, corporate research laboratories, and government labs. (These also serve as mechanisms for diffusion.)

Although infrastructural activities are often as challenging as basic research, they attract less attention in universities and national laboratories. The U.S. academic community gives little emphasis to fields such as solid-state chemistry, polymer chemistry, and electrochemistry, which are major priorities in Japanese and German laboratories. Federal agencies do support such work, but primarily in relation to their own missions, which do not cover commercially relevant technologies in a comprehensive way. Their narrow mission orientations mean that agencies give little attention to the potential generality of this work or to diffusion of the resulting know-how. Duplication coexists with gaps. While some duplication of R&D can be productive, stimulating competition and verifying results, too much is wasteful—especially when other fields are starved for support. Industry, too, suffers from unnecessary duplication and lack of communication, conducting infrastructural R&D in a fragmentary fashion, often without adequate documentation and with little awareness of what is going on elsewhere. Falling into a gray area between proprietary and nonproprietary, much of this work does not become public knowledge. For maximum benefit to the economy, it should.

Government involvement in infrastructural technology goes back decades. The idea is implicit in the mission of agencies including the National Bureau of Standards (now NIST), the U.S. Geological Survey, and the old National Advisory Committee for Aeronautics. In 1979, President Jimmy Carter sent a special message to Congress proposing initiatives "to significantly enhance our nation's industrial innovative capacity and thereby help to revitalize America's industrial base," including "a program to cooperate with industry in the advancement of generic technologies that underlie the operations of several industrial sectors."[13] While many other recommendations of President Carter's Domestic Policy Review of Industrial Innovation have since been implemented in one form or another, the incoming Reagan administration ignored those dealing with cooperative, generic R&D. More recently, however, President Bush's administration has signaled its approval of public support for generic technology, a concept encompassing some of the attributes we include under the infrastructural category.

Public investments in infrastructural technology lead to improved productivity in two ways. First, they compensate for the understandable reluctance of private investors to pay for technology that benefits many, thus providing industry a broad array of knowledge, facilities, and capabilities that would otherwise not be available. Second, if managed properly, they could lessen duplication by aiming at R&D results of maximum generality, placing them in the public domain, and emphasizing their dissemination. Infrastructural technology represents a relatively safe policy from a political point of view. If effectively coupled to industry, it is likely to be useful and unlikely to attract unfavorable attention through expensive failures. The marginal returns of increased spending on projects dealing, say, with preventing corrosion or improving highway durability should be large and fairly predictable, even though such projects are unlikely to produce quick, visible results in terms of improved industrial competitiveness.

Although some part of infrastructural work will consist of large, lumpy investments such as synchrotron light sources and national computer networks, most projects will be decentralized "small science" efforts, covering a broad spectrum of R&D that can be performed in an equally broad spectrum of institutions. There is no reason to fear that opportunities to participate, or the rewards from doing so, will go to a small number of geographically concentrated organizations. On the other hand, some of this work may resemble R&D performed by for-profit consulting firms that may complain about "unfair competition." Government agencies should avoid conscious overlap with such work, but might in some cases compensate the performers in return for broader diffusion—for example, through publications and software.

---

[13] *Public Papers of the Presidents of the United States: Jimmy Carter 1979*, Book II, June 23–December 31, 1979 (Washington, DC: U.S. GPO, 1980), pp. 2068–2974. The recommendation resulted from 18 months of study by an interagency task force headed by Jordan Baruch, assistant secretary of commerce for Science and Technology, and supported by a panoply of private sector advisory committees. See also *Advisory Committee on Industrial Innovation: Final Report* (Washington, DC: Department of Commerce, September 1979).

Not all infrastructural technologies require direct support. Software packages for computer-aided engineering, measuring instruments, or chemical analytical services often pay back commercial development without public assistance. Where the market for such tools might be uncertain, government could stimulate demand indirectly by subsidizing potential users if productivity would benefit. Here, government action would stimulate technological learning, as in many of the large defense investments of the past.

A federal government program of infrastructural investments would operate differently from those for pathbreaking or strategic technologies. Given the broad but diffuse benefits, we would not necessarily expect particular constituencies to petition for specific projects. Nor, given the diversity of topics, mechanisms, and motivations for such work, would we expect a single set of criteria to apply to all (beyond the overall goals of technical excellence and potential for improving productivity or quality). Instead, a program to support infrastructural technology should rely on joint guidance from government and industry experts, factoring in technical issues and business needs about equally to set priorities. The most difficult management challenge will be ensuring work of consistently high quality and usefulness—setting standards that are comparable with the best basic research, while choosing topics to meet the needs of productivity and competitiveness rather than to fill conceptual gaps in the edifice of science. Chapter 10 gave an example, recommending a national program in manufacturing science and technology that would involve a variety of institutions including universities and national laboratories.

A More Hazardous Policy Domain: Strategic Technology

We define strategic technology by association with the competitive needs of firms in specific industry sectors that are of high importance to the U.S. economy, generally because of strong upstream or downstream linkages to other sectors, including defense. A strategic technology may account for a tiny fraction of GNP or employment, yet have broad ramifications; machine tool production, for example, accounts for less than 1 percent of manufacturing employment in the United States, but about 25 percent of manufacturing employment makes use of machine tools.[14] If government-industry partnerships to develop

[14] Maryellen R. Kelley and Harvey Brooks, *The State of Computerized Automation in U.S. Manufacturing,* John F. Kennedy School of Government, Harvard University, 1988.

strategic technologies are successful, economic returns may be clear and prompt. Nevertheless, because of their association with particular industries or groups of firms, strategic technology programs come closest, of the three types of policy initiative we discuss, to the political hazards of industrial policy.

There will be many cases in which the critical requirement for competitive success is some factor other than technology. These are not candidates for strategic technology projects. As described here, strategic projects are restricted to cases in which essential technological capabilities cannot be sustained by the industry alone—either because the pace of progress and scale of investment are beyond the reach of firms, or because the industry faces serious challenges in mitigating negative externalities (e.g., environmental impacts).

In some cases, firms will be able to collectively finance strategic technology investments. Government might participate on a "seeding" basis, helping to initiate and legitimize cooperation through consortia or trade associations and alleviating fears of antitrust exposure. Alternatively, an industry consortium with its own funds might work in partnership with a publicly funded government or university lab. Those who follow Japanese technology policy will find this category familiar. MITI is famous for strategic technology projects aimed at increasing Japanese shares of world markets: VLSI microelectronics and civil aviation, to name two examples, one successful and the other not.

In contrast to pathbreaking technologies, which explore possibilities from which *new* industries may arise, strategic projects arise from the needs of an *existing* sector. Strategic projects do not have the high *technical* risk associated with pathbreaking programs; it is high *business* risk that will normally explain the shortfall in private investment. Decisions to embark on strategic technology projects must be based largely on economic grounds, along with a determination that the industry they support is critical to the nation's future—inevitably a political decision. There have been two widely debated recent U.S. examples of strategic technology proposals, both seeking defense support: semiconductor process technology, culminating in formation of the Sematech consortium (see Chapter 8), and high-definition video displays, which generated a great deal of controversy when linked with one set of possible applications—high-definition television (HDTV).

Some form of industry association or consortium will almost always be involved in a strategic technology project as a means for assembling the private sector's share of the financial support, providing an organizational structure, preventing discrimination among firms, and distributing the benefits throughout the industry in a fair way. In the case of Sematech, formation of a consortium preceded, and was an important factor in, the federal decision to grant support. In contrast, the push to create a federally funded program for HDTV did not proceed to formation of a consortium. With no consensus on business strategy, advocates failed to muster political support.

Those who might be doubtful about strategic technology projects are entitled to ask why the government should be involved at all if an industry consortium is required in any event. In some cases, government involvement is not necessary. Many industry associations operate their own laboratories or sponsor R&D in the interest of their memberships.[15] But there are several reasons why federal participation might be appropriate:

1. Government may have a stake in the outcome—for example, if the industry in question has special importance for defense, or if the program was intended to mitigate negative externalities such as environmental hazards.

2. With business remaining nervous over antitrust enforcement, government participation (although not necessarily funding) may be required before industry is willing to establish the consortium. Thus the founders of Microelectronics and Computer Technology Corporation (MCC), which did not seek federal financing, lobbied for the National Cooperative Research Act of 1984[16] (which eliminated application to registered R&D consortia of the treble damages provision of federal antitrust law) while planning their joint venture.

3. Success or failure of a relatively small industry may have critical implications for a much larger portion of the economy. This is particularly true of projects built on vertical relationships in industry. Sematech, for example, was established by semiconductor manufacturers concerned about the ability of their suppliers to keep up with technical advances driven by Japanese as well as American firms. For such industries, in which world

---

[15] A good example is the American Newspaper Publishers Association, which has been a major force in the development and adoption of computer technology and other innovations. See Anthony Smith, *Goodbye Gutenberg: The Newspaper Revolution of the 1980s* (New York: Oxford University Press, 1980), especially pp. 77–78.

[16] Public Law 98-462, 15 U.S.C. 4301–4305.

demand is characterized by wide fluctuations between shortage and overcapacity, government involvement might help mitigate the impact of periods of shortage if, for example, U.S. manufacturers could not obtain the latest products from foreign-based suppliers.

Strategic projects start with identification of an industry seeking some form of federal partnership. There is no way to avoid the political reality of picking a "winner." This offends not only those committed to exclusive reliance on market mechanisms, but also those who fear that private firms, finding federal help addictive, will lose the sense of responsibility for their own competitiveness, or that federal resources will flow to the politically well connected instead of the strategically important, or that selection of one technical approach may crowd out others. Strategic projects must therefore be clearly differentiated from pathbreaking and infrastructural projects so that the latter two, which should be less controversial, are not embroiled in the questions that inherently surround strategic technologies. Strategic technology investments will always demand special scrutiny.

Box 12-B presents sample criteria for evaluating strategic technology investments. The appropriate sources of advice and evaluation for such proposals would be business managers and technical experts, with decisions based primarily on business judgments. Technical knowledge generated via a strategic technology consortium would typically be shared among consortium members, only later being diffused more widely.

How about the participation of foreign-owned firms? Given the prevalence of strategic alliances and partial foreign equity ownership, the only answer is: it depends. If the purpose of Sematech, for example, is to assure the viability of a U.S.-based semiconductor equipment industry, that industry should be encouraged to sell to Japanese as well as U.S. chipmakers to broaden its market base, even if that means equipping foreign competitors with tools developed with U.S. government support. On the other hand, if the purpose is to ensure the viability of a U.S.-based chip industry, Sematech might want to broaden the base of technology available to that industry by welcoming the participation of U.S. subsidiaries of Japanese-owned equipment manufacturers. Such questions had little visibility in the debate over Sematech—a failing our proposals are meant to overcome.

Terminating a project will always be a challenge. A government with the courage to risk selecting particular industries to help must

BOX 12-B: SAMPLE CRITERIA FOR STRATEGIC TECHNOLOGY INVESTMENTS

The benefits expected from strategic technology investments are nearer-term than those from pathbreaking investments, and the ability to make quantitative estimates should be that much better.

1. Can a persuasive case be made that the industry to be assisted is sufficiently critical to a wide and important enough segment of the economy (high-value-added production, a source of high-wage employment) or to national security? Would the competitive health of many other linked industries be jeopardized if the domestic base of the applicant industry were replaced by import dependence?
2. Is accelerated development, acquisition, and workforce mastery of technology really the key to maintaining competitiveness ? Or are factors such as industrial structure, government regulation, unfair trade practices of competitors, obsolete management strategies, or inadequate workforce training more important?
3. Does the applicant industry appear to have an adequate strategic plan, with defined goals and milestones that would result in sustained competitiveness, taking into account the likely response of its foreign rivals, and assuming adequate U.S. government policy response with respect to any nontechnological factors identified in question 2?
4. Will the combination of government interventions and industry actions be sufficient to enable the industry to become self-sustaining so that it can acquire the follow-on generations of technology without government assistance?
5. Assuming that strategic technology is supported through an industry consortium or government-industry partnership (perhaps involving universities or federal laboratories), will the parallel investments in the member companies suffice to ensure timely commercialization and meet the criterion of self-sustainability in question 4?
6. What are the estimated benefits and penalties of including affiliates of foreign-based multinationals as members of the consortium or partnership eligible for U.S. government assistance? What, if any, criteria in the way of codes of conduct, structural characteristics, reciprocal national treatment by home government of foreign affiliates, and so forth, should be set for membership in consortia seeking U.S. government support?
7. What is the necessary composition of a group for study and evaluation that will provide public credibility, industry confi-

dence, and political legitimacy for a proposed investment in strategic technology? Are there industries, interest groups, or other stakeholders that are likely to be adversely affected by government support of a proposed project? How should these interests be represented in the decision process?

8. How, by whom, and how frequently should progress toward meeting the goals of a strategic technology investment be assessed? What relative weight should be given to the assisted industry, independent experts, government officials, and Congress in assessing progress?

have equal courage to terminate that assistance when it is no longer needed, or when it becomes clear that success cannot be achieved. Commitment to a public/private partnership should be associated from the outset with criteria for a clear end to the project, if not a specified termination date.

A New Policy Challenge: Diffusion

For all the difficulties the nation faces in agreeing on new policies for generating technology, we face as much or more trouble in applying the results of R&D effectively. Indeed, many American companies do a poor job of using existing knowledge, much less the latest outputs of the laboratory. This imposed few penalties in the days of unchallenged technological supremacy. But in competition against swift-moving, sophisticated competitors, American firms must be able to find and exploit narrow margins of superiority in technical expertise, margins that will be constantly at risk. The strong emphasis we have placed throughout this book on diffusion and absorption of existing technology in addition to generation of new knowledge reflects our belief that policy should move toward demand-side strategies that will help U.S. enterprises find, adapt, and put to use the best technology available.

While the results of publicly funded R&D must find their way to commercial firms if they are to be of economic benefit, it is the diffusion of *all* nonproprietary knowledge—not just government R&D—that is essential to U.S. competitiveness. Indeed, increasingly that knowledge will be found overseas. The very slow spread of advanced quality control practices through U.S. industry following their reimport from

Japan illustrates the problem; these methods have little to do with research, everything to do with shopfloor know-how. In addition, the acceptability of policies aimed at generating industrially relevant technology depends on an explicit diffusion strategy that guarantees to Americans that they will reap their share of the benefits. Since our economic competitors already have much more effective diffusion policies, the United States cannot capture the benefits of its own investments without an equally aggressive diffusion strategy.

At the heart of the problem lies the ongoing explosion of knowledge. With technical output doubling every 10 to 15 years, not even the largest firms can keep up across the board. The sheer size of the technology base makes it hard for companies to locate and tap the know-how needed to solve their particular problems. In the absence of dynamic technological communities and high labor mobility such as those found in California's Silicon Valley or Boston's Route 128 (and sometimes even in their presence), the traditional unwillingness of American firms to cooperate with one another hinders diffusion.

Smaller firms have special difficulty recognizing proven design and analysis methods, production processes, and technical tools that could help them. With government paying for so much knowledge creation, it makes sense for government to facilitate its application. The net effect is to spread the economic benefits and increase competition—not by picking winners, but by increasing access to innovative capacity.

Policy design requires attention to the diversity of channels through which technology flows in a modern industrial economy. Chapters 2 and 3 describe many of the forms of technical knowledge and the myriad paths through which it circulates in the economy, suggesting the need for five types of diffusion policy:

1. Systematic acquisition and dissemination of foreign technical knowledge to individuals and firms;
2. Evaluation, adaptation, and dissemination of technical information within the United States;
3. Technical services and industrial extension, particularly to small and medium-sized firms;
4. Collaborative technical activities among firms; and
5. Investments in human resources.

All five should focus on downstream processes of commercialization and technology absorption rather than on R&D itself.

Although diffusion-oriented policies have a long and honored history in agriculture, and these policy areas have extensive precedents, the record of support for commercial manufacturing has been poor. NIST's new mission, legislated in the 1988 Omnibus Trade and Competitiveness Act, includes two diffusion-oriented programs in addition to financing of "pre-competitive generic" projects through the Advanced Technology Program. One is a technology extension program to help smaller manufacturers improve their productivity; the other is the establishment of manufacturing technology centers in cooperation with the states. These new and relatively modest initiatives provide a basis for the steps we outline in the following paragraphs.

Acquisition and diffusion of foreign technical knowledge.   The starting point of a diffusion strategy must be to change U.S. attitudes toward the achievements of others. In comparison with Japanese companies, Americans suffer from a "not-invented-here" syndrome—in part a by-product of national policies focused on prestige. These attitudes are costly in both time and dollars. A good diffusion strategy would give as much emphasis to importing and adapting knowledge as to growing it at home. Collection and evaluation of information from abroad, and the acquisition of new insights through joint projects with Japan and the European Community, can help achieve this goal.

The National Technical Information Service (NTIS), part of the Department of Commerce, serves as a central repository and distribution source for reports on the results of government-funded R&D. NTIS remains small, underfunded, and not especially responsive to the needs of the technical community, in part as a consequence of the Reagan administration's unsuccessful attempt at privatization. Congress not only blocked that attempt, but through the Japanese Technical Literature Act of 1986 (P.L. 99-382) gave Commerce a new mission: to monitor technical developments in Japan, translating and distributing information on Japanese technology. A similar effort should be made to improve U.S. access to European technical information. The Department of State should, at the same time, upgrade both the number and the technical qualifications of the science counselors in our embassies overseas. Other industrial nations have much greater technical capability in their embassies, and can engage in technical scanning and evaluation of U.S. capabilities that our embassies cannot match.

Evaluation, adaptation, and dissemination of technical information within the United States.   It has been recognized for 30 years that quality control, adaptation to user needs, and dissemination are critical for enhancing the appropriability of the results of both public and private

R&D.[17] At one time the White House Office of Science and Technology (now the Office of Science and Technology Policy, OSTP) encouraged all federal R&D agencies to establish centers in university, nonprofit, and national laboratories for data evaluation, compilation, and dissemination.[18] The agencies also sponsored critical reviews of technical fields and the consolidation of technical knowledge in engineering handbooks. As the government's emphasis on large, mission-oriented projects grew, these diffusion-oriented efforts languished.

The development of computer networks and electronically accessible data collections makes scientific and technical information (STI) services easier to use. Professional societies are beginning to experiment with electronic journals. The government's investment in the National Research and Education Network (NREN)—a central part of the strategy to improve the nation's information infrastructure—will make expanded STI services available to thousands of laboratories in universities, industry, and government, and should attract investment by private information vendors as well. At the same time, OSTP should work closely with the Office of Management and Budget to reexamine and strengthen the guidelines for agency STI policies embodied in OMB Circular A-130. This document sets policy for agency obligations to distribute information to the public. It is essential that policy encourage dissemination of reliable information that is adapted to user needs.

Technical services and industrial extension. More than 40 states now provide assistance to companies with technical or business problems ranging from shopfloor reorganization to debugging new products and planning new investments. Many of these programs are new and relatively small. But taken together, the states are spending over half a billion dollars annually.[19] In addition, NIST has begun a program of

---

[17] President's Science Advisory Committee, *Science, Government, and Information: The Responsibilities of the Technical Community and the Government in the Transfer of Information*, prepared by the Panel on Scientific Information chaired by Alvin Weinberg (Washington, DC: U.S. GPO, 1963).

[18] The Federal Council for Science and Technology established its Committee on Scientific and Technical Information in 1963. It coordinated federal scientific and technical information policy until its transfer to NSF in 1972 and subsequent demise. Charles R. McClure and Peter Hernon, *United States Scientific and Technical Information Policies* (Norwood, NJ: Ablex, 1989), p. 41.

[19] In 1988, 44 states invested $550 million in programs to promote technological innovation. Industrial extension programs get perhaps 10 percent of the state total. See Megan Jones, "Helping States Help Themselves," *Issues in Science and Technology*, vol. 6, no. 1 (Fall 1989), p. 56; David Osborne, "Refining State Technology Programs," *Issues in Science and*

Manufacturing Technology Centers, five of which had been established by 1991. NIST is also experimenting with other ways to enhance downstream performance, through industrial extension.

Several hundred thousand small and medium-sized firms stand to benefit from extension programs. Few of these companies engage in R&D; some do not employ even a single engineer. Given this, the federal government might take two further steps. First, NIST could rapidly expand its manufacturing centers, with the goal of perhaps three dozen by the end of the decade (and funding totaling, say, $100 million annually). Second, the federal government could share costs equally with qualifying state extension programs. Many of these enjoy a few years of success, only to die when a change in state government accompanies a recession year, as happened recently in Massachusetts. The federal government should help stabilize the most effective state initiatives and give them better access to the technical resources of federally funded laboratories. In addition, NIST should be directed to evaluate the effectiveness of extension efforts, whether funded by states or by the federal government, and to disseminate the results and use those results in improving its own programs.

Collaborative technical activities.   Cooperative R&D serves to diffuse technology as well as generate new knowledge. The well-known projects sponsored by Japan's Ministry of International Trade and Industry, or the Framework Program of the European Community, probably have greater impact through diffusion than through technology generation. After all, the most effective way to understand technologies is to apply them to some practical end. Joint R&D brings people together to attack common problems and provides a window on technologies that companies might not learn of otherwise. Some companies also detail technical staff to universities to share in research and gain access to skilled graduates and faculty expertise; this is a primary motive for industry support of engineering research centers created under sponsorship of the National Science Foundation. Although joint research is nothing new, the past decade has seen a steady growth in U.S. R&D consortia. The demonstrated effectiveness of consortia in speeding technology diffusion justifies higher levels of federal cost sharing than has been common in the United States.

For more than a decade, policymakers have also sought better ways

---

Technology, vol. 6, no. 4 (Summer 1990), p. 55; Philip Shapira, *Modernizing Manufacturing: New Policies to Build Industrial Extension Services* (Washington, DC: Economic Policy Institute, 1990), p. 24.

of exploiting technologies resident in national laboratories or otherwise stemming from federally funded R&D (Chapter 3). If cooperative arrangements with industry are to flourish, the laboratories will have to change in operating style and culture. Personnel exchanges could help speed the process of change. It is now relatively easy for scientists and engineers from industry to spend time at a federal laboratory working on cooperative projects or making use of specialized facilities. It should be just as easy for laboratory employees to take temporary assignments in industry. This would help them understand the needs of business and the imperatives of economic competition.

Demonstration projects mix knowledge generation by government with an explicit effort to attract industry to the new technology. Such projects take federal agencies further from research collaboration, deeper into commercialization. In the absence of market forces that could help shape technical objectives and discipline managers, many such projects—for example, DOE's ceramic engine work and some of its solar energy R&D—have suffered from too much "technology push." On the other hand, NASA's civil aviation programs—many of which emphasize validation of emerging technologies, collecting test data and other engineering information, and making it available to industry—can boast a relatively impressive track record.[20] Industry typically shares in the costs of this work. Demonstrations will always suffer from an inherent dilemma: they have high visibility and must therefore be seen to "succeed." This forces decision makers to be conservative in their technical choices. In civil aviation, with safety paramount, conservative choices will often be appropriate. But aviation is the exception, and many other technology demonstrations have proven inefficient and expensive relative to their benefits.

Investments in human resources.   This is the most powerful diffusion strategy in the long run. Three issues stand out. First, the dismal state of precollege math and science education will require major commitments from the Department of Education and the National Science Foundation. In particular, these agencies need to develop mechanisms to stimulate innovations in teaching and learning and to diffuse the best of them throughout the country. As with new technologies in our

---

[20] National Research Council, U.S. Civil Aviation Manufacturing Industry Panel, *The Competitive Status of the U.S. Civil Aviation Manufacturing Industry: A Study of the Influences of Technology in Determining Competitive Advantage* (Washington, DC: National Academy Press, 1985).

factories, the rate of adoption of new ideas in U.S. schools is very low.[21]

The second issue is the future of American colleges and universities, which perform a critical role in the creation and diffusion of knowledge. If universities are to promote innovation, they must be able to attract students to technical careers. Reversing the trend of recent years, the federal government should significantly expand aid to students and universities to keep tuition within a tolerable range and ease financial pressures on the universities. For their part, the universities need to reform their engineering curricula. The bias toward research must be balanced with attention to engineering design, manufacturing systems, and process development.

Finally, the Departments of Education and Labor, together with the states and private industry, must articulate a strategy for preparing young people to make the transition from school to work. In Germany, for instance, three-quarters of noncollege-bound high school students receive three years of apprenticeship training, combining classroom instruction with work experience. American schools need to take more responsibility for preparing and placing students; at the same time, American industry must take a long-term approach to improving workers' skills and opportunities.[22]

Other mechanisms for education and training now function with little or no government support. Professional and technical societies, for instance, plan and conduct short courses in specialized subjects, especially newer topics that may be unfamiliar to engineers and scientists 10 or 15 years out of school. Universities likewise put on more short courses and summer institutes today than two decades ago, sometimes in cooperation with technical societies or consultants and consulting firms. But relatively few companies actively encourage their employees to take advantage of these opportunities, while federal support for continuing education of working engineers and scientists (in contrast to postdoctoral research fellows) has been negligible. Indeed, the federal government does little to support education and training for those already employed, regardless of occupation. A knowledge-based

---

[21] Task Force on K-12 Science and Math Education, *In the National Interest: The Federal Government in the Reform of K-12 Science and Math Education* (New York: Carnegie Commission on Science, Technology and Government, 1991).

[22] For policies to improve the skills of blue- and gray-collar workers, see Office of Technology Assessment, *Worker Training: Competing in the New International Economy* (Washington, DC: U.S. GPO, September 1990), pp. 37–70.

economy requires much more attention to continuing education and training.

HOW WILL THE NEW TECHNOLOGY POLICIES BE IMPLEMENTED?

Most federal agencies should be eager to participate in the broad new technology policies described in the preceding pages. However, all will have to overcome shortcomings rooted in structure, history, and political experience. Institutional innovations will be required to implement the policies we have described. These have two components:

- The R&D management component: development of new capabilities in agencies asked to manage investments in pathbreaking, infrastructural, or strategic technologies; and
- The diffusion component: creation of new relationships between the government and those elements of the private sector that transmute R&D—including federal R&D—into social returns.

Defense agencies, civil mission agencies, and basic research agencies will all face problems in establishing new working relationships with industry. Each has its strengths: DARPA in pathbreaking exploration of emerging technologies, NIST in practical, productivity-enhancing infrastructural technology. But it seems certain that old agencies will have to learn new ways of operating. If necessary, new agencies will have to be created. In either case, progress will be slow, and expectations must be realistic. Therefore, it is time to get started.

Defense Agencies

Most recent attention to DoD's contributions has focused on DARPA. The reason is clear: DARPA has an excellent reputation for investments in emerging technology of potential military importance, and many such technologies are at least potentially dual use. DARPA's track record rests on experience in managing technical risk and in demonstrating the potential of new technology to military customers. To apply DARPA's expertise in developing pathbreaking technology in areas further from its military mission, the agency could be reconstituted as the *National* Advanced Research Projects Agency, retaining its defense mission but adding to it a wider set of national needs, when requested and supported by nondefense agencies.[23] Even if DARPA sticks to defense, however, the agency will need to develop collaborative relationships on dual-use technology with commercial firms to ensure military access.[24]

But DARPA is probably not the best agency for other new types of federal investment. It does not have much expertise in developing or ensuring the application of the incremental and widely applicable innovations characterizing infrastructural technology. Moreover, DARPA's mode of operation—highly autonomous, with decisions protected from challenge by the shield of national security—and its tight, top-down management style will not work well for strategic technology projects, which should be open to public scrutiny and managed in partnership with the industry. Shared decision processes will challenge defense agencies accustomed to contractor-vendor relationships.[25] (The negotiations for DARPA support of Sematech, for example, revealed tensions over the extent to which the consortium would be subject to DARPA technical direction.) Policymakers should be skeptical about DARPA's ability to handle the political complexities inherent in collaboration with commercial industry.

Civil Agencies

The Department of Commerce seems the appropriate home for the infrastructural technology mission, having already been assigned this responsibility by Congress in the 1988 Trade Act. Within Commerce, NIST should take the lead. The primary repository for federal government experience in everyday industrial technology, NIST has wide experience in diffusing know-how from its own laboratories and a long and honorable history of resisting political interference. Yet NIST is by tradition a research organization, not an administrative agency. If given responsibilities for handling both intramural and extramural R&D, it

---

[23] Task Force on Science, Technology, and Economic Performance, *Technology and Economic Performance: Organizing the Executive Branch for a Stronger National Technology Base* (New York: Carnegie Commission on Science, Technology, and Government, 1991).

[24] Lewis M. Branscomb, testimony to the Joint Economic Committee, U.S. Congress, September 12, 1991.

[25] The government already has the ability (31 U.S.C. 6301 *et seq.*) to enter into cooperative agreements with industry in which the objectives are jointly defined by both parties. However, this mechanism is rarely used in a defense R&D or procurement context, which relies primarily on government-directed contracts.

risks damaging both its relationships with industrial partners and the vitality of its internal programs. Similar conflicts between the performers of internal and extramural research can be observed in NASA centers. However, other federal agencies—NIH and NOAA, for example—have developed policies and organizational mechanisms to limit these conflicts.

Outside NIST and NOAA, the Commerce Department has little experience in R&D management. Its primary constituency—the business community—remains ambivalent about the department's role in the technical life of the nation. Policymakers skeptical about Commerce's ability to manage large projects are unlikely to award it a budget big enough to make a difference.

Looking beyond Commerce, NASA and DOE both have extensive experience in contract R&D, and both operate large national laboratories with high levels of technical skill. Both can make important contributions in the identification and pursuit of pathbreaking technology consistent with their missions. Each, however, has drawbacks: the Department of Energy because of its tradition of closely held, in-house projects; NASA because of its recent history of technically troubled projects. Skepticism about redirecting these two agencies must also arise because of their past reliance on spinoff. Given their culture, NASA and DOE will have difficulty developing appropriately collegial and collaborative relationships with contractors while retaining administrative accountability for public funds. Both are more accustomed to carrying development through to end products required for their missions. However, as these agencies gain greater experience with Cooperative Research and Development Agreements and other nontraditional mechanisms, they should be able to take on greater responsibilities, particularly in pathbreaking projects.

NSF, NIH, and other civil sponsors of basic research could in principle expand their scope. Agency managers and constituents, however, would see such a move, even if accompanied by budget increases, as threatening funding for academic science, already hard pressed for resources.

In recent years, NSF has increased its support for engineering research and its programs of interdisciplinary centers, many of which involve industry participation. But NSF's organizational culture reflects the norms of university science, so it would be relatively difficult to graft a more applied culture onto the agency. Either the graft would not take, or, if it did, NSF's traditional mission would suffer. In any case, the foundation's academic constituency would strongly oppose any actual or potential threat to its research emphasis, including any weakening of the scientific community's control (exercised largely through peer review) over project selection and quality. An industrially oriented program calling for greater agency discretion in selecting and managing projects would properly be seen as incompatible with the tradition of unsolicited proposals evaluated by disciplinary peers.

However, there is no clear line between the engineering research already funded by NSF and infrastructural technology. Much of NSF's current funding for engineering, materials, and computing programs, which constitutes almost 30 percent of the agency's budget, could be classified as support for research into infrastructural technologies. This work will continue to make a valuable contribution to the commercial technical base. The same conclusion holds for much of the engineering and more applied scientific research supported by the mission agencies. In addition to our recommendation in Chapter 10 that NSF establish a manufacturing sciences directorate, we suggest that the foundation's engineering research budget increase somewhat faster than overall NSF spending. We also urge that NSF's support of interdisciplinary centers be expanded, and that centers be permitted to compete for continuing federal funding (rather than expected eventually to become self-supporting—an unlikely prospect).

Coordination

There remains the question of how the nation's new technology strategy should be guided. Today the Office of Science and Technology Policy and the Office of Management and Budget rely heavily on the interagency Federal Coordinating Council on Science, Engineering, and Technology (FCCSET) and its working groups to assemble budget information and to coordinate agency activities. When policy issues, such as high-performance computing, cut across jurisdictions, one agency may be assigned lead responsibility. It would be possible to emulate this practice by assigning particular pathbreaking or strategic technology projects to individual agencies. But at present, OSTP does not have the staff capability (or budget) to assess the technological condition of U.S. industry, as called for earlier in this chapter, and to translate that assessment into a portfolio of pathbreaking, infrastructural, and strategic investments. Nor does it have the necessary power to broker deals between agencies and enforce agreements. FCCSET has traditionally functioned simply as an information-sharing body. In the 1991 budget Congress added $5 million to the OSTP appropriation for a Critical Technologies Institute, which might in principle have been used to begin strengthening the office's analytical capabilities. Such an outcome, however, seems unlikely absent a major shift in administration policy.

At present, the agencies have a great deal of autonomy in managing their R&D budgets. Overall spending levels and big science projects draw scrutiny, but programs viewed as routine do not. The agencies will understandably guard their autonomy, yet a more effective technology policy means setting priorities across and within agencies and enforcing them. There is no need for OSTP to act as czar. There is a need for OSTP to strengthen its own staff capabilities, breaking free of dependence on people detailed to the office from elsewhere in government. OSTP must also continue working to turn FCCSET into a decision-making body—one able to take action. Too often in the past, FCCSET has stalled action rather than initiated it.

Private Sector Responsibilities

Putting the government's house in order is only half the job, since the policies and processes we recommend depend critically on industry participation. Institutional innovation in the private sector will be needed to provide mechanisms for creating consensus on R&D objectives and fostering public/private working relationships. The Computer Systems Policy Project, with membership consisting of the chief executive officer and the chief technical officer of each of the leading computer manufacturers headquartered in the United States, has been effective in transmitting the industry's technical and business views to government officials without coming to be seen as simply a lobbying effort. More such experiments will be needed. Business leaders will have to begin viewing Washington as an ally over the long run. Today, business too often lobbies government on a case-by-case basis, seeking favors such as tax breaks or trade protection. Trade associations could help create consensus on technology goals if they could move past lowest-common-denominator positions. The Aerospace Industries Association has made significant contributions already, through its efforts to define technology roadmaps. Other business and trade associations could learn from this example. So could technical societies.

MOVING INTO A NEW ERA

In this book, we have sought to picture a national science and technology enterprise that is

- more farsighted, because of a structured approach to support for pathbreaking technologies;
- more productive, because of broader and more coherent defense and civil investments in infrastructural technology;
- better prepared to compete, because of a stronger base of strategic technology, created in a less politicized and more disciplined way;
- more competent, because of increased emphasis on diffusion of technical knowledge and skills; and
- less isolated in defense and civil sectors, because of more effective access by defense to the commercial industrial base.

We envision the Defense Department not as an enclave following its own rules, but as a partner with a more competitive set of industries. Increased attention to diffusion of knowledge and skills, based on a more realistic understanding of industrial innovation, would replace naive reliance on spinoff. This will require a fundamental change in the way Congress and the executive think about technology policy.

The U.S. government should put aside its fear of moving beyond the support of science to embrace engineering and technology. We have attempted to provide criteria that can distinguish appropriate modes of support from those that a democratic government in a free-market society should not attempt. Americans cannot afford the luxury of being either so cautious or so overconfident that we fail to earn our place as first among equals in a competitive world.

Leaders in industry should call for government support of the portions of the knowledge base that generate widespread benefits but do not draw private investment. Federal programs in these areas must be developed in collaboration with technical experts and managers in private industry, who must share responsibility for disciplined decisions.

Americans are coming to recognize that the technological fruits of investments in national defense will provide less nourishment in the future. New policies are needed for a world where the United States holds no technological monopoly, and the nation should get on with the job of creating them.

Index

411

era of economic transition and, 372–73
on federal technology transfer, 75–81
fostering diffusion of technology as, 370,
380, 398–405 (*see also* Diffusion of
technology)
in France, 212–15
in Germany, 227–31
historical lack of explicit articulation of,
45–46
implementation of, 371, 405–10
industrial policy and, 45, 46–50
infrastructural technology investments as,
370, 380–81, 382–83(table), 384,
390–93
international comparisons of, 209–10
in Japan, 236–40, 394, 402
in manufacturing, 335–49
in microelectronics industry, 279–82
need for new federal, 24–26, 52–53,
359–60, 369–70
overview of, 369–72
pathbreaking technology investments as,
370, 380–81, 382–83(table), 384–90
pitfalls of, 374–78
pitfalls of, avoiding, 378–79
private sector responsibilities and public,
373–74
recommendations for changes in defense
sector, 160–61, 363–65, 367–68
segregation of defense and commercial
sectors because of, 133–34, 142–54
in software industry, 306–12
strategic technology investments as, 370,
380–81, 382–83(table), 384
in Sweden, 219–22
Technology reserve, to preserve defense
technology base, 162, 362
Technology sharing, 185–88
long-term research as form of, 186–87
as requirement for success, 175
at Rockwell International, 181–82
through vertical integration, 187–88
Technology transfer. *See also* Defense-
commercial technology linkages;
Spinoff technology
agricultural, 60 n.7
federal policies on, 75–81
within firms (*see* Technology sharing)
vertical and horizontal, 55
Teflon, 56, 57–58

Texas Instruments (TI), 259
Third-tier firms, 168–69
Thomson firm, France, 216–17, 244
Toyota, 32
Trade associations, 410

United States
defense-related research and development in, vs. select countries, 210,
211(table), 243–47
economic competitiveness of (*see* Economic competitiveness, U.S.)
technology investments in, overview of,
91–95
UNIX operating system, 305
Upstream activities
in technological innovation, 19, 20, 21,
162, 246, 373
*Use of Commercial Components in Military
Equipment,* 156

VAX computers, 73
Velcro, 56, 57
Very High Speed Integrated Circuits
(VHSIC) program, 73, 75, 184, 256,
268–72
lessons of, 272
origins and goals of, 269–70
planning and management of, 270–72
planning successor to, 280
Very-Large-Scale Integration (VLSI) microelectronics, 51, 259, 394

Watson, Thomas J., Jr., 65–66
Weapons procurement. *See* Defense procurement
Workforce. *See also* Education and training
of workforce; Human resources
cross-cultural differences in, 51, 114–15,
247, 323
German, 51, 323 n.9
impact of defense sector on technical,
115–18
Japanese, 114–15, 323 n.9
U.S. technical, 114–15, 323
World War II, relationship of federal government to science and technology following, 12 n.5

Xerox Corporation, 318

About the Authors

JOHN A. ALIC

John Alic has been on the staff of the congressional Office of Technology Assessment since 1979, where he directed projects beginning with *U.S. Industrial Competitiveness: A Comparison of Steel, Electronics, and Automobiles* (1981), and including *International Competitiveness in Electronics* (1983), *International Competition in Services* (1987), and *Commercializing High-Temperature Superconductivity* (1988). Alic has also contributed to OTA assessments dealing with labor market issues, most recently, *Worker Training: Competing in the New International Economy* (1990), and to studies of automobile fuel economy. He is currently directing an assessment of economic links between the United States and Mexico.

A graduate of Cornell, Stanford, and the University of Maryland, Alic was a member of the engineering faculties of Wichita State University (1972–1978) and the University of Maryland (1966–1972) prior to joining OTA. He has published extensively on international competitiveness, industrial and technology policies, engineering design, fatigue and fracture of materials, and technical education.

LEWIS M. BRANSCOMB

Dr. Lewis M. Branscomb is Albert Pratt Public Service Professor at Harvard University's John F. Kennedy School of Government and directs the school's Science, Technology, and Public Policy Program in the Center for Science and International Affairs.

A research physicist at the U.S. National Bureau of Standards (now the National Institute of Standards and Technology) from 1951 to 1969, he was appointed director of NBS by President Nixon. In 1972, Dr. Branscomb was named vice president and chief scientist of IBM

Corporation and served on the IBM Corporate Management Board. While at IBM, he was appointed by President Carter to the National Science Board and in 1980 was elected chairman, serving in that capacity until May 1984.

Dr. Branscomb was graduated from Duke University *summa cum laude* in 1945, and was awarded a Ph.D. degree in physics by Harvard University in 1949. Until his appointment to the Harvard faculty in 1986, he was an elected overseer of Harvard University. He holds honorary doctor of science degrees from fourteen colleges and universities.

Formerly a director of the IBM Europe, Middle East, Africa Corporation and of General Foods Corporation, he is a director of Mobil, MITRE, and Lord Corporations, and of the C.S. Draper Laboratories, Inc.

Among his other presidential appointments, Branscomb was named to President Johnson's Science Advisory Committee, and by President Reagan to the National Commission on Productivity. He is a member of the National Academy of Engineering, the National Academy of Sciences, and the National Academy of Public Administration.

He serves on the Technology Assessment Advisory Committee to the Technology Assessment Board of the U.S. Congress, and in 1991 was appointed to Massachusetts Governor William Weld's Council on Economic Growth and Technology. Professor Branscomb has written extensively on atomic physics, information technology, science and technology policy, and management of technology.

HARVEY BROOKS

Harvard Brooks is Benjamin Peirce Professor of Technology and Public Policy (Emeritus) and former dean of Engineering and Applied Physics at Harvard University (1957–1975). Before coming to Harvard as professor of Applied Physics in 1950 he had been associate laboratory head of the Knolls Atomic Power Laboratory of General Electric in Schenectady from 1946 to 1950. Trained originally as a theoretical physicist at Yale, Cambridge, and Harvard Universities, he is a member of the National Academy of Sciences, the National Academy of Engineering, and is a former president of the American Academy of Arts and Sciences (1970–1975). From 1975 to 1986, he headed the Science, Technology, and Public Policy Program of the Kennedy School of Government at Harvard.

Brooks was a member of the President's Science Advisory Committee from 1959 to 1964, of the National Science Board from 1962 to

1974, and chairman of the Committee on Science and Public Policy of the National Academy of Sciences from 1966 to 1971. Most recently, he has been a member of several task forces of the Carnegie Commission on Science, Technology, and Government. He has served as a consultant to the OECD, as a member of the UN Advisory Committee on Science and Technology for Development, and on numerous other government and international advisory bodies. He was the first chairman of the board of the German Marshall Fund of the United States (1972–1978) and is a director of Raytheon Corporation.

ASHTON B. CARTER

Ashton Carter is director of the Center for Science and International Affairs, and Ford Foundation Professor of Science and International Affairs at Harvard's Kennedy School of Government. He received his doctorate in theoretical physics from Oxford University, where he was a Rhodes Scholar. He subsequently held positions at Rockefeller University, the congressional Office of Technology Assessment, the Office of the Secretary of Defense, and MIT.

In addition to writing numerous scientific publications and government studies, Carter co-edited and co-authored *Ballistic Missile Defense, Managing Nuclear Operations,* and *Soviet Nuclear Fission: Control of the Nuclear Arsenal in a Disintegrating Soviet Union,* as well as wrote OTA's *Directed Energy Missile Defense in Space.* He is a member of advisory bodies to the White House Office of Science and Technology Policy, the Office of the Secretary of Defense, OTA, the National Academy of Sciences, the American Association for the Advancement of Science, the American Academy of Arts and Sciences, and the Carnegie Commission on Science, Technology, and Government. Carter is also a member of the Defense Science Board, the Council on Foreign Relations, the American Physical Society, and the International Institute for Strategic Studies, and he is a trustee of The MITRE Corporation.

GERALD L. EPSTEIN

Dr. Epstein directed the Dual-Use Technologies Project of the Science, Technology, and Public Policy Program at Harvard University's John F. Kennedy School of Government, which he joined in 1989.

He received S.B. degrees in physics and in electrical engineering from MIT. He was awarded a Fannie and John Hertz Foundation Fellowship for graduate study, which he used at the University of California at Berkeley. Doing research in experimental astrophysics at the Lawrence Berkeley Laboratory, he received his Ph.D. in physics in 1984.

In 1983, Dr. Epstein was selected as a Congressional Fellow by the Office of Telchnology Assessment, where he was an analyst and project director in the International Security and Commerce Program and the Energy and Materials Program. At OTA, he was an author of *Arms Control in Space: Workshop Proceedings* (1984), *Ballistic Missile Defense Technologies* (1985), *Starpower: The U.S. and the International Quest for Fusion Energy* (1987), and *Holding the Edge: Maintaining the Defense Technology Base* (1989).